AF327358

Nanostructure Science and Technology

Series Editor:
David J. Lockwood, FRSC
National Research Council of Canada
Ottawa, Ontario, Canada

For further volumes:
http://www.springer.com/series/6331

Qian Liu • Xuanming Duan • Changsi Peng

Novel Optical Technologies for Nanofabrication

 Springer

Qian Liu
National Center for Nanoscience
  and Technology
Beijing, China, People's Republic

Xuanming Duan
Technical Institute of Physics and Chemistry
Chinese Academy of Sciences
Beijing, China, People's Republic

Changsi Peng
Institute of Information
  Optical Engineering
Soochow University
Suzhou, China, People's Republic

ISSN 1571-5744
ISBN 978-3-642-40386-6   ISBN 978-3-642-40387-3 (eBook)
DOI 10.1007/978-3-642-40387-3
Springer Heidelberg New York Dordrecht London

Library of Congress Control Number: 2013951106

Preface

Novel optical technologies have promoted great advances in nanofabrication. This book focuses on the novel and promising techniques in super-resolution laser nanofabrication, demonstrating that they now play an important role in enhancing nanofabrication and offer vast potential in opening up new frontiers. It provides readers essential information on the newest laser techniques used for nanofabrication, and some seemingly impossible directions for further development. It will also familiarize readers including engineers, scientific researchers, and graduate students in nanoscience and nanotechnology with nanofabrication advances and trends based on novel laser techniques.

Because laser techniques have been developing at a rapid pace and are now being applied numerous fields, the subject is extremely broad and it is very difficult to cover everything in a single book. Therefore, this book focuses on some recently developed and emerging laser nanofabrication techniques, and further techniques potentially suitable for mass production. Every chapter in the book is essentially self-contained so as to provide in-depth coverage for those readers who are only interested in specific techniques. At the same time, the chapters are interlinked and form an organic whole on the latest advances and future development in laser nanofabrication.

This book is the result of a joint effort. Prof. Qian Liu, Dr. Taifeng Yu, and Dr. Chuanfei Guo from National Center for Nanoscience and Technology contributed Chaps. 1, 2, and 3; Prof. Xuanming Duan from Technical Institute of Physics and Chemistry, Chinese Academy of Sciences, focused on Chap. 4; Prof. Changsi Peng, Prof. Chinhua Wang, Dr. Su Shen, Prof. Linsen Chen, and Dr. Fuyang Xu from Soochow University were the lead authors of Chaps. 5, 6, and 7. The research work covered in this book was partly supported by National Basic Research Program. And we are grateful to our family members for their understanding. We especially thank Springer editor of the book, June Tang, for her patience and support throughout its preparation.

June 2013
Beijing

Qian Liu
Xuanming Duan
Changsi Peng

Contents

Chapter 1
Introduction

Laser is another important invention of mankind in the twentieth century, after the atomic energy, computer, and semiconductor. Since it was found in 1960, laser has been developed rapidly, not only giving the traditional optical science a new life but also leading to the rise of a new industry. Up to now, laser has been widely used in medicine, beauty, manufacturing industry, weapon system, and other fields, because of its characteristics of higher power, less divergence angle, better monochromaticity, shorter light pulse, and continuously tunable spectrum output compared to other light sources. Because laser technique has been developing at a rapid pace and is being applied to more and more fields, a full description of laser technique is very difficult and even impossible. Therefore, this book will focus on some recently developed and emerging laser nanofabrication techniques.

Obviously, laser processing technique is one of the earliest and most matured laser techniques. After 50 years of development, it has been applied to cutting, welding, surface treatment, punching, micromachining, surface structure fabrication, etc. With a decrease in machining dimension, higher and higher fabrication resolution is desired, especially in microelectronics and semiconductor industry. On the other hand, rapid development of nanoscience and nanotechnology leads to a large demand in the field of nanofabrication too. However, usual laser fabrication is still limited to micrometer scale because of optical diffraction limit of laser system, gravely impeding the applications of laser in nanoscale fabrication. Recently, some super-resolution laser fabrication methods have been developed to meet the requirement of smaller and smaller fabrication size. The fabrication techniques beyond the optical diffraction limit will greatly push laser technique forward. Another issue, which is considered by scientific and industrial communities, is how to further develop the applications of laser in industrial production on a large scale, especially to develop some novel laser techniques suitable for fabrication of micro-/nanostructures.

Q. Liu et al., *Novel Optical Technologies for Nanofabrication*, Nanostructure Science and Technology, DOI 10.1007/978-3-642-40387-3_1,
© Springer-Verlag Berlin Heidelberg 2014

1.1   Super-Resolution Laser Fabrication Techniques

With the development of modern sciences and technologies, we are getting into a nanoscale world. In the past several decades, the minimum node size in integrated circuits (ICs) based on photolithography has been reduced at a rate of about 30 % every 3 years according to Moore's law and has reached 22 nm so far. Besides, micro- and nanoscale components based on laser fabrication also are required in the fields of microelectronics, micro/nano diffractive optical elements (DOEs), microelectromechanical system (MEMS), optomagnetic storage, biochips, nanoscale electronic circuits, photonic crystal and optical communications, and so forth. To meet the requirement of smaller and smaller fabrication size, it is very necessary to research and develop new laser techniques with fabrication resolution beyond diffraction limit.

The super-resolution laser fabrication technique refers to those with a fabrication or processing size smaller than optical diffraction limit of a laser system. Breaking through the diffraction limit of a laser system is not only very important for optical theory, but also opens a new route in decreasing laser processing size. Recently, the emerging super-resolution laser techniques are developing rapidly, including super-resolution laser direct writing technique, femtosecond laser fabrication technique based on two-photon absorption, and laser fabrication technique based on surface plasmonic polaritons (SPPs).

Laser direct writing (LDW) is a kind of well-known technique to fabricate microstructures, which is based on interaction between laser beam and acceptor material. It is often used to fabricate photomask for photolithograph in IC industry and other micro-components, and main acceptor material is organic photoresist now. With the rapid development of large-scale integrated circuit (LSIC) technology and nanofabrication technique, higher and higher fabrication resolution is required for LDW technique. For this purpose, a novel super-resolution LDW technique has been developed based on nonlinear interaction between acceptor materials and laser beam. By means of the nonlinear effect, the technique successfully broke through the diffraction limit of the optical system, opening a new technical route to improve the resolution of LDW technique, which is much different from the traditional methods relying on increasing NA and/or shortening the incident light wavelength. More importantly, this technique expands the acceptor material range from organic photoresist to inorganic materials, metallic materials, semiconductor materials, and so on, greatly widening LDW technique's fabrication capability in various material-based micro-/nanostructures and devices.

Two-photon absorption fabrication technique based on femtosecond laser has been developed as a super-resolution laser technique suitable for fabricating macro-/nanostructures. Two-photon absorption means that an organic molecule can absorb simultaneously two photons with the same frequency or different frequencies and be excited from a lower energy state to a higher energy state. The energy difference between the two states of the molecule is equal to the energy sum of the two photons. When a femtosecond laser beam is focused on one point in organic liquid, two photons will be absorbed by the organic molecules at the point and result in a fast-curing dot with size breaking through the diffraction limit. By scanning with such kind of focused laser beam according to designed solid loci, corresponding three-dimensional structure induced by two-photon polymerization can be fabricated easily. Two-photon polymerization is a popular technique with the capacity to directly fabricate 3D micro-/nanostructures and has been successfully applied to fabrication of a variety of photonic and micromechanical devices.

Photo-nanolithography based on SPPs is another super-resolution laser technique, capable of making nanopatterns with resolution beyond the optical diffraction limit. The SPPs-based photolithography is currently a fast growing research area in which new ideas and improvements emerge constantly. It is believed that the technique will have significant potential for the next-generation photolithography technique with characteristics of nanoscale, large area, fast speed, and cost-effective fabrication.

1.2 Laser Fabrication Techniques for Mass Production

Laser interference lithography (LIL) is one promising technique for mass production because of its low cost, high fabrication resolution, and high processing rate. LIL is a high-efficiency writing technique because it can easily fabricate periodic structures by using interference pattern on the target material in one step. LIL, as a featured technique, is particularly suitable for fabrication of photonic crystals, high-density templates, as well as periodic structures for MEMS and bioscience and has attracted wide attention in nanoscale manufacturing of periodic and quasiperiodic features in recent years. At the present, a challenge of LIL technique is how to get smaller interference structures.

Another promising nanofabrication technique for mass production is nanoimprint lithography (NIL). The conventional NIL process, on the order of a few minutes per wafer, is still far from meeting the demands of many practical applications, especially in photonics, biotechnology, organic optoelectronics, etc. To increase processing efficiency, the concept of roller nanoimprinting is being pursued by many investigators as a means to increase throughput with high-resolution feature. Several types of flexible roll-to-roll nanoimprinting techniques have been developed. Nowadays, the major roadblock for large-format NIL is how to obtain the imprinting master mold, although the next-generation lithography approaches have demonstrated great capability in fabricating large-scale planar micro- and nanostructures. NIL is likely to become a key method to fabricate surface nanorelief for mass production of large-format micro-/nanostructure.

1.3 Main Contents and Brief Introduction

This book focuses on the latest progresses in laser nanofabrication, mainly dealing with some emerging techniques such as super-resolution laser direct writing technique, two-photon absorption technique based on femtosecond laser, super-resolution fabrication technique based on SPPs effect, and novel laser fabrication based on wrinkle, as well as the most promising laser techniques suitable for mass production. This book consists of seven chapters and the brief introduction is given as follows:

In Chap. 1, background, target, and main contents of this book are described. Meanwhile, a brief review of laser development is also given.

In Chap. 2, a new conceptual laser direct writing (LDW) technique with super-resolution fabrication ability is introduced. Principle and acceptor materials of the super-resolution LDW technique and its applications in grayscale photomask, ordered nanoribbons, and nanorelief are described in detail. An outlook on further development of the technique and potential applications in the future is also provided.

In Chap. 3, one type of novel micro-/nanostructure fabrication method is presented based on highly ordered, designable, and defect-free wrinkling fabricated with laser path-guided technique, providing a cheap and easy way to realize large-area fabrication of the surface complex structures. The mechanism dealt with this method is studied deeply, and various existing and potential applications are also proposed and discussed.

In Chap. 4, the 3D micro-/nanostructure fabrications based on two-photon absorption of femtosecond laser are introduced. The various functional materials used for 3D micro-/nanostructures, such as luminescent polymers, nanocomposites, metal, and glass, are also described in detail. The prospect of micro-/nanomanufacturing based on femtosecond laser is also discussed.

In Chap. 5, laser interference lithography (LIL) technique is discussed, including different interference patterns obtained by modifying the parameters of the LIL setup (incident configuration, polarization, phase, and intensity), graded index photonic crystal lens fabricated by using LIL patterns, patterns fabricated by direct writing with LIL technology, and in situ patterned semiconductor quantum dots (QDs) prepared by using LIL patterns.

In Chap. 6, progress on the different implementation schemes of the SPPs-based photolithography is reviewed, including prism-coupled SPPs nanolithography, grating-coupled SPPs nanolithography, and superlens imaging nanolithography techniques.

In Chap. 7, two techniques suitable for fabrication of large-format master mold are introduced. Several types of flexible roll-to-roll nanoimprinting, such as UV, thermal-embossing, and roll-to-roll seamless nanoimprinting lithography, are discussed in detail. Finally, potential applications of the techniques are briefly reviewed.

Chapter 2
Super-Resolution Laser Direct Writing and Its Applications

2.1 Introduction

Nowadays, various lasers for different performance goals have been used in almost all corners in our world from military weapon to civil DVD, due to its pure color, excellent coherence, good directivity, high power density, and large power range. Laser as a powerful tool has greatly pushed optics forward since it emerged in the 1960s, resulting in some new subordinate subjects and applications in optics such as laser holographic optics, laser metrology, and laser manufacturing.

As early as in the 1970s, laser direct writing (LDW) technique had been already used for drilling and cutting purposes because of the simple procedure, high precision, and low cost. In 1980s, LDW microfabrication techniques were improved greatly and developed rapidly. In 1983, Gale et al. [1] made microlens array on photoresist using LDW technique in line raster scanning mode. Koronkevich et al. [2] fabricated masks and kinoform optical elements in 1984 by using LDW technique in polar coordinates scanning mode. In 1989, Heidelberg University [3] designed a two-dimension laser direct writer; its scanning accuracy and the fabrication resolution were better than 1 μm. In 1990, Haruma et al. [4] fabricated micro-Fresnel lens with the resolution even better than those made by e-beam lithography. Correspondingly, LDW instruments have already been developed also, e.g., a commercial laser writer [5] can achieve feature size of 0.35 μm and alignment accuracy of 50 nm in a 200 $\times$ 200 mm scanning area; another writer [6] used for industry can write patterns on an area of 1,600 $\times$ 1,400 mm with the feature size as small as 1 μm. Now LDW is not only one of the main methods for fabricating photomasks in IC industry, but also becomes an important technique suitable for high-precision single-piece manufacture or low-volume production.

As an optical system, the spatial resolution of the laser direct writer is subjected to the optical diffraction limit. We know that in a laser manufacturing system, the resolution (R) and depth of focus (DOF) are two of the most important parameters,

Q. Liu et al., *Novel Optical Technologies for Nanofabrication*, Nanostructure Science and Technology, DOI 10.1007/978-3-642-40387-3_2,

which mainly relate to wavelength of incident light (λ) and numerical aperture (NA), and can be described as:

$$R = k_1 \frac{\lambda}{NA} = k_1 \frac{\lambda}{n \sin \theta}.$$

(2.1)

$$DOF = k_2 \frac{\lambda}{(NA)^2} = k_2 \frac{\lambda}{(n \sin \theta)^2}$$

(2.2)

where k_1 and k_2 are related to working conditions, such as exposure process and properties of acceptor material; n denotes refraction index of light transferring medium (air or liquid for immersion lens); and θ represents half-angular aperture of objective lens.

From Eq. 2.1, shortening the wavelength of incident light, increasing the numerical aperture (NA) of objective lens, as well as decreasing the parameter k_1 by optimizing the process conditions are effective routes to raise the fabrication resolution of a laser direct system. Among them, shortening wavelength is the most common method for increasing resolution so far, but it also has to face some unavoidable challenges. For example, the wavelength of the incident light could not be decreased infinitely, the shorter the wavelength is, the more sophisticated facilities and more expensive techniques are required. More basically, optical diffraction limit in principle cannot be broken through by reducing wavelength only. On the contrary, a longer wavelength is desired in LDW technique for obtaining long depth of focus (DOF), which can avoid the possible focus error and ensure the writing depth. According to Eqs. 2.1 and 2.2, longer wavelength will lead to a longer DOF but a lower resolution R, resulting in a dilemma in wavelength selection for a longer DOF or for a higher R.

Note that the acceptor materials applied to LDW technique are extremely important for the final products. Currently, main acceptor materials used in LDW are organic photoresists, the same as the materials used for optical projection lithography. A typical photoresist solution is usually composed of resin, photoactive component, and solvent. Resin is the base material, while photoactive component and solvent are used to control the physical properties of the resist [7]. Owing to the photoresist preparation on a substrate is by a wet spin-coating method, contamination and edge bend are hardly avoided. Additionally, whenever a laser with a new wavelength is used in LDW, a new photoresist has to be developed correspondingly because it is only sensitive to a definite wavelength. Due to these shortages of organic resists, exploring new types of acceptor materials including inorganic materials, metallic materials, and even semiconductor materials is becoming more and more important for the development and application of LDW techniques.

On one hand, higher and higher fabrication resolution is required for an LDW system, with the rapid development of large-scale integrated circuit (LSIC) technology and nanometer-scale fabrication. However, the current LDW equipments can hardly achieve very high spatial resolution because of the optical diffraction limit. On the other hand, most of the laser direct writers use photoresists as the acceptor materials, hindering its wider applications in surface structure fabrication. To overcome the difficulties mentioned above, LDW technique should be improved at least in two aspects: (1) enabling the writing on various materials and (2) raising the spatial resolution even beyond the diffraction limit.

To increase the spatial resolution and expand the range of optional acceptor materials, we developed a new conceptional LDW technique with super-resolution fabrication capability. By means of smartly utilizing the nonlinear interaction between laser and acceptor material in the technique, the diffraction limit of the optical system has been successfully broken through, opening a new technical route to improve the resolution of LDW, which is very different from the traditional methods relying on increasing NA and/or shortening the incident light wavelength. More importantly, this technique expands the acceptor material range from organic photoresist to inorganic materials, metallic materials, and semiconductor materials, greatly widening the application of LDW technique in developing new types of devices and new micro-/nanoscale fabrication methods.

In this chapter, we will introduce super-resolution mechanism, the laser direct writing system, and suitable acceptor materials and discuss some new applications of the super-resolution LDW in continuous grayscale photomask, nanorelief, ordered nanoribbon, and so forth.

2.2   Principle of Super-Resolution LDW

The diffraction limit of an LDW system can be broken by utilizing the nonlinear interaction effect between incident laser beam and acceptor materials. Figure 2.1 shows an example of a hole array made by the super-resolution LDW. Note that the diameter of the holes is about 40 nm far beyond the diffraction limit of the LDW system, which adopts a 532 nm laser with objective lens of 0.9 *NA*.

Several explanations have been presented on the phenomenon of beyond diffraction limit of LDW technique [8–11]. Here we also try to explain this effect from our viewpoint based on a simple and feasible model [12].

In fact, the interaction between laser and the irradiated material is a very complex process. When laser beam hits the material, the corresponding results induced by the laser–matter interaction could lead to multiple effects including photochemical reaction, thermal effect, ablation effect, melting, and phase transition and oxidation of the material, resulting in complex physical and chemical changes in the acceptor material. Generally, because the energy profile of the laser beam is Gaussian type, the temperature pattern on the exposed spot area will approximately be a circular symmetry Gaussian distribution. Therefore, the temperature of central part of the spot is much higher than that of the surround part, as shown in Fig. 2.2. Therefore different responses of the material to the laser beam occur at different regions in the laser-irradiated spot. By selecting suitable laser power to match thermal thresholds of the material, the resolution of LDW can be effectively increased [9].

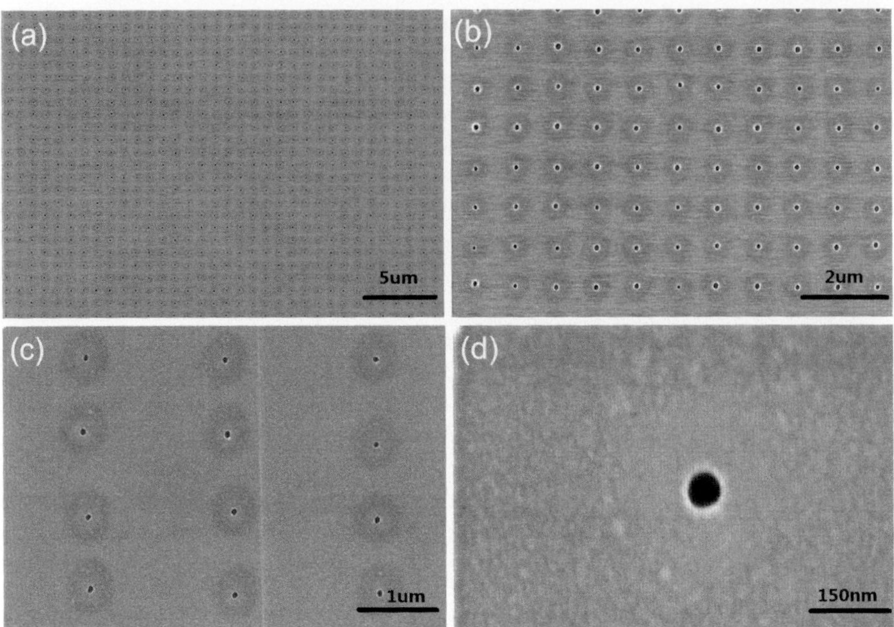

Fig. 2.1 SEM images at different magnifications of hole array fabricated by LDW on a 40 nm thick Ti film, each hole with a diameter of 40 nm

Fig. 2.2 The schematic image of the temperature profile inside a laser spot of Gaussian profile

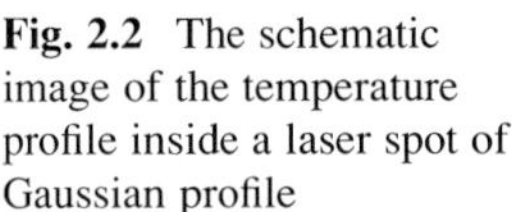
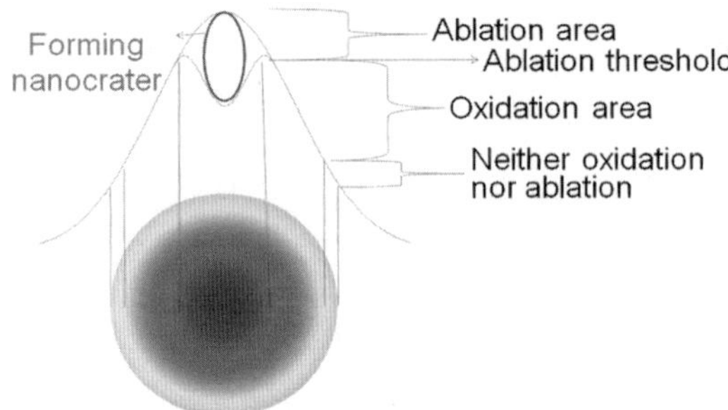

Here, we explain it in detail by taking a titanium thin film with a thickness of 40 nm as an example. When the laser beam reaches the surface of the titanium film, the Gaussian distribution laser beam absorbed by the film converts to a Gaussian distribution temperature field. In the thermal process, we are concerned about two temperature thresholds. One is ablation temperature of titanium, and the other one is the temperature for activating oxidation [13], the former is much higher than the latter. When a low power (<2 mW) for a pulse width of 200 ns was adopted in our

LDW, no changes was observed in the Ti films because the temperature is too low to cause neither oxidation nor ablation. By using medium powers (2–10 mW) with a 200 ns pulse width, the laser-induced temperature in the titanium films could be higher than the oxidation threshold but lower than the laser ablation threshold; bulge structures found in this case are shown in Fig. 2.2. While at a higher power ($\geq$10 mW, 200 ns pulse width), the temperature at the spot center is higher than the ablation threshold of titanium, leading to gasification in the central part. But at the surrounding part of the spot, the temperature is lower than ablating threshold value because of the Gaussian temperature distribution, making the size of the ablated hole much smaller than the size of the laser beam, even beyond the diffraction limit (see Fig. 2.2). It can be seen from here that the breaking through diffraction limit in the super-resolution LDW stems from the nonlinear interaction of laser beam and the acceptor material. In micro-/nanofabrication field, the critical size of the product is as small as submicrometers or even several tens of nanometers; thus, LDW writer used in this field must have very high accuracy, stability, and applicable acceptor materials.

2.3   Laser Direct Writer

For breaking through diffraction limit, a high-quality laser direct writer is essential. Figure 2.3 shows the principle of a home-built laser direct writer for super-resolution fabrication.

The LDW system is composed of three main parts, laser moderation part (the main part) for controlling the power and pulse of the laser beam; focusing detection part (monitor module in Fig. 2.3) for keeping the light beam focused on the desired position; and sample scanning part (scanning stage). A computer is used to control the whole system, including the operations of all parts in the system, such as adjusting the power and pulse frequency of the laser beam, uploading parameters of the designed pattern, and controlling and monitoring the writing process.

1. *Laser Moderation*

This writer adopts a solid laser as the writing light source with the instability less than 1 %. The laser beam passes through an acousto-optic modulator (AOM), a rotary mirror, and an expander and finally focused on the surface of the sample by an objective lens. The objective lens is fixed on a PZT nano-grade positioner which can move along Z-axis (parallel to the light beam) to adjust the distance between the objective lens and the sample. The sample is mounted on a two-dimensional PZT scanning stage with a step of 1 nm.

2. *Focusing Detection*

Because of the unevenness and slope on the sample surface, as well as the external vibration and errors of the scanning stage, the laser beam could not always keep exactly focusing on the sample surface during the process of writing. To ensure that the laser beam is invariably focused on the surface of the sample, a focus servo

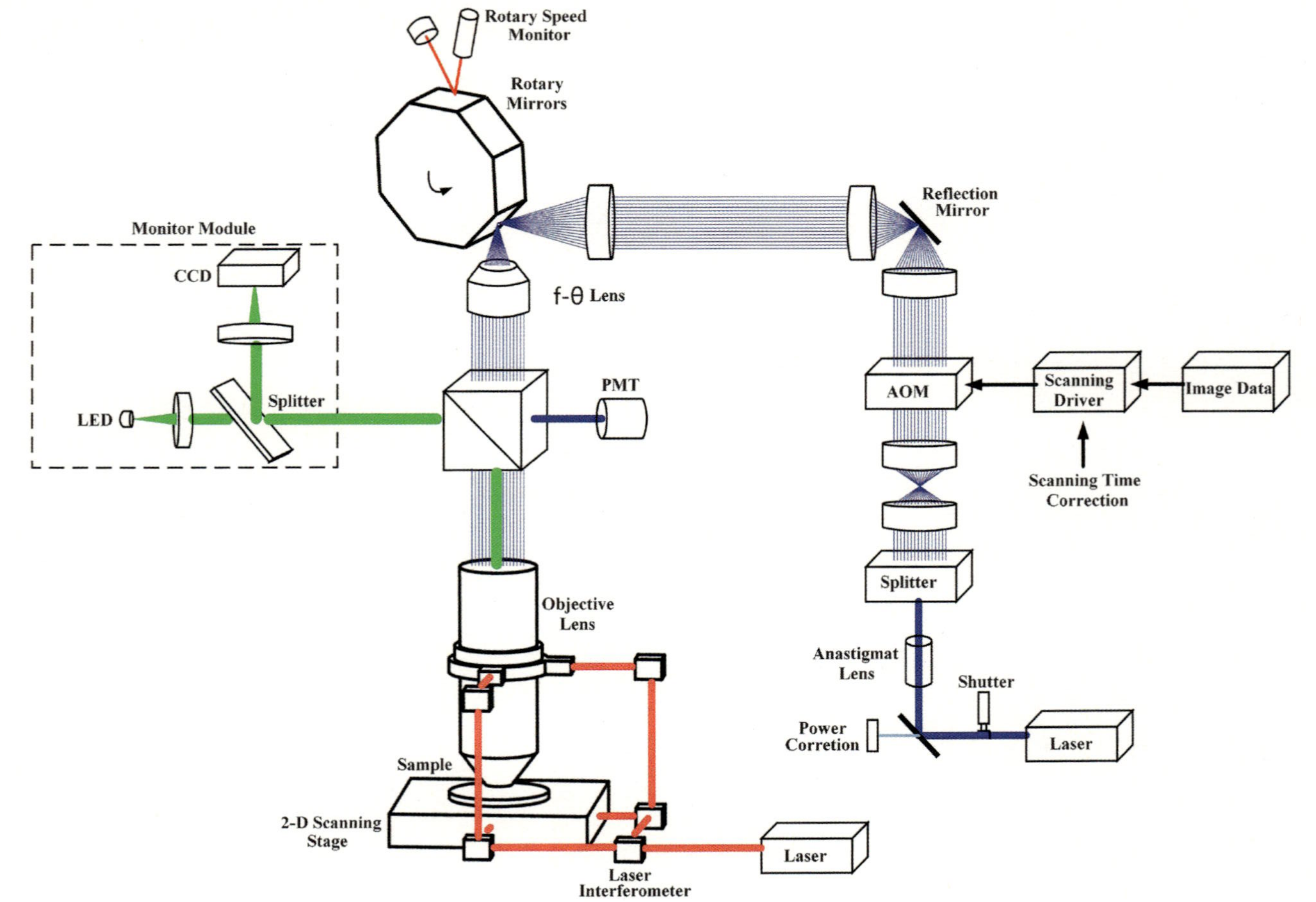

Fig. 2.3 Sketch of a home-built LDW for super-resolution fabrication

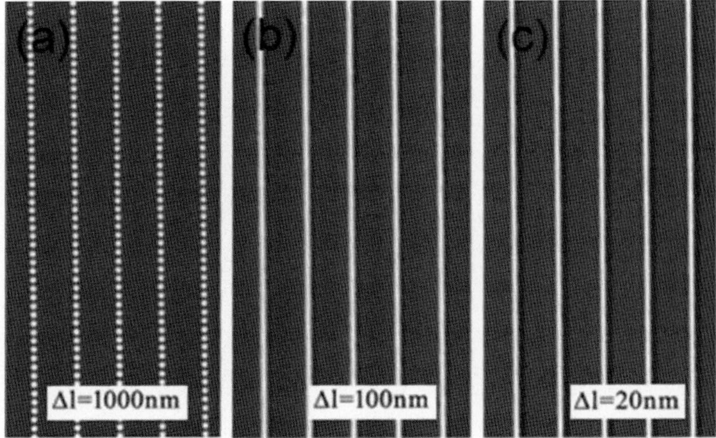

Fig. 2.4 *Dots* and *lines* written on the samples by LDW; Δl is a moving step length in line direction

module is employed. In the focus servo module, another laser beam with a different wavelength is used to track focus. The reflective light from the sample surface is collected by a quadrant detector. The signal of the detector can indicate the defocus distant. Then the objective lens moves up or down according to the feedback signals to correct the focus error automatically.

3. *Scanning Stage*

A high accurate x-y scan stage is used in the system. In the writing process, the sample is fixed on the stage and can move in x-y plane with the scan stage. The scan process is controlled by using the PZT. Since the LDW instrument is used for micro-/nanoscale fabrication, the scanning stage must be of high stability, high alignment, high repeatability, and high accuracy (at nanoscale).

Laser direct writer has the advantages of easy operation, direct writing, and friendly working environment. It is a convenient tool for research and development purpose. In IC industry and micro-/nanofabrication, LDW system is mainly used in fabrication of photomasks, MEMS, integrated optics elements, writing logos and labels on electronic packaging, and other nano-/micro surface structures.

This laser direct writer adopts the pulse laser writing and raster scanning mode. All the working parameters, such as the power and pulse width of the laser beam, as well as the scanning step length, can be controlled and adjusted. Laser duration in one pulse will "write" one "dot" on the surface of the sample. Then the sample moves one step to a new position and finishes the next writing action. Any pattern or feature written by LDW system on the surface of the acceptor material is composed of a series of the dots. If adopting too small step length in writing process, the written dots may overlap partially, and the dots form lines or other 2D patterns on the surface (Fig. 2.4). Varying laser power and/or pulse width can generate various patterns on the acceptor material (Fig. 2.5), such as grayscale mask, nanorelief, micro/nano diffractive optical elements (DOE), microelectromechanical systems (MEMS), and even more complex patterns.

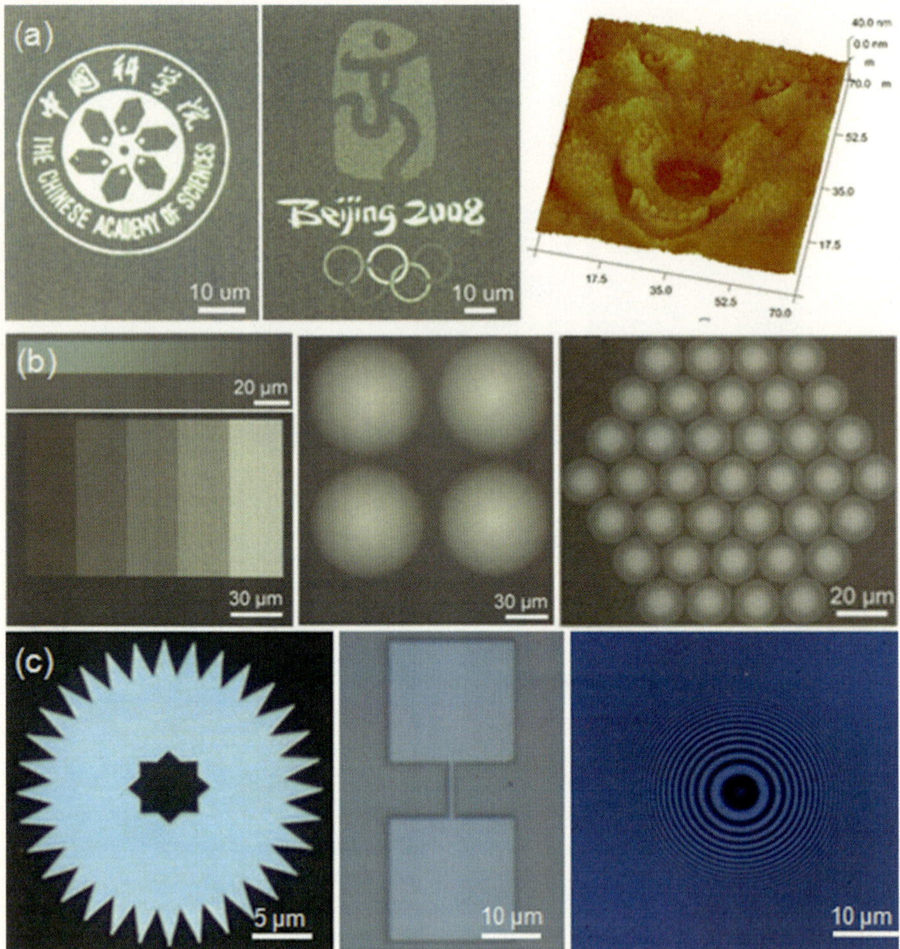

Fig. 2.5 (**a**) Nanoscale reliefs, (**b**) structures of grayscale masks, and (**c**) MEMS and DOEs

2.4 Acceptor Materials for Laser Direct Writing

So far, in the field of photomask making, the main acceptor materials used in LDW technique are organic photoresists. When the photoresists are exposed under fixed wavelength of laser beam, their physical and/or chemical properties will be modified at the irradiated areas, and the designed patterns can be remained after developed. Photoresists can be divided into positive resists and negative ones. The differences between them are that the exposed portions for a positive photoresist become soluble and the unexposed portions remain insoluble, while the situation is on the contrary for a negative resist [7].

With rapid development of IC industry and micro-/nanofabrication, the integration level of electronic elements is becoming higher and higher, and corresponding size of electronic elements is becoming smaller and smaller. Generally, adopting a laser with a shorter wavelength can efficiently solve this resolution challenge, just like the adopted technical route by using shorter and shorter light wavelengths (436 nm → 365 nm → 248 nm → 193 nm → 13.5 nm) in microelectronics and semiconductor industry. However, this can only postpone the deadline of reaching the physical limit of resolution, and corresponding LDW equipment will also be more and more expensive and finally reach an unacceptable price. In addition, developing new types of photoresists to match a new shorter-wavelength laser is not an easy issue, and corresponding photoresist needs a very long R&D period and a high investment. Hence new fabrication methods and different acceptor materials are necessary for the extending applications. In order to fundamentally overcome the difficulties mentioned above, breaking through diffraction limit of the LDW is a promising way. For this purpose, development of new types of acceptor materials in a wider range is necessary and will greatly push forward the development of super-resolution LDW and expand its applications.

Theoretically, any material might "be written" by laser. Acceptor materials with nonlinear feature can be inorganic materials, organic materials, metallic materials, even semiconductors, and so on. In this section, we will introduce a couple of nonlinear acceptor materials.

2.4.1 Metallic Materials

The transparent metallic oxides such as In_2O_3 and SnO_2 are star materials in the photoelectrical industry owing to their applications in the field of transparent conductive films. For example, ITO film (tin-doped indium oxide) has been widely used [14]. In recent years, the oxides of In, Sn, and Zn are in the spotlight because of the diverse morphologies and desired photoelectrical properties of the In_2O_3, SnO_2, and ZnO nanostructures [15]. These oxides are also good candidates for gas sensors [16]. Based on the metal-transparent-metallic-oxide(s) (MTMO) systems, we present a new application of the metals as acceptor materials in the field of micro-/ nanofabrication, and we have found that continuously variable transmittance (T) is available for many systems consisting of a metal and its oxides, e.g., Sn and tin oxides, and In and indium oxide as well.

2.4.1.1 Sn and Tin Oxides

Metal tin has three relatively stable oxides, i.e., tetragonal t-SnO, orthorhombic o-SnO$_2$ and t-SnO$_2$ [15–24]. Among these oxides, t-SnO$_2$ is the most transparent, followed by o-SnO$_2$. The t-SnO is an undesired material for many applications such as transparent conducting films due to its adverse impact on transparency. But this characteristic is helpful for fabricating grayscale masks because it may bring richer gray levels. Besides these oxides, we have reported another transparent and stable phase: amorphous SnO_x [25]. The different optical properties of Sn and tin oxides provide the possibility to adjust the transparency of a Sn film because of controllable phase compositions at various temperatures, so that it might be used as a potential grayscale material. Moreover, the system consisting of the Sn and its oxides has an applicable wavelength range in the visible and near ultraviolet region down to 350 nm, and the transmittance of SnO_2 is very high in this region [26].

Figure 2.6a [27] shows XRD spectra of 20 nm Sn films heated at different temperatures for 5 min, and Fig. 2.6b shows the optical density (OD) spectra of the films from 350 to 700 nm. The phase evolution, which is similar to the reported result [24], is known from the XRD spectra and the near ultraviolet–visible (NUV-V) OD spectra that the t-SnO_2 (the sample was annealed at 600 °C) is the most transparent, the OD value of the t-SnO_2 is below 0.04 ($T > 91$ %) even at 350 nm. The t-SnO is the main phase of the sample (annealed at 400 °C), which has an OD below 0.24 ($T = 58$ %) from 350 to 700 nm. The as-deposited Sn film is almost opaque and has an OD of 0.65 ($T = 22$ %) at 350 nm. The distinction of the OD can be larger by using a thicker film. The results verify that the Sn–SnO_x system is suitable for photolithography in the wavelength range from visible to NUV region down to 350 nm, covering some commonly used wavelengths in photolithography such as I-line and G-line. The nanostructures fabricated by Sn thin film have been found to be able to reach high resolution (~200 nm), which is smaller than diffraction limit of the 532 nm laser system.

2.4.1.2 Indium and Indium Oxide

There is difference in the case of indium–indium oxide system compared with Sn–tin oxide system. In–InO system has only one stable oxide, In_2O_3, with the body-centered cubic (BCC) structure (some researchers reported other structures and the phase of In–O [28, 29]). The In–In_2O_3 system is simpler than the Sn–SnO_x system. Figure 2.7 shows that the annealed In film (In_2O_3) is very transparent in the visible region. Although it becomes less transparent with the decreasing of wavelength in NUV region, the optical density of the In_2O_3 film is still small. For example, a 20 nm In film can have an OD ranging from 1.10 to 0.10 (T from 7.7 % to 79 %) at 365 nm in an oxidation process, and the OD range of a 30 nm In film is from 1.55 to 0.15 (T from 2.8 % to 71 %).

Indium, as nonlinear grayscale mask material, has both a large OD value and large OD range, in the same time it has low energy consumption in mask fabrication and considerable stability. Like tin metal, indium also has low melting temperature, transparent and stable oxides, and suits for being nonlinear acceptor material.

2.4.1.3 Comparison of In and Sn

1. Indium has a larger OD than tin at the same thickness. From Figs. 2.6 and 2.7, for
   a 20 nm thickness, the OD of an In thin film is around 1.1, while at the same

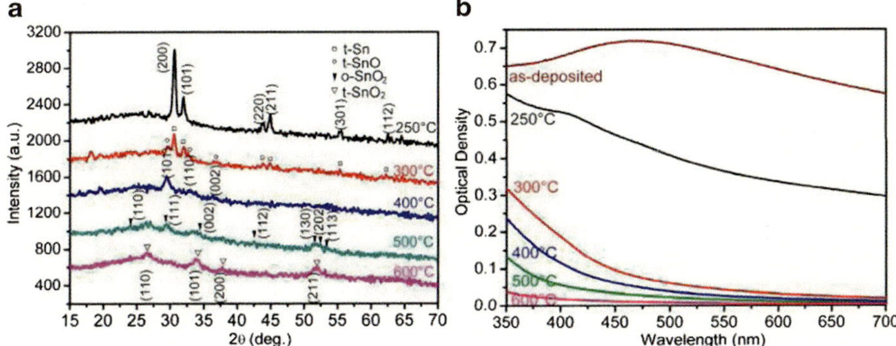

Fig. 2.6 (**a**) XRD spectra of 20 nm Sn films heated at different temperatures for 5 min, showing the phase evolution from Sn to SnO, orthorhombic (o-) SnO_2, and finally to tetragonal (t-)SnO_2. (**b**) NUV-vis spectra of the films annealed at different temperatures; the t-SnO_2 film has an OD value less than 0.04 from 350 to 700 nm (Reproduced from Ref. [27] by permission of Optical Society of America)

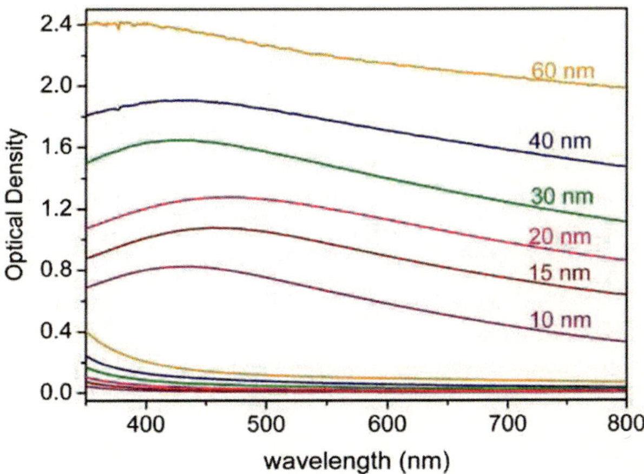

Fig. 2.7 NUV-vis spectra of In films with different thicknesses (*upper* part) and those of the films annealed at 350 °C for 1 h (*lower* part). Spectra of an In film and the corresponding In_2O_3 film are in the same color (Reproduced from Ref. [27] by permission of Optical Society of America)

thickness, the Sn film has an OD of only 0.65. After well annealed, OD of the original 20 nm metallic films changes to In (In_2O_3) (~0.12) and Sn (SnO_2) (~0.08). Therefore, indium is a better material for grayscale mask making than tin, although SnO_2 is slightly more transparent compared with In_2O_3.

2. According to our experience, Sn and a-SnO_x system can be well controlled only when the thickness of the film is less than 20 nm. And it needs to use a long pulse exposure (>10 μs) to prepare the Sn–SnO–SnO_2 system, and a typical writing laser dose is 15–4,000 J/cm^2. While to obtain the In–In_2O_3 system a 100 ns pulse width or even shorter is enough, corresponding energy density is only around 1.5 J/cm^2. Compared with the two material systems for grayscale mask, In–In_2O_3 has obvious advantages to the Sn–SnO_x system, lower energy consumption and a higher writing speed. The different pulse width requirement for

the two material systems may be because forming of t-SnO$_2$ phase needs intermediate oxides, but In$_2$O$_3$ can be directly oxidized from In.

3. Experimental results [29] show that the grayscale masks made by both materials are stable. The gray levels of the masks are not changed after long time exposure in air ambience or UV light. The analyses for metallic thin films show that each metallic Sn grain is coated by a very thin a-SnO$_x$ layer, while the surface of In has a thin In$_2$O$_3$ layer formed in air to prevent further oxidation, therefore both of them are stable. In addition, for the optical properties of the masks fabricated by LDW, after heating at 100 °C for hours, no change was found in Sn film [30], but In film became slightly more transparent (which can hardly be observed by naked eyes). We think that for In film, a protective layer might be adopted to avoid the further oxidization.

2.4.2 Bimetallic Materials

The working principle of bimetallic thermal resist is that two different metals may form a eutectic alloy, according to the explanation in the literature [31]. The melting temperature of the new eutectic alloy is lower than the melting point of either of the component metals. A bimetallic thermal alloying resist film consists of three distinct layers: two thin layers with different metals as the imaging layers and a thick layer as the underlay layer. The two different metals should be selected in accordance with the requirement that the melting temperature of their eutectic alloy is less than 300 °C. At the interface of the two metallic layers, the atoms of the metals permeate each other to form an alloy thin layer. This thin layer of alloy has a lower melting temperature. When a laser beam with a suitable power hits the imaging layer, the metals melt beginning from the interface between the two metals. After the end of the laser irradiation, the heated spot cools down and forms eutectic alloy, while the unirradiated parts of the film keep unchanged. The surface of the thin film now composes of three kinds of materials, two original metals (unexposed part) and eutectic alloy (exposed areas). Due to the different chemical properties of the original metals and the eutectic alloy, pattern transferring becomes possible when using a proper solution and provides us the basics for fabricating micro-/nanoscale structures and masks.

The bismuth–indium, bismuth–tin, and indium–tin bimetallic thermal resists have been studied [31, 32], and HCl:H$_2$O$_2$:H$_2$O solution has been used as the developer. From the results of the literature [33], the Bi–In bimetallic alloying resist can be used to produce grayscale masks (Fig. 2.8). Figure 2.9 shows the optical density (OD) of two samples with different exposure intensities. Bi–In–O material has a good result; the values of OD change from 3.45 to 0.66 at 365 nm when the laser power varies from 0 to 0.4 W.

OD values of the samples and the transmission of the materials change accordingly with the intensity change of the laser beam in fabrication process. Obviously, the material transforms from opaque to transparent. Here we may be confused by

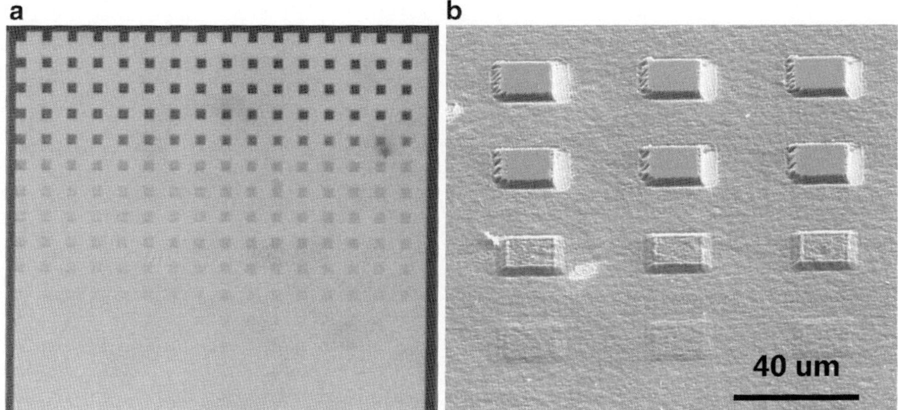

Fig. 2.8 (**a**) A grayscale mask was created on Bi/In/O film (back-lit microscope image, 20×20 µm for each gray square). (**b**) SEM picture of squares with different heights created on Shipley photoresist with the grayscale mask (Reproduced from [33] by permission of SPIE and thanks to Prof. G. Chapman for giving the reprint permission)

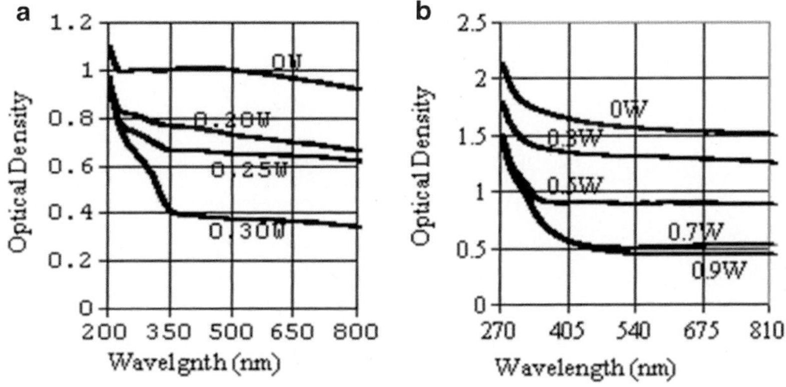

Fig. 2.9 Absorption through (**a**) 15/15 nm BiIn and (**b**) 48/48 nm BiIn exposed with the argon laser at different intensities (Reproduced from [33] by permission of SPIE and thanks to Prof. G. Chapman for giving the reprint permission)

the mechanism of the bimetallic alloying resists discussed above. In fact, it is suspicious that the two different metals can form eutectic alloy by the laser irradiation. As we have known, due to the high conductivity, metals are good conductors and good electromagnetic shielding materials. So that metals are not transparent in the vast range of electromagnetic waves, including the range from ultraviolet to visible light. Eutectic alloys, including Bi/In, Bi/Sn, and In/Sn, are still metallic materials. They should keep opaque in the ultraviolet-visible light range, neither transparent nor grayscale, unless there are opposite experiment data. Hence the interpretation of eutectic alloy forming from the bimetallic alloying resists and becoming transparent is not reasonable. Please notice that in Fig. 2.9, the material is not a bimetallic alloy, but Bi–In–O; the added oxygen must be the key factor. This relationship between OD and laser beam intensity is attributed to the metallic oxidation. The material should be transformed from opaque metal to transparent metallic oxide.

Further investigation of the transparency conversion mechanism of bimetallic Bi/In has been made [34]. Bi/In thin films were annealed in the oxygen environment with different temperatures to simulate the conditions of long pulse laser exposure. Structural and optical properties analyses indicate that oxidation is regarded as the reason of transparency conversion induced by heat treatment and long-pulsed laser exposure, while laser ablation is demonstrated to be the main reason of transparency conversion induced by short-pulsed (~7 ns) laser.

For a bimetallic film, the oxidation is a complex process from the surface and inward. The oxidation should start from outside surface of the film. Oxygen atoms diffuse from the air to the first metallic layer, and gradually approach the interface of the bimetal, and then the second layer. Theoretically, only after the first layer is completely oxidized, then the second layer begins the oxidation process. If we assume that the metallic oxide is transparent, but metal itself is not, it seems that the gray levels of the film come from the second layer or only one metallic layer may be enough. Therefore, single metallic element and its oxides seem to be simpler and have clear mechanism, compared with the bimetallic alloying resists.

2.4.3   Semiconductor Phase-Change Materials

As promising inorganic photoresists, phase-change materials have attracted much attention recently [35–38]. Their optothermal responses have broad-spectrum features [39], indicating their application prospect. Semiconductor phase-change materials as inorganic photoresist can be easily deposited onto substrates to produce uniform, large-area resist film (both planar and non-planar) in vacuum [39–41], which is compatible to full vacuum processing technical requisition in next-generation microelectronic industry. In addition, this kind photoresist will greatly simplify the production procedures because they need neither to select particular light source nor to be used in a special environment and can completely eliminate the pre-baking and post-baking steps required for organic photoresists [42].

Ge–Sb–Te (GST) is a kind of typical phase-change semiconductor material. Its optical reflectivity and electrical resistance have distinct differences at amorphous and crystalline states. Based on these properties, GST has been widely used in optical data storage [43–45] and phase-change random access memory (PRAM) [46–49]. Recently, some works on using GST materials as inorganic photoresists have been reported based on the different chemical and selective etching properties in two phases [39, 50, 51]. Generally, as a phase-change memory material, it has to meet the requirements of high spatial resolution, fast response speed, low energy dissipation in phase transition, and low activation energy. These demands for phase-change memory material are similar to the requests of photoresist. If using focused laser beam under a suitable exposure condition to irradiate the metal chalcogenide semiconductors, phase transition from amorphous to a crystal phase in the region of the exposed spot occurs. Employing selective etching property or other different chemical and physical properties between its amorphous and crystalline phases, GST material may be considered as a candidate of photoresists used in IC and microscale fabrications. However, the GST phase-change materials are far from good inorganic photoresist candidates due to its poor surface roughness and higher energy consumption. In this section, an improved GST-based material, $Ge_2Sb_{1.5}Bi_{0.5}Te_5$ (GSBT), will be introduced as an inorganic photoresist [52].

GSBT thin film with thickness of 100 nm was deposited onto Si substrates by RF magnetron sputtering under working pressure of 0.1 Pa in argon flow. The substrates were not heated, and kept at ambient temperature during the deposition period. Figure 2.10a shows the surface of the as-deposited GSBT film, which is composed of superfine granules without apparent crystalline grains. Obviously, the as-deposited GSBT film is smooth and uniform (2 nm roughness for 100 nm thick) which is suitable for super-resolution usage. After annealed in vacuum ambient at 150 °C for 15 min, the surface (Fig. 2.10b) becomes rougher, and the granules of the annealed surface are much larger than that of as-deposited one. The nature of the GSBT films is shown by the X-ray diffraction (XRD) experiments as in Fig. 2.10c. The XRD curve below indicates that the as-deposited film is in amorphous state, while the upper XRD curve on annealed film exhibits several distinct peaks corresponding to a face-centered cubic (FCC) crystalline structure, revealing the phase of the GSBT film from amorphous to polycrystalline state. Obviously, the phase change can also be generated by optothermal effect induced by laser dose in LDW technique.

The phase transition of GSBT film indicates that this material has suitable thermal threshold for thermal mode lithography. As a resist, smooth and uniform surface, suitable thermal threshold are only the basic conditions. Its photo-thermal properties are more important for photolithography. As a photothermal inorganic photoresist, the material must have a good thermal absorption property in a broad light spectrum range. Figure 2.10d shows the UV-visible absorption spectra of the amorphous state GSBT film. It is shown obviously in the figure, amorphous GSBT film can effectively absorb light in a wide range of wavelength, especially in UV band which partially covers the wavelength range used in microelectronics industry. It is obvious that the acceptor material GSBT possesses great potential as an inorganic resist. Additionally, a good etching selection between amorphous and polycrystalline states is also necessary for a practical resist GSBT.

Further study indicates that Bi-doped phase-change material $Ge_2Sb_{1.5}Bi_{0.5}Te_5$ (GSBT) is a smart material and can be either negative or positive resist depending on the developing solutions used [53].

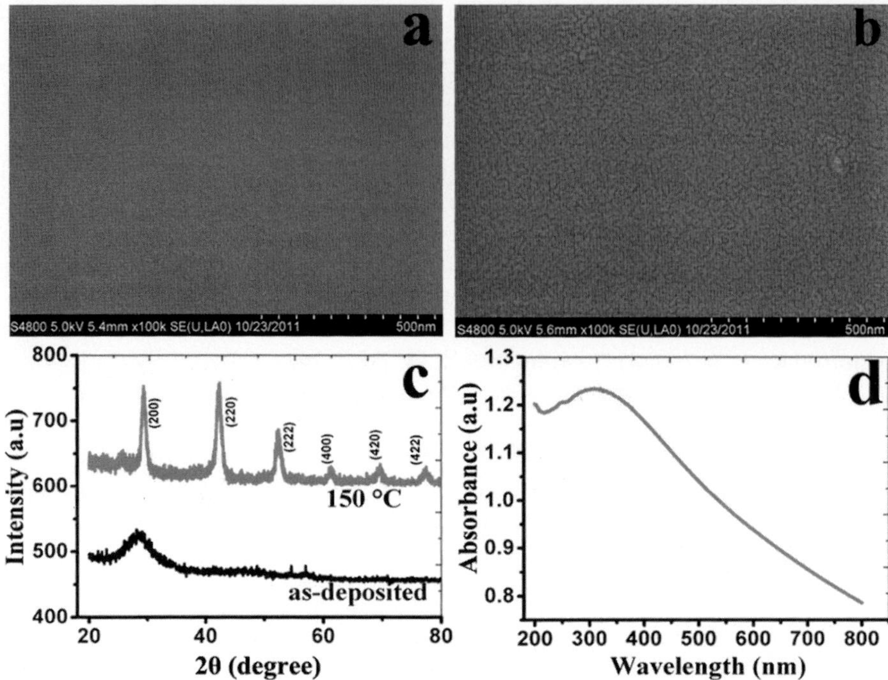

Fig. 2.10 SEM images of 100 nm thick GSBT films. (**a**) As-deposited. (**b**) Annealed in vacuum at 150 °C for 15 min. (**c**) XRD patterns of the two films. (**d**) UV-visible absorption spectra of GSBT (Reproduced from Ref. [52] by permission of Optical Society of America)

2.5   Applications Based on Super-Resolution LDW

LDW technique with super-resolution can obtain a fabrication resolution beyond diffraction limit, greatly expanding its application scope and providing many new chances to explore new techniques and new methods. In this section, we will discuss some applications with super-resolution fabrication feature based on the super-resolution LDW techniques.

2.5.1   MTMO Grayscale Photomasks

The fast-developing modern technologies, such as telecommunication, computer, information display, and imaging, create the increasing demands for the integration of electronic and photonic systems. Besides the densification of IC chips, the miniaturization of optical devices has been another essential task for us. Micro-/ nanoscale optical devices and microelectromechanical systems (MEMS) not only need planar designs for the high-quality surfaces topography but also require the 3D structures with the third dimension in vertical orientation. The grayscale lithography technique, which is simple, direct, and cost-effective, has been considered to be one kind of ideal fabrication method for high-efficiency micro optical devices and three-dimension (3D) MEMS [54–60]. In the technique, grayscale mask fabrication is one of the key points, which will deal with both the LDW technique and the mask materials.

Two mask fabrication techniques have been commonly used so far, one is to fabricate the mask on a chromium film on glass substrate (COG) [61, 62] by using microfabrication technique and the other is to fabricate the mask on high-energy-beam sensitive (HEBS) [63] glass by using electron beam. Compared the mask fabrication processes and the properties of the two materials, although COG material is very simple, the mask fabrication procedure needs many steps such as film deposition, lithography, etching, and resist striping. More fatally, the mask based on COG cannot achieve high resolution. These demerits limit its further development in grayscale mask fabrication, while HEBS glass technique shows its advantages in making true grayscale masks. The shortage of HEBS glass is that it is a very complicated material system, and its working process depends on the expensive high-energy electron beam system. The common problem in both techniques is too costly. To overcome the difficulties mentioned above, we developed a novel metal-transparent-metallic-oxides (MTMO) grayscale mask based on our LDW technique and successfully realized super-resolution fabrication. Compared with two foregoing techniques, this technique is simpler in manufacturing process and cheaper in manufacturing cost. In this section, we will introduce the fabrication, mechanism, and applications of the new grayscale mask and briefly discuss its super-resolution feature.

2.5.1.1 Metal/Oxides Grayscale Masks

By means of laser direct writing on metal films, we present here a simple, cheap, and stable grayscale photomask based on the metal-transparent-metallic-oxides (MTMO) systems, very different from techniques based on COG and HEBS glass [63]. In principle, metals are opaque materials, but many of them can be transformed into transparent metallic oxides by laser irradiation on the metallic films. If the transmittance of the transparent-metallic oxides is variable and controllable, corresponding metals, basically, can be used to make grayscale masks. The MTMO grayscale mask can be fabricated with only two simple steps: metallic film deposition and laser direct writing. In the following, we will take Sn/tin oxides and In/indium oxide grayscale mask as examples [27] to illustrate the fabrication process, and briefly compare In with Sn mask.

Sn (In) films (10–60 nm) were deposited on glass substrates by a radio-frequency magnetron sputtering with a power of 30 W and pressure of 0.57 Pa. Note that we adopted our patented technique [64] to prepare the films so as to obtain smooth surface composed of fine grains.

Generally speaking, the Sn (In) films prepared by common physical vapor deposition methods can only obtain rough surfaces with big grains when the thickness of Sn (In) layer exceeds 15 nm. Such inhomogeneous morphology is incapable of making high-resolution grayscale pattern. Ostwald ripening effect is the main reason of deterioration of the films, because in the ripening process of the deposition, small grains tend to be vanished while big ones grow even bigger, according to Gibbs–Thomson interpretation in thermodynamics [65]. In order to prepare films with fine granular surface, the ripening effect should be suppressed, i.e., preventing the growth of big grains in the nucleation process. Our measure is to adopt a multilayer deposition process, in which the thickness of each layer is thinner than 15 nm so as to keep ultrafine grains. It must be mentioned that the multilayer deposition is not the reason of the refinement. Smooth surfaces with fine grains will not be obtained if the films are simply deposited only with several layers. In fact, a thin oxide coating formed on every metal layer plays a crucial role in preventing the re-sublimation and growth of the metal grains, as shown in Fig. 2.11

Figure 2.12 shows the differences of the films with various deposition layers and with or without oxidation. Figure 2.12a–c shows that the grain sizes reduce with the increase of the number of layers, oxidized in the deposition intervals. Figure 2.12d–f shows the 20 nm In film prepared by one time deposition, four times of deposition (without surface oxidation in the deposition intervals), and four times of deposition (5 nm $\times$ 4, with surface oxidation for each layer), respectively. It is obvious that the grain size does not change after four times of deposition (Fig. 2.12e), unless the film was exposed to oxygen ambience in every deposition intervals (Fig. 2.12f).

The grayscale mask fabrication is performed by a home-built laser direct writer, which adopts an objective lens (NA 0.90), a 532 nm laser (Spectra Physics, Millennia Pro 2i) with the repetition rate of 250 Hz, and the scanning step length of 200 nm, and the focused laser spot size of around 350 nm. The laser pulse width and the laser power can be adjusted from 30 ns to 1 ms and from 1 to 15 mW, respectively. The writing conditions of the laser beam are controlled by an acousto-optic modulator. The whole operation process and all parameters are controlled by a computer, including the image file with all writing parameters and writing path data. Figure 2.13 shows an example of a 10 bit bitmap image written by the LDW on a Sn thin film. It is easy to see differences in the grayscale pattern images written on refined (Fig. 2.13a) and rough (Fig. 2.13b) surfaces of Sn films. Obviously, image quality, gray levels, and resolution based on refined surface in Fig. 2.13a are much better than that based on the rough surface in Fig. 2.13b. More importantly, the resolution of the mask in Fig. 2.13a reaches 200 nm, beyond the diffraction limit of our LDW system [30].

The indium and its oxide In_2O_3 can be also used to fabricate grayscale mask [27], and the process is similar as that of Sn films described above. As MTMO material, In/In_2O_3 system has been also proved to be a good choice to make grayscale mask (Fig. 2.14).

Figure 2.15a, b demonstrates that various 3D microstructures are created by using the MTMO grayscale masks. Focusing effect of the microlens array is shown

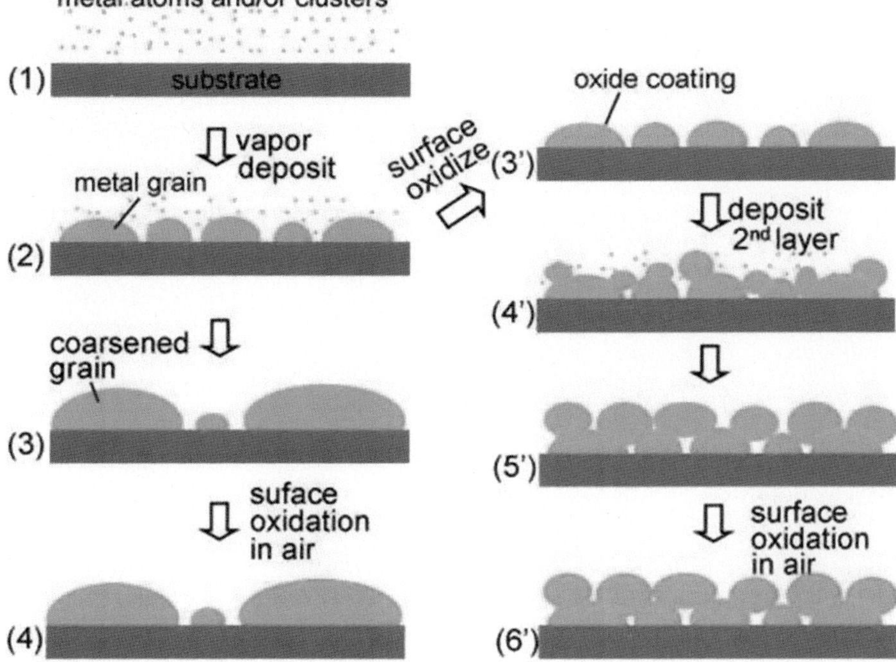

Fig. 2.11 Schematic illustration of the metallic films with the same nominal thickness prepared by routine method (route 1-2-3-4) and the refinement (route 1-2-3'-4'-5'-6'). The latter is composed of two layers of refined grains as a result of the interdiction of homoepitaxy caused by the oxide coating on the metal surface (Reproduced from Ref. [27] by permission of Optical Society of America)

in Fig. 2.15c, d. Besides, the MTMO systems can also be used as amplitude gratings or other microoptics with variable amplitude (Fig. 2.15d).

To check the practical usability, a 3×3 SiO_2 microlens array was fabricated by our Sn-based MTMO grayscale mask with 30 μm diameter each lens. The results of focusing and imaging effect of this DOE are shown in Fig. 2.16.

It can be seen from Fig. 2.16 that the DOEs have a good performance in surface profiles. Likewise, the focusing and imaging effect of the single Fresnel lens is workable as shown in Fig. 2.16d–f, indicating that the DOEs made by MTMO grayscale masks are practicable [66].

2.5.1.2   Mechanism of the Grayscale Features of the MTMO Systems

The mechanism of the grayscale feature of the MTMO grayscale masks can be ascribed to the coexistence of the opaque metal and the corresponding transparent metallic oxide(s). But the details for the Sn/tin oxides and In/indium oxide systems are different. Experiments have indicated that [25, 27], under a short pulse exposure

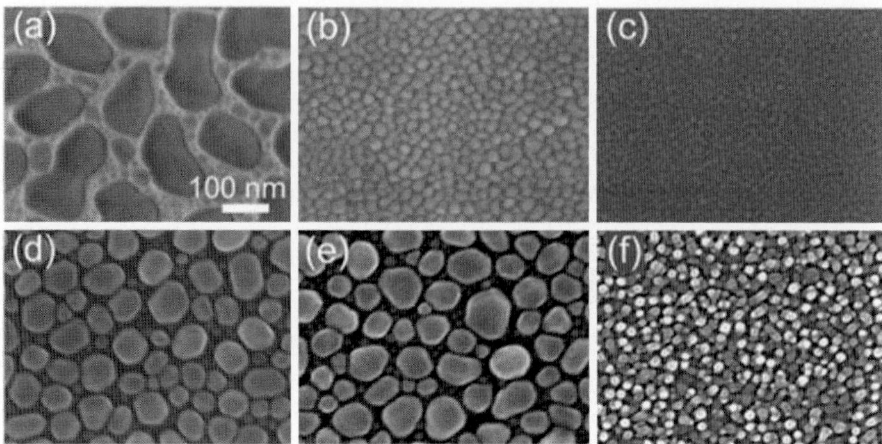

Fig. 2.12 SEM images of different Sn and In films with the same thickness of 20 nm. (**a–c**) 20 nm Sn films deposited in one time (20 nm), twice (10 nm × 2), and four times (5 nm × 4), respectively. (**d–f**) 20 nm In film prepared by one time deposition, four times of deposition (without surface oxidation in the deposition intervals), and four times of deposition (5 nm × 4, with surface oxidation for each layer), respectively (Reproduced from Ref. [27] by permission of Optical Society of America)

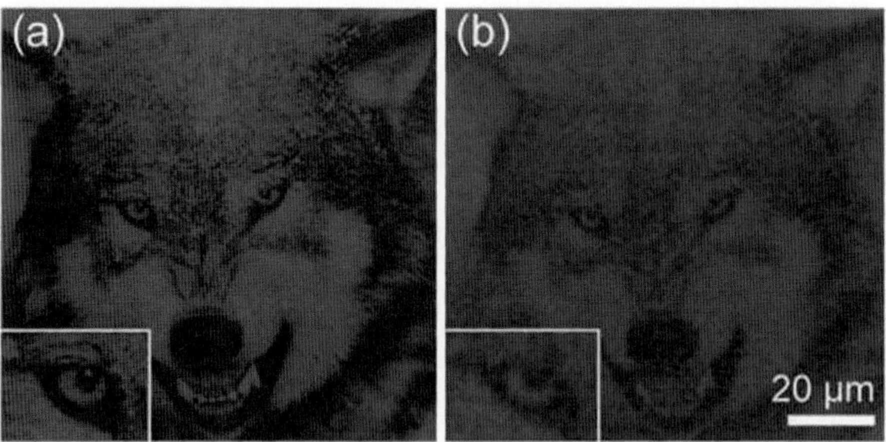

Fig. 2.13 (**a, b**) Grayscale patterns written on the 20 nm Sn films with refined and rough surface, respectively. The latter does not show fine structures of the wolf. The *insets* are magnified images of the wolf's eye, clearly showing that the refined film can achieve much better gray levels and finer features

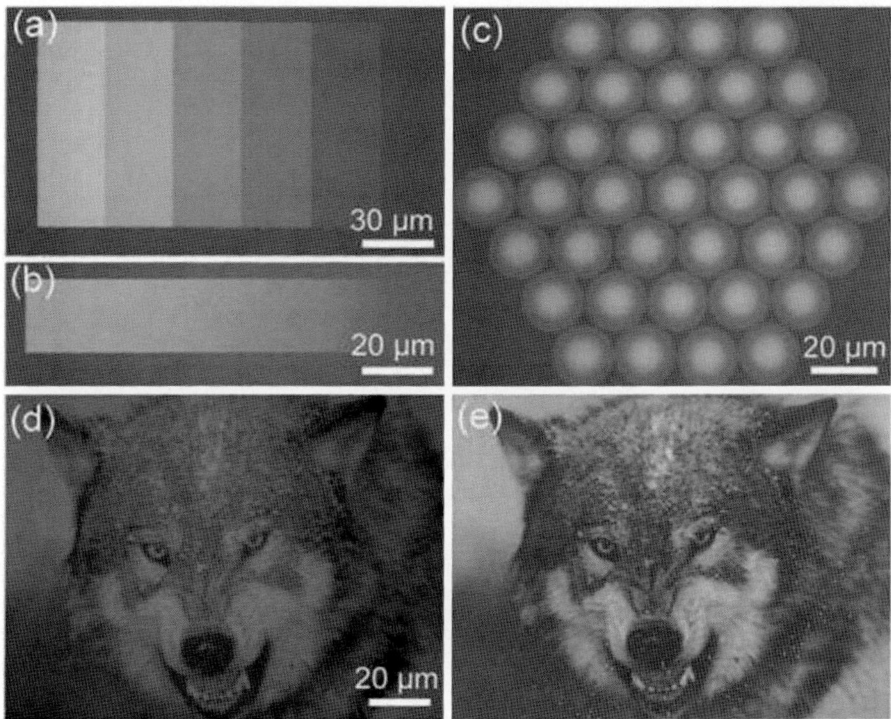

Fig. 2.14 (**a–e**) Grayscale patterns written on refined In films. (**d**) and (**e**) Complex grayscale patterns written under 2.6–10 mW 1.0 μs pulse width and 1.5–8 mW 1.0 ms pulse exposure, respectively. Image (**e**) is color-inverted (Reproduced from Ref. [27] by permission of Optical Society of America)

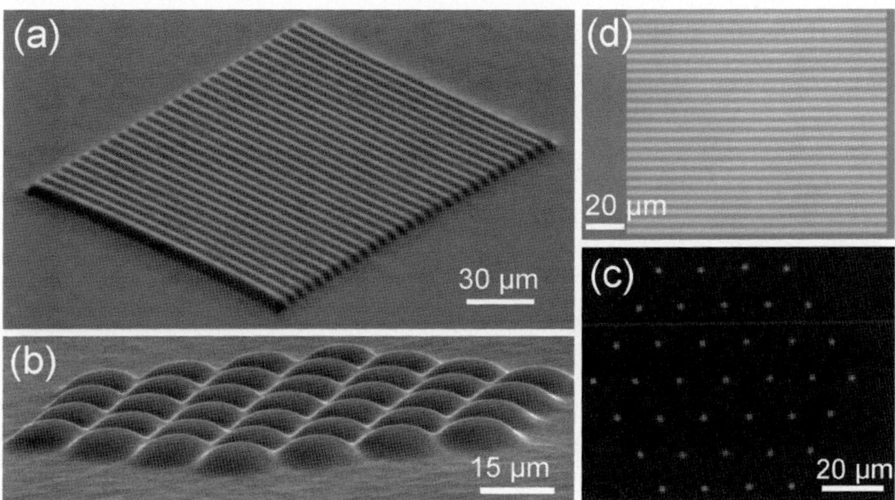

Fig. 2.15 SEM images of (**a**) surface relief phase grating with a period of 5 μm and (**b**) microlens array fabricated in an SU-8 photoresist by using MTMO grayscale masks. (**c**) Focusing effect of the microlens array. (**d**) Optical image of a MTMO grayscale pattern, which can be used as an amplitude grating or a mask for fabricating surface relief grating (Reproduced from Ref. [27] by permission of Optical Society of America)

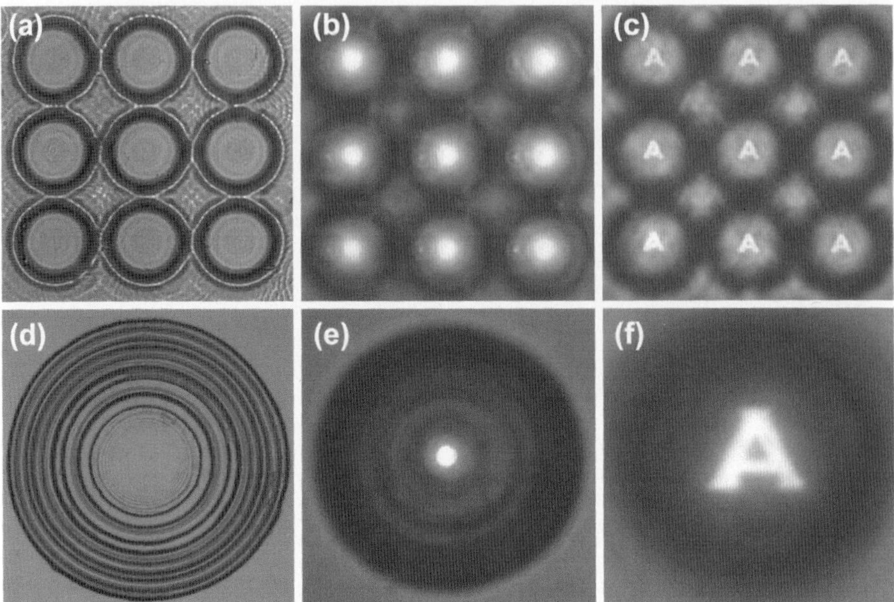

Fig. 2.16 Optical microscopy images of SiO_2 microlens array and single Fresnel lens. (**a**) Morphology of the DOE, (**b**) focusing image of the lens array, and (**c**) letter "A" imaged through the microlens array. (**d–f**) are corresponding profile, imaging, and focusing of single Fresnel lens, respectively (Reproduced from Ref. [67] by permission of Optical Society of America)

($<$1,000 ns) of laser, the outside surfaces of Sn grains are oxidized to form the shell/core structures of a-SnO_x/Sn (Fig. 2.17). Obviously, the gray level, i.e., the transmittance, of the shell/core structure is determined by the thickness of the transparent a-SnO_x shell. Therefore, longer pulse width produces (oxidation) thicker a-SnO_x layers (transparent), resulting in more transparency of the film. In this case, the Sn films written by LDW can be regarded as the layered oxidation, as shown in Fig. 2.18a [27]. But this model is only suitable for relatively thin Sn films, typically no more than 20 nm. The minimum optical density (OD) of Sn film under short pulse exposure is not as small as that exposed under long pulse, e.g., OD changes from 0.73 to 0.20 for a 20 nm Sn film under 200 ns pulse, while the OD range is from 0.73 to 0.08 under a pulse width of 1.0 ms [30]. When Sn film is exposed under a much longer pulse ($>$10,000 ns), the Sn grains are decomposed to subgrains with a scale of ~5 nm. These subgrains can be in different phase structures, i.e., Sn, t-SnO, o-SnO_2, and t-SnO_2, with different optical properties. As we have known, t-SnO_2 is the most transparent oxide among the tin oxide group. The gray levels of the multiphase system, the Sn-SnO-SnO_2 system, are determined by the mixture ratio of different phases. When the Sn completely transformed to t-SnO_2, OD of the film reaches its minimum. Schematic illustration of this model is shown in Fig. 2.18b.

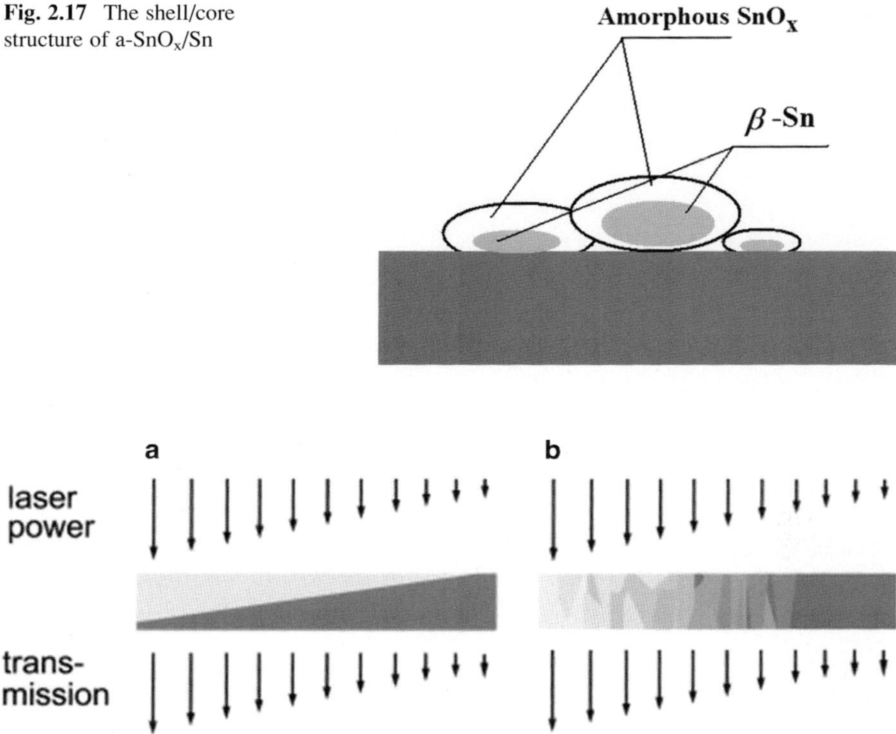

Fig. 2.17 The shell/core structure of a-SnO$_x$/Sn

Fig. 2.18 Schematic illustration of (**a**) layered oxidation model and (**b**) grain model for explaining the grayscale features of MTMO systems (Reproduced from Ref. [27] by permission of Optical Society of America)

Indium and its oxide are different from Sn/tin oxide. Indium does not transform to amorphous oxide under short pulse exposure, In_2O_3 is its sole stable oxide. Therefore, the OD values are simply determined by the thickness of the In_2O_3. The grayscale feature of In/In_2O_3 can be explained by the model shown in Fig. 2.18a, i.e., layered oxidation model. Simulation of the OD of the In_2O_3/In bilayer (In film is 20 nm thick before oxidation) at wavelengths of 365 and 532 nm is shown in Fig. 2.19, from which we know the film's transmission achieves its maximum before being completely oxidized. This trend is especially remarkable at a short wavelength, at which In_2O_3 is less transparent. The explanation is simple: the OD of the In_2O_3 film is oscillated with the thickness (in fact, OD of In_2O_3 is decreased with the thickness in our experimental range), while that of the In is almost linear to the film thickness (see the inset in Fig. 2.19). In is even more transparent than In_2O_3 when it is thinner than 4 nm, so that In_2O_3 contributes the majority of the film's OD. According to this result, the film does not need to be fully oxidized in laser writing.

In and Sn, as the low melting point metals, have very transparent and stable oxides, that's why we choose them as the media to fabricate grayscale masks.

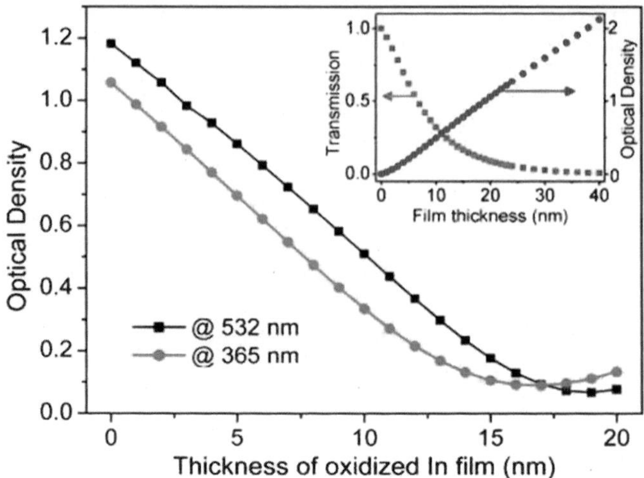

Fig. 2.19 Simulation of the In$_2$O$_3$/In bilayer's OD at wavelengths of 365 and 532 nm. In film is 20 nm thick and has been oxidized previously. It is noted that minimum OD occurs before the film is completely oxidized. The *inset* shows the OD or T of the In film versus film thickness (Reproduced from Ref. [27] by permission of Optical Society of America)

Experimental results have shown both grayscale masks are stable (gray levels are not changed) after long time exposure in air ambience or UV light, demonstrating that the MTMO masks are competent for ordinary gray-tone lithography in laboratory. Additionally, a thin SiO$_2$ protective layer deposited on the mask can significantly enhance its stability in more rigorous conditions. Regardless of In or Sn film, the gray level of the exposed area is a function of laser power, so that the LDW technique is adequate for fabrication of grayscale masks using In or Sn films. Performances of the two films, however, are not exactly the same. For example, In has a larger OD than Sn with the same thickness and the In–In$_2$O$_3$ system can be obtained by using short pulse exposure, but Sn/SnO$_x$ system is much richer in grayscale levels because of more phases of tin oxides. It should be noted that MTMO is suitable for many metals besides In and Sn.

2.5.2   Ordered TiO$_2$ Nanoribbons and Devices

Due to its high refractive index, self-cleaning property, high dielectric constant, biocompatibility, and chemical stability, TiO$_2$ has been used in etching masks in the fabrication processes of indium-doped tin oxides (ITO) and silicon MEMS devices [67] as well as blazed gratings for realizing multilevel DOE fabrications [68]. In addition, TiO$_2$ film can be as a cleaning layer coated on the surface of solid immersion lens for effectively removing organic contaminants based on the self-cleaning effect [69]. Recently, TiO$_2$ micro-/nanostructures have also attracted more and more attentions in dye-sensitized solar cells [70], photocatalysts [71], electrochemical sensors [72], biomedical implants [73], and so on.

One-dimensional (1D) TiO_2 nanostructures such as nanotube, nanowire, and nanobelt are important members of the TiO_2 nanostructure family. Usually natural state of 1D TiO_2 exists as single crystal, having a higher surface-to-volume ratio, a lower charge carrier recombination rate compared with the Ti nanoparticle [74, 75]. Different from nanorod or nanotube, free TiO2 nanoribbon is easy to twist and becomes disordered because of the large aspect ratio and flexibility. Obviously, such disordered nanoribbons are unable to satisfy the application requirements. Some progress has been made in fabricating ordered TiO2 nanoribbons. For example, Park and coworkers fabricated ordered nanoribbons via combing polymer templates with physical vapor deposition [76]. By employing sol–gel deposition, defined polymer templates, or complex surface relief gratings, highly ordered TiO_2 nanostructures and nanoscale array with different shapes have been successfully fabricated [68, 77]. However, these techniques are usually complex and costly because of dependency on templates and masks. So far, it is not an easy issue yet for fabrication of the refined and arbitrarily shaped nanostructures.

Here we propose a simple, maskless, path-directed, and low-cost method to make highly ordered TiO_2 nanoribbons [78]. Combining LDW with wet etching technique enables arbitrary-shaped and ordered TiO_2 nanoribbon structures to be made. The aspect ratio and length of the nanoribbons could be controlled easily. The lengths of the TiO_2 nanoribbons obtained by this method can reach several centimeters, while the widths of the ribbons are adjustable from 150 nm to several micrometers. As shown in Fig. 2.20, the fabrication process needs only three steps: (1) a Ti film is deposited on a glass substrate (see Fig. 2.20a); (2) patterns are written on the Ti film by using a laser direct writer, as shown in Fig. 2.20b and d; and (3) TiO2 nanoribbons are obtained by using wet etching technique (Fig. 2.20c and e).

The principle of this fabrication method could be described as below. When the laser beam hits the surface of the Ti film during the laser direct writing process, the energy of the laser is absorbed by the film and converts to heat, causing metal oxidation in the areas exposed to laser. As a result, stable TiO2 phase forms [8, 12, 30], and both of the volume and height of the oxidized regions are changed [8]. After nano-sized TiO_2 strips (or other patterns) form on the surface of the film, a wet etching process is carried out at room temperature. Dilute fluorhydric acid (HF, the volume concentration: 4.5 %) is used as the etching solution. TiO_2 has a lower solubility in HF than that of metallic Ti. Hence, in the HF solution, the TiO_2 nanoribbons (laser-written part) remain unchanged, while the Ti film (unwritten part) is removed due to the different etching ratio between TiO_2 and metallic Ti. After the wet etching process, the highly ordered TiO_2 ribbons are left on the surface of the substrate, as shown in Fig. 2.21. It should be point out specially that by using our laser direct writer (NanoLDW-I, a 532 nm laser and 0.90 NA objective lens), ribbons with 150 nm far beyond the diffraction limit of the optical system (about 300 nm) could be obtained. By controlling the writing power of the laser beam, we can fabricate various micro-/nano features. In fact, narrower nanoribbons could be fabricated by optimizing the matching laser power with thin film process.

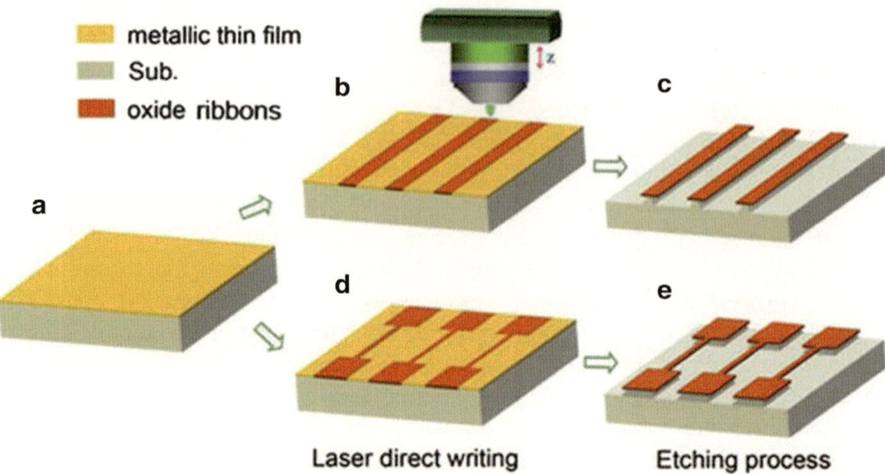

Fig. 2.20 Schematic diagram illustrates the process of path-directed fabrication of suspended TiO_2 nanoribbons (**a**) Ti film on the glass substrate. (**b**) and (**d**) patterns fabricated by laser direct writing. The structures obtained after the etching process: (**c**) arrays of partially suspended nanoribbons, (**e**) arrays of completely suspended nanoribbon beams (*thin* part) (Reproduced from Ref. [79] by permission of The Royal Society of Chemistry)

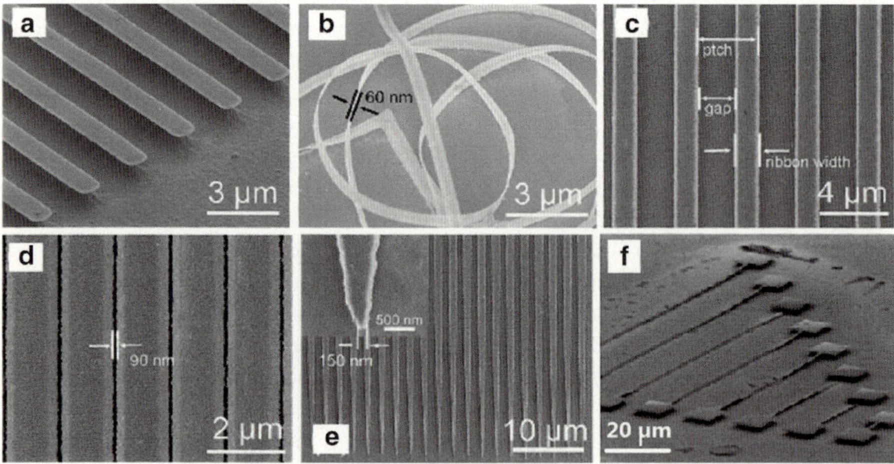

Fig. 2.21 Typical SEM images of arrays of suspended structures fabricated by LDW and wet etching. (**a**) Partially suspended nanoribbons etched in dilute HF solution for 3 min. (**b**) Curling nanoribbons after etching for about 5 min. (**c**) and (**d**) Nanoribbons with gap widths of 2 mm and 90 nm, respectively. (**e**) Arrays of nanoribbons with a gradually changed width, the end size of the tip reach about 150 nm, corresponding to 20 J/mm^2 laser energy density. (**f**) The bridge-like structures with different suspension lengths (Reproduced from Ref. [79] by permission of The Royal Society of Chemistry)

Considering glass is a good substrate material and has higher etching rate (about 230 nm min^{-1}) in dilute HF solution, if the etching time in HF is long enough, the glass substrate starts to be etched after the metallic Ti surface is removed. In other words, the TiO_2 nanoribbons can be partially (suspended) or completely peeled away from the glass substrate by extending the etching time so as to obtain the freely or half freely high-ordered nanoribbons. Adopting this processing method, various types of TiO_2 nanoribbon structures, such as suspended belts, bridge-like structure, and curling nanoribbons, can be fabricated easily on glass substrate. Of course, partially suspended and bridge-like structure nanoribbon beam can be directly fabricated on an underlay such as Si and SiO_2, by fitting partial nanoribbon still on the substrate. Especially, the nanoribbons supported partly on the substrate remind us that it might be a more reasonable method for building devices based on nanoribbons. Compared with the totally free ribbons, such nanoribbons have higher reliability and orderliness as well as more anti-bending and anti-torsion, thus avoiding degradation of performance.

It should be noted that (a) after wet etching, the width of the ribbons is a little different from the width written by laser. We think the width difference comes from the oxidation difference between the central and edge parts because the two parts absorb different amounts of energy according to the Gaussian distribution of laser beam. The insufficient oxidation makes that the Ti material at the edge part is removed by the HF solution. (b) The cross section of the ribbons made by this method is not a rectangle but thicker center and thinner edges because of Gaussian distribution of laser beam.

Various nanoribbon arrays, like squares, triangles, rings, and grids, can also be fabricated using the super-resolution LDW combined with wet etching method, either substrate supported or suspended on substrate (Fig. 2.22). All the structures are constructed by 60 nm thick, 1.6 μm wide TiO_2 ribbons. These seamless-connected structures indicate that LDW combined with wet etching method cannot only be used to obtain high-quality nanoribbons but also can be used to fabricate complex geometric shapes constructed by ribbons. The fabrication of complex geometric shapes proves the advantages of this fabrication method, such as accurately controlled size and arbitrarily shapes, which are helpful for creating various practical devices with complex structures.

In order to understand the nature of nanoribbon, the crystal structure of the obtained nanoribbons was analyzed by using TEM. Figure 2.23 shows the enlarged TEM image of the ribbon structure and the selected area electron diffraction (SAED) patterns of the nanoribbon. It should be noted that the samples used for TEM analysis are completely suspended and transferred from the glass substrate to TEM support (as shown in Fig. 2.23a, b), indicating that the nanoribbons made by this method have good transferability. Figure 2.23c further shows an enlarged image of one side of the square ribbon structure. Brightness differences of the SAES patterns corresponding to upper half and lower half in Fig. 2.23d indicate that the crystalline quality of the nanoribbon in the central part is obviously better than that at the edge, and the SAED results show us that the nanoribbon structure induced by LDW is polycrystalline rutile TiO_2.

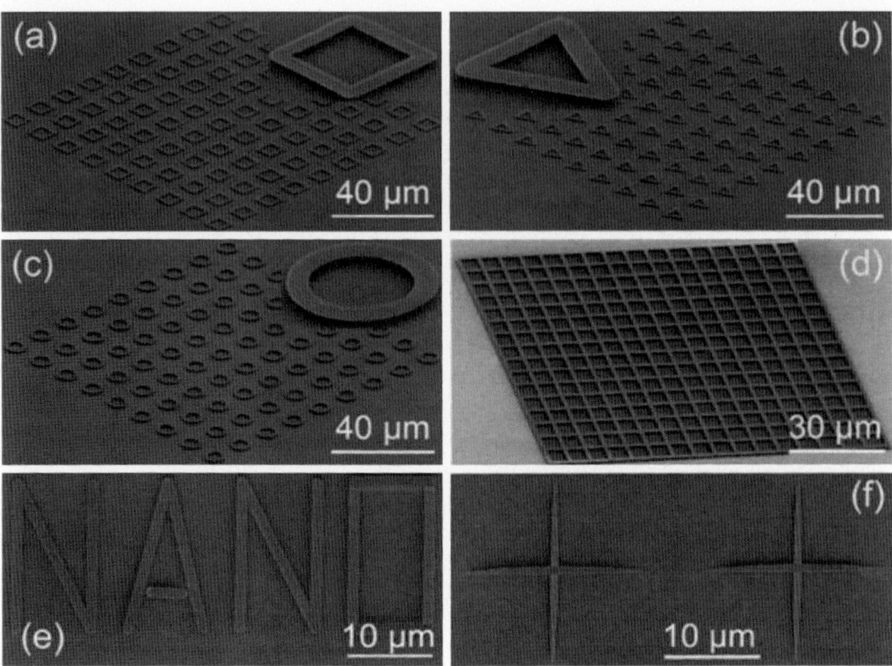

Fig. 2.22 SEM images of various complex-shaped structures made up of nanoribbons: (**a**) square array; (**b**) triangle array; (**c**) circular array; (**d**) micronetworks; (**e**) letters of "NANO"; (**f**) cross-like structures with a gradually changed width (Reproduced from Ref. [79] by permission of The Royal Society of Chemistry)

To build reliable MEMS and nanodevices, the mechanical properties of the building materials should be investigated. Young's modulus (E) of the nanoribbon was tested directly using atomic force microscopy (AFM) (see Fig. 2.24). The test sample is a bridge-like suspended nanoribbon beam as shown in Fig. 2.24b. Figure 2.24c shows the dynamic evolution of the load force (F_{load}) versus time, the stress of the probe suddenly vanished when F_{load} reached the critical value (fracture stress = 4.3 μN) indicates the TiO_2 suspended nanoribbon beam broke under this force (i.e., it means the fracture strength of this nanoribbon is about 250 GPa). The inset in Fig. 2.24c is the AFM image, which further confirms that the fracture of the beam has happened. Figure 2.24d plots out the beam deflection (d_{beam}) along the nanoribbon beam (X direction, as shown in Fig. 2.24b) at three fixed load forces. The plots show the maximum F_{load} is 2.48 μN to ensure our study on Young's modulus E in the range of elastic deformation. Based on the formula [79] of $d_{beam} = 4F_{load}L^3(x/L)^3(1-x/L)^3/ET^3W$, where L, W, T, and E are the length, width, thickness, and Young's modulus of the nano-beam, respectively, we calculated the Young's modulus $E = 120$ GPa, which is consistent with the reported results [80]. The experimental results demonstrate that the mechanical property of the TiO_2 nanoribbon beam is acceptable for use in device fabrication.

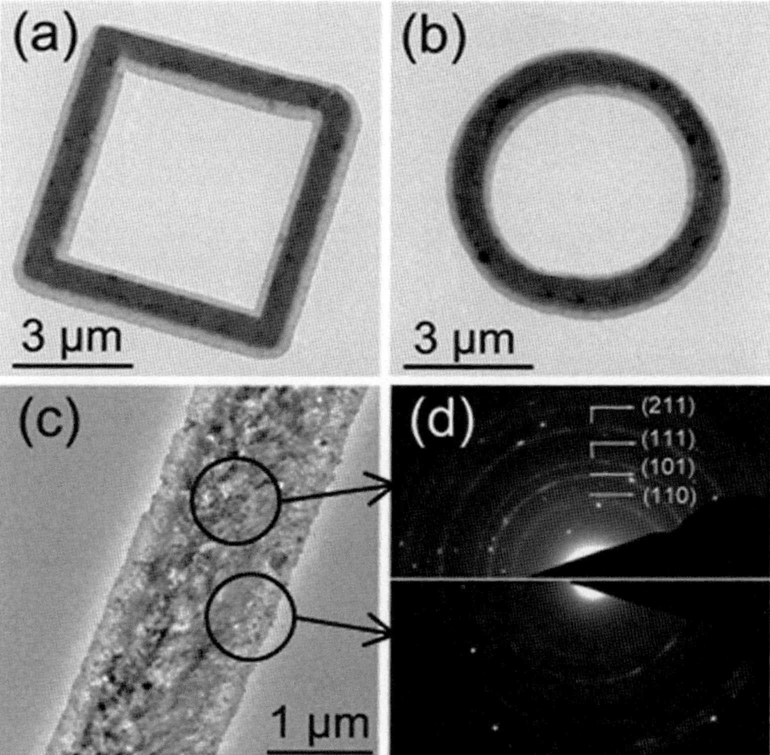

Fig. 2.23 TEM morphologies of nanoribbons transferred to the TEM grid. (a) and (b) Square and circle nanoribbon structure on the TEM support. (c) An enlarged image of one side of the square ribbon structure.(d) Selected area electron diffraction (SAED) patterns, showing that the nanoribbons are composed of rutile TiO_2 (Reproduced from Ref. [79] by permission of The Royal Society of Chemistry)

In the laser direct writing process, the gap width between the nanoribbons can be down to tens of nanometers (Fig. 2.21d), which is much smaller than the 532 nm wavelength of the laser, and this may provide us with an indirect nanofabrication technique. In fact, not only the nanoribbon width can be controlled precisely, but also the thickness of the nanoribbon can be controlled simply by adjusting the thickness of Ti film. That means aspect ratio of the ribbon can be controlled in both length to width and thickness to width. The method also provides us one realizable route of creating nanoribbon based array-devices without transfer process.

2.5.3 Nanorelief Fabrication by LDW

With the rapid development of micro-/nanofabrication technologies, the micro-/ nanoreliefs show their importance due to their potential applications in micro artworks, image storage, anti-counterfeiting, micro-/nano diffractive optical elements

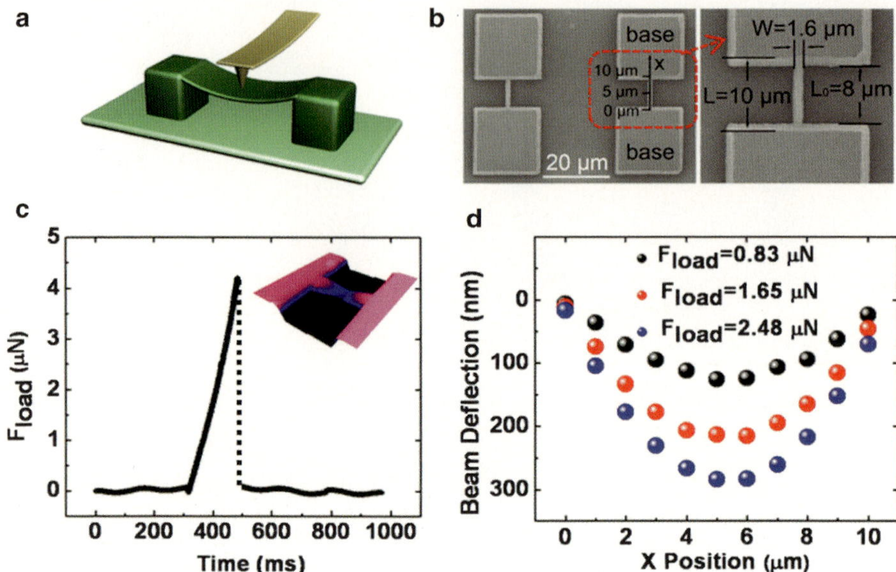

Fig. 2.24 (**a**) Schematic construction of testing a suspended TiO_2 nanoribbon beam by an AFM. (**b**) SEM images of the bridge-like TiO_2 nanoribbon beams. (**c**) F_{load} varies versus time. (**d**) Deflection along X direction of the suspended nanoribbon beam at three fixed load forces (Reproduced from Ref. [79] by permission of The Royal Society of Chemistry)

(DOEs), and micro-/nanoelectromechanical system devices. Here micro-/nanorelief means a quasi three-dimensional micro/nanostructure. At present, most microlens array, DOEs, and MEMS are reliefs with the critical heights of micrometers and nanometers.

There are many techniques that can fabricate micro-/nanostructures, such as e-beam lithography (EBL), nanoimprint lithography (NIL), and focused ion beam (FIB) milling [55, 81–83]. But these mainstream techniques are either too costly or too complicated for fabricating high-resolution relief structures. Generally, the microreliefs are fabricated by using grayscale photolithography [61, 84, 85], but the resolution hardly reaches sub-500 nm. Two-photon absorption (TPA) is an attractive technique to break the limit. TPA has been used in making complicated 3D structures, e.g., micro/nanobulls, photonic crystals, and DOEs, and the resolution can reach sub-200 nm [86–88]. However, only transparent polymers can be used as working media, limiting its application.

The advantages of LDW are simple and maskless, and not requiring rigorous working conditions. LDW can be applied to many types of materials, including metals, organic and inorganic materials, and semiconductor materials, whether transparent or not. The resolution of the LDW decides it is suitable for fabricating sub-micrometer structures on metallic films [8].

Using LDW, complex and high-resolution nanoreliefs can be directly fabricated in metallic Sn films [25]. In this experiment, a 12 nm thick Sn thin film was sputtered on

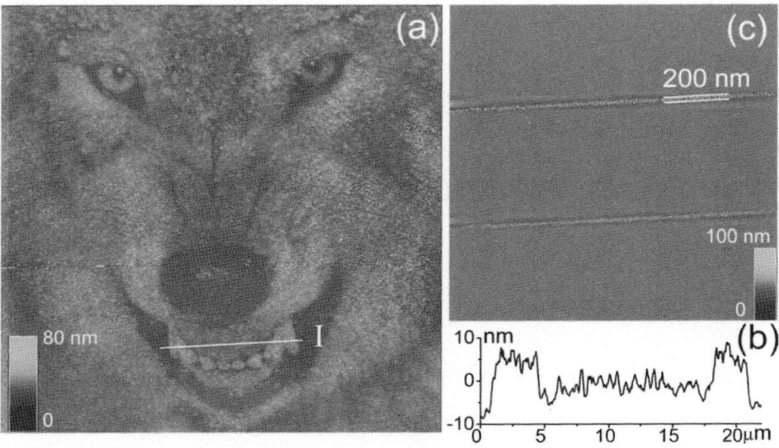

Fig. 2.25 (a) AFM topographic image of the nanorelief (70 × 70 µm) and (b) section analysis along the line "I". (c) AFM image (20 × 20 µm) of raster scanned lines with resolution of around 200 nm (Reproduced from Ref. [25] by permission of Optical Society of America)

a glass substrate, and the roughness of the film surface was 2 nm; a bitmap file of a wolf head was drawn on the Sn (similar to Fig. 2.13) film by a laser direct writer. The grayscale of the picture was realized by controlling the laser power, where a high power made a high transmittance and low reflectance, and vice versa. A micrometer-sized bas-relief of the wolf head was fabricated as shown in Fig. 2.25a.

AFM analysis (Fig. 2.25b) shows that the image written by the LDW is not limited in a 2D plane and the third dimension is varied in height. The experimental data indicate that the varied height Δh and the optical density of the Sn film are almost linear to the writing laser power. This relationship provides great advantages in fabrication of fine optical elements and complicated micro-/nanoreliefs. LDW cannot only be used to create perfect microscale bitmap image as expected but also used to control the height in the Z direction. The resolution of the fine structures is ~200 nm, much smaller than the diffraction limit of 532 nm laser according to AFM topographic image shown in Fig. 2.25a. From the section analysis along marked "I" line, we can see the height varies clearly (Fig. 2.25b). The different heights correspond to the different gray levels of the pixels in this image, in other words, following the changes of the writing laser power. From the AFM analysis, the maximum change of the height is around 12 nm, equal to the deposited thickness d of the Sn film, i.e., $\Delta h/d \sim 100$ %. The results demonstrate that LDW technique can continuously control the transmittance, reflectance, and height of the reliefs fabricated on Sn metallic films. LDW is a suitable tool for fabricating metallic nanoreliefs with super-resolution fine structures and a continuously variable height.

For the reason of varying height, one of the explanations could be the oxidation of the material [8] when exposed to the laser beam. Oxygen atoms enter into the film and cause the thickness increase. Theoretically, if the Sn is oxidized completely from white tin (β-Sn) to SnO_2, the volume increases ~34.5 %. If the height change of the Sn

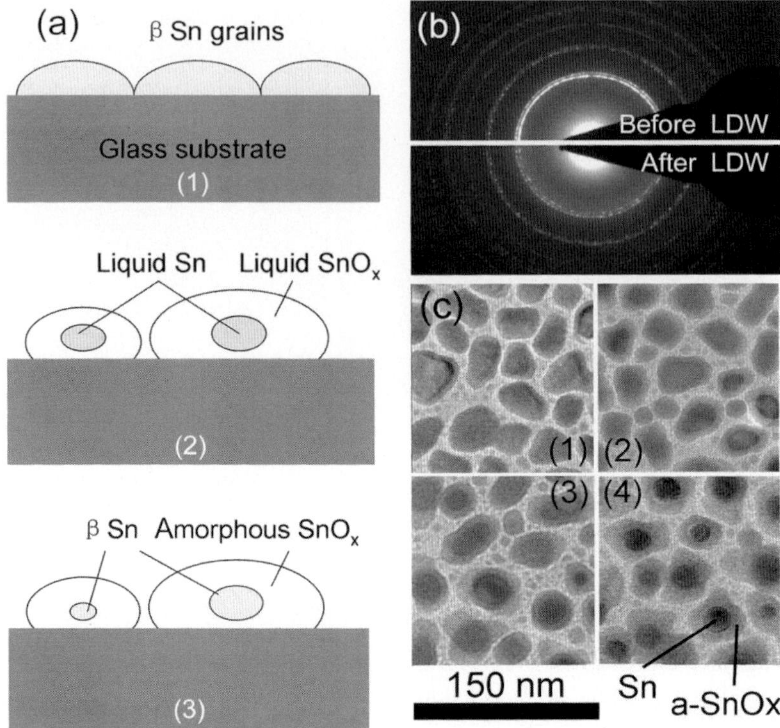

Fig. 2.26 Mechanism of height change in the nanoreliefs. (**a**) The model: (*1*) as-deposited Sn film with flat grains, (*2*) liquid Sn/SnO$_x$ grains induced by LDW, and (*3*) after cooling down. (**b**) SAED results verify the formed Sn/a-SnO$_x$ core/shell structure. (**c**) TEM images show the film morphologies of Sn/a-SnO$_x$ core/shell structures: (*1–4*) corresponding to different laser powers from low to high (Reproduced from Ref. [25] by permission of Optical Society of America)

film is only from the volume change caused by oxidation, then the maximum value is only about 4 nm. Therefore the oxidation could not be the only reason for the height change (~12 nm, Δh/d ~ 100 %) after the laser irradiation.

The β-Sn is found to be transformed to amorphous (a-)SnO$_x$ instead of crystal oxides such as SnO or SnO$_2$ after laser exposure.

$$Sn + (x/2)O_2 \xrightarrow{\text{pulsed laser}} SnO_x$$

The as-deposited Sn film prepared for LDW is composed of fine grains which are coated with thin a-SnOx shells (thickness ~2 nm) with flat morphology (Fig. 2.26a-1) [24, 89]. Due to the low melting point, ~232 °C, the β-Sn grains easily melt under laser exposure. Therefore when irradiated by laser beam with proper power, the Sn melts to liquid Sn/SnO$_x$ structures [22]. Because the surface tension exists and the melting liquid does not wet the underlying glass substrate, the Sn/SnO$_x$ structures form ball-like shapes (Fig. 2.26a-2). As a result, the spherical morphology retains after cooling down (Fig. 2.26a-3) and increases the grain height efficiently.

With a higher laser power, several fine grains which are originally adjacent melt and aggregate together to form larger liquid Sn/SnO_x grains. The aggregation will cause roughness, which brings additional height change. Thus it can be seen that laser power can be adjusted to control the temperature of the exposure spot, so as to control the grain height. The existence of β-Sn/a-SnO$_x$ structure has already been directly confirmed by selected area electron diffraction (SAED) patterns and transmission electron microscopy (TEM) images (Fig. 2.26b, c).

2.6 Outlook

LDW is a simple, direct, and controllable fabrication technique and has been widely used in field of microfabrication. With the rapid development of nanoscience and nanotechnology, LDW technique also faces the challenge of the smaller and smaller structure size. The development of the super-resolution LDW makes it possible to fabricate structures from microscale to nanoscale, therefore expands the applications of LDW technique.

By means of nonlinear interaction between laser beam and the acceptor material, we propose a new concept of LDW technique with super-resolution fabrication capacity based on a local miniature effect to laser beam size. This super-resolution LDW provides a new route to increase fabrication resolution, different from the traditional methods of increasing numeral aperture of objective lens and/or shortening the wavelength of incident laser beam.

More importantly, the super-resolution LDW broke the limit of acceptor materials, and many different materials can be used now, such as inorganic material, metallic material, and semiconducting material. And the super-resolution LDW technique has been applied to fabricate grayscale mask, nanorelief, MEMS, microoptical devices, and so on.

In the future, besides its applications in nanostructure and nanodevice fabrications, the super-resolution LDW technique is also expected to be applied in microelectronics and semiconductor industries.

References

1. Gale MT, Karl K (1983) The fabrication of fine lens arrays by laser beam writing. Proc SPIE 398:347–353
2. Koronkevich VP, Kiriyanov VP, Kokoulin FI (1984) Fabrication of kinoform optical elements. Optik 67(3):257–266
3. Rensch C, Nell S, Schickfus M (1989) Laser scanner for direct writing lithography. Appl Opt 28(17):3754–3758
4. Haruna M, Wakahayashi K, Nishihara H (1990) Laser beam lithographed micro-Fresnel lenses. Appl Opt 29:5120–5126
5. SVG Optronics Co. www.SVGoptronics.com

6. Heidelberg Instruments. http://www.himt.de/en/products/vpg1600.php
7. Campbell SA (2008) Fabrication engineering at the micro- and nanoscale, 3rd edn. Oxford University Press, Oxford
8. Gorbunov AA, Eichler H, Pompe W, Huey B (1996) Lateral self limitation in the laser-induced oxidation of ultrathin metal films. Appl Phys Lett 69(19):2816–2828
9. Kurihara K, Nakano T, Ujiie M, Tominaga J (2008) High-speed fabrication of large-area nanostructured optical devices. Microelectron Eng 85:1197–1201
10. Kurihara K, Yamakawa Y, Nakano T, Tominaga J (2006) High-speed optical nanofabrication by platinum oxide nano-explosion. J Opt A-Pure Appl Opt 8:139–142
11. Kuwahara M, Mihalcea C, Atoda N, Tominaga J, Fuji H, Kikukawa T (2002) Thermal lithography for 0.1 μm pattern fabrication. Microelectron Eng 61–62:415–419
12. Wang YS, Guo CF, Cao SH, Miao JJ, Ren TL, Liu Q (2010) Controllable fabrication of super-resolution nanocrater arrays by laser direct writing. J Nanosci Nanotechnol 10:7134–7137
13. Tosto S, Di Bartolomeo A, Di Lazzaro P (1996) Surface ablation by excimer laser irradiation of Ti and Ti6Al4V alloy. Appl Phys A 63:385–389
14. Granqvist CG, Hultaker A (2002) Transparent and conducting ITO films: new developments and applications. Thin Solid Films 411(1):1–5
15. Pan ZW, Dai ZR, Wang ZL (2001) Nanobelts of semiconducting oxides. Science 291 (5510):1947–1949
16. Comini E (2006) Metal oxide nano-crystals for gas sensing. Anal Chim Acta 568(1–2):28–40
17. Wang ZL, Pan Z (2002) Junctions and networks of SnO nanoribbons. Adv Mater 14 (15):1029–1032
18. Lee AF, Lambert RM (1998) Oxidation of Sn overlayers and the structure and stability of Sn oxide films on Pd (111). Phys Rev B 58(7):4156–4165
19. Pan XQ, Fu L (2001) Oxidation and phase transitions of epitaxial tin oxide thin films on (10–12) sapphire. J Appl Phys 89(11):6048–6050
20. Arbiol J, Comini E, Faglia G, Sberveglieri G, Morante JR (2008) Orthorhombic Pbcn SnO_2 nanowires for gas sensing applications. J Cryst Growth 310(1):253–260
21. Chen YX, Campbell LJ, Zhou WL (2004) Self-catalytic branch growth of SnO_2 nanowire junctions. J Cryst Growth 270(3–4):505–510
22. Kolmakov A, Zhang Y, Moskovits M (2003) Topotactic thermal oxidation of Sn nanowires: intermediate suboxides and core-shell metastable structures. Nano Lett 3(8):1125–1129
23. Batzill M, Diebold U (2005) The surface and materials science of tin oxide. Prog Surf Sci 79 (2–4):47–54
24. Domashevskaya EP, Chuvenkova OA, Kashkarov VM, Kushev SB, Ryabtsev SV, Turishchev SY, Yurakov YA (2006) TEM and XANES investigations and optical properties of SnO nanolayers. Surf Interface Anal 38(4):514–517
25. Guo CF, Zhang Z, Cao S, Liu Q (2009) Laser direct writing of nanoreliefs in Sn nanofilms. Opt Lett 34(18):2820–2822
26. Zhao J, Huo LH, Gao S, Zhao H, Zhao JG (2006) Alcohols and acetone sensing properties of SnO2 thin films deposited by dip-coating. Sens Actuators B Chem 115(1):460–464
27. Wang CY, Cimalla V, Romanus H, Kups T, Ecke G, Stauden T, Ali M, Lebedev V, Pezoldt J, Ambacher O (2006) Phase selective growth and properties of rhombohedral and cubic indium oxide. Appl Phys Lett 89:011904–011906
28. Okamoto H (2007) In-O (indium-oxygen). J Phase Equilib Diff 28(6):591–592
29. Guo CF, Cao S, Jiang P, Fang Y, Zhang J, Fan Y, Wang Y, Xu W, Zhao Z, Liu Q (2009) Grayscale photomask fabricated by laser direct writing in metallic nano-films. Opt Express 17 (22):19981–19987
30. Chapman G, Tu Y, Sarunic M, Dhaliwal J (2001) BiIn: a sensitive bimetallic thermal resist. Proc SPIE 4345:557–568
31. Tu Y, Chapman G (2003) Bi/In as patterning and masking layers for alkaline-based Si anisotropic etching. Proc SPIE 4979:87–98

32. Chapman GH, Tu Y, Choo C, Wang J, Poon DK, Chang M (2006) Laser-induced oxidation of metallic thin films as a method for creating grayscale photomasks. Proc SPIE 6153:61534G
33. Cao SH, Guo CF, Wang YS, Miao JJ, Zhang ZW, Liu Q (2008) Transparency conversion mechanism and laser induced fast response of bimetallic Bi/In thin film. Proc SPIE 7269:726910–726913
34. Lyubin V, Arsh A, Klebanov M, Dror R, Sfez B (2008) Nonlinear photoresists for maskless photolithography on the basis of Ag-doped As2S3 glassy films. Appl Phys Lett 92 (1):011118–011120
35. Lyubin V, Klebanov M, Bar I, Rosenwaks S, Eisenberg NP, Manevich M (1997) Novel effects in inorganic As50Se50 photoresists and their application in micro-optics. J Vac Sci Technol B 15(4):823–827
36. Min'ko VI, Shepeliavyi PE, Indutnyy IZ, Litvin OS (2007) Fabrication of silicon grating structures using interference lithography and chalcogenide inorganic photoresist. Semicond Phys Quantum Electron Optoelectron 10(1):40–44
37. Chiu KP, Lai KF, Yen SC, Tsai DP (2009) Surface plasmon polariton coupling between nano recording marks and their effect on optical read-out signal. Opt Rev 16(3):326–331
38. Shintani T, Anzai Y, Minemura H, Miyamoto H, Ushiyama J (2004) Nanosize fabrication using etching of phase-change recording films. Appl Phys Lett 85(4):639–641
39. Choi BJ, Choi S, Eom T, Rha SH, Kim KM, Hwang CS (2010) Phase change memory cell using Ge2Sb2Te5 and softly broken-down TiO2 films for multilevel operation. Appl Phys Lett 97(13):132107–132109
40. Risk WP, Rettner CT, Raoux S (2009) Thermal conductivities and phase transition temperatures of various phase-change materials measured by the 3ω method. Appl Phys Lett 94(10):101906–101908
41. Jain H, Vlcek M (2008) Glasses for lithography. J Non-Cryst Solids 354(12–13):1401–1406
42. Kolobov AV, Fons P, Frenkel AI, Ankudinov AL, Tominaga J, Uruga T (2004) Understanding the phase-change mechanism of rewritable optical media. Nat Mater 3(10):703–708
43. Lee ML, Yong KT, Gan CL, Ting LH, Muhamad Daud SB, Shi LP (2008) Crystalline and thermal stability of Sn-doped Ge2Sb2Te5 phase change material. J Phys D: Appl Phys 41 (21):215402–215405
44. Sun ZM, Zhou J, Ahuja R (2007) Unique melting behavior in phase-change materials for rewritable data storage. Phys Rev Lett 98(5):055505
45. Nakayama K, Takata M, Kasai T, Kitagawa A, Akita J (2007) Pulse number control of electrical resistance for multi-level storage based on phase change. J Phys D: Appl Phys 40 (17):5061–5065
46. Kim C, Kang DM, Lee TY, Kim KHP, Kang YS, Lee J, Nam SW, Kim KB, Khang Y (2009) Direct evidence of phase separation in Ge2Sb2Te5 in phase change memory devices. Appl Phys Lett 94(19):193504–193506
47. Lee J, Choi S, Lee C, Kang Y, Kim D (2007) GeSbTe deposition for the PRAM application. Appl Surf Sci 253(8):3969–3976
48. Park SJ, Kim IS, Kim SK, Yoon SM, Yu BG, Choi SY (2008) Phase transition characteristics and device performance of Si-doped Ge2Sb2Te5. Semicond Sci Technol 23 (10):105006–105011
49. Chu CH, Shiue CD, Cheng HW, Tseng ML, Chiang H-P, Mansuripur M, Tsai DP (2010) Laser induced phase transitions of Ge2Sb2Te5 thin films used in optical and electronic data storage and in thermal lithography. Opt Express 18(17):18383–18393
50. Kim JH (2008) Effects of a metal layer on selective etching of a Ge5Sb75Te20 phase-change film. Semicond Sci Technol 23(10):105009–105015
51. Xi HZ, Liu Q, Tian Y, Wang YS, Guo SM, Chu MY (2012) Ge2Sb1.5Bi0.5Te5 thin film as inorganic photoresist. Opt Mater Expr 2(4):461–467
52. Xi HZ, Liu Q, Guo SM (2012) Phase change material Ge2Sb1.5Bi0.5Te5 possessed of both positive and negative photoresist characteristics. Mater Lett 80:72–74

53. Rogers JD, Kärkkäinen AHO, Tkaczyk T, Rantala JT, Descour MR (2004) Realization of refractive microoptics through grayscale lithographic patterning of photosensitive hybrid glass. Opt Express 12(7):1294–1303

54. Reimer K, Quenzer HJ, Jürss M, Wagner B (1997) Micro-optic fabrication using one-level gray-tone lithography. Proc SPIE 3008:279–288

55. Jiang H, Yuan X, Yun Z, Chan YC, Lam YL (2001) Fabrication of microlens in photosensitive hybrid sol–gel films using a gray scale mask. Mater Sci Eng C 99:16–22

56. Christophersen M, Phlips BF (2008) Gray-tone lithography using an optical diffuser and a contact aligner. Appl Phys Lett 92(19):194102–194104

57. Gimkiewicz C, Hagedorn D, Jahns J, Kley E-B, Thoma F (1999) Fabrication of microprisms for planar optical interconnections by use of analog gray-scale lithography with high energy beam sensitive glass. Appl Opt 38(14):2986–2990

58. Waits CM, Morgan B, Kastantin M, Ghodssi R (2005) Microfabrication of 3D silicon MEMS structures using gray-scale lithography and deep reactive ion etching. Sens Actuator A Phys 119(1):245–253

59. Waits CM, Modafe A, Ghodssi R (2003) Investigation of gray-scale technology for large area 3D silicon MEMS structures. J Micromech Microeng 13(2):170–177

60. Reimer K, Hofmann U, Juerss M, Pilz W, Quenzer HJ, Wagner B (1997) Fabrication of microrelief surfaces using a one-step lithography process. Proc SPIE 3226:2–6

61. Gal G (1994) Method for fabricating microlenses. US Patent 5,310,623, 10

62. Chen C, Hirdes D, Folch A (2003) Gray-scale photolithography using microfluidic photomasks. Proc Natl Acad Sci U S A 100(4):1499–1504

63. Zhang JM, Guo CF, Liu Q (2010) A superfine crystalline metal or alloy thin film and its fabrication method. China Patent, CN201010033715.9

64. Krishnamachari B, McLean J, Cooper B, Sethna J (1996) Gibbs-Thomson formula for small island sizes: corrections for high vapor densities. Phys Rev B 54(12):8899–8907

65. Zhang JM, Guo CF, Wang YS, Miao JJ, Tian Y, Liu Q (2012) Micro-optical elements fabricated by metal-transparent-metallic-oxides grayscale photomasks. Appl Optics 51 (27):6606–6611

66. Kim SS, Chun C, Hong JC, Kim DY (2006) Well-ordered TiO_2 nanostructures fabricated using surface relief gratings on polymer films. J Mater Chem 16:370–375

67. Cheong WC, Yuan L, Koudriachov V, Yu WX (2002) High sensitive SiO2/TiO2 hybrid sol–gel material for fabrication of 3 dimensional continuous surface relief diffractive optical elements by electron-beam lithography. Opt Express 10(14):586–590

68. Hong HG, Kim YJ (2008) Self-cleaning effect of solid immersion lens using photocatalyst TiO_2 film for near-field recording. Jpn J Appl Phys 47(7):5939–5943

69. Feng X, Shankar K, Varghese OK, Paulose M, Latempa TJ, Grimes CA (2008) Vertically aligned single crystal TiO_2 nanowire arrays grown directly on transparent conducting oxide coated glass: synthesis details and applications. Nano Lett 8(11):3781–3786

70. Chen JIL, Ozin GA (2009) Heterogeneous photocatalysis with inverse titania opals: probing structural and photonic effects. J Mater Chem 19:2675–2678

71. Shankar K, Bandara J, Paulose M, Wietasch H, Varghese OK, Mor GK, Latempa TJ, Thelakkat M, Grimes CA (2008) Highly efficient solar cells using TiO_2 nanotube arrays sensitized with a donor-antenna dye. Nano Lett 8(6):1654–1659

72. Popat KC, Leoni L, Grimes CA, Desai TA (2007) Influence of engineered titania nanotubular surfaces on bone cells. Biomaterials 28(21):3188–3197

73. Wu NQ, Wang J, Tafen DN, Wang H, Zheng JG, Lewis JP, Liu XG, Leonard SS, Manivannan A (2010) Shape-enhanced photocatalytic activity of single-crystalline anatase TiO_2 (101) nanobelts. J Am Chem Soc 132(19):6679–6685

74. Chong SV, Suresh N, Xia J, Salim NA, Idriss H (2007) TiO_2 nanobelts/CdSSe quantum dots nanocomposite. J Phys Chem C 111(28):10389–10393

75. Park JM, Nalwa KS, Leung W, Constant K, Chaudhary S, Ho KM (2010) Fabrication of metallic nanowires and nanoribbons using laser interference lithography and shadow lithography. Nanotechnology 21:215301–215305
76. Xia DY, Jiang YB, He X, Brueck SRJ (2010) Titania nanostructure arrays from lithographically defined templates. Appl Phys Lett 97(22):223106–223108
77. Wang YS, Wang R, Guo CF, Miao JJ, Tian Y, Ren TL, Liu Q (2012) Path-directed and maskless fabrication of ordered TiO2 nanoribbons. Nanoscale 4:1545–1548
78. Sader JE (1995) Parallel beam approximation for V-shaped atomic force, microscope cantilevers. Rev Sci Instrum 66(9):4583–4587
79. Wu KR, Ting CH, Wang JJ, Liu WC, Liu CH (2006) Characteristics of graded TiO_2 and TiO2/ITO films prepared by twin DC magnetron sputtering technique. Surf Coat Technol 200:6030–6036
80. Fujita T, Nishihara H, Koyama J (1982) Blazed gratings and Fresnel lenses fabricated by electron-beam lithography. Opt Lett 7:578–580
81. Chou SY, Krauss PR, Renstrom P (1995) Imprint of sub-25 nm vias and trenches in polymers. Appl Phys Lett 67:3114–3116
82. Tseng AA (2005) Recent developments in nanofabrication using focused ion beams. Small 1:924–929
83. Su J, Du J, Yao J, Gao F, Guo Y, Cui Z (1999) New method to design halftone mask for the fabrication of continuous microrelief structure. Proc SPIE 3680:879–883
84. Yu W, Yuan X, Ngo N, Que W, Cheong W, Koudriachov V (2002) Single-step fabrication of continuous surface relief micro-optical elements in hybrid sol–gel glass by laser direct writing. Opt Express 10:443–448
85. Kawata S, Sun H, Tanaka T, Takada K (2001) Finer features for functional microdevices. Nature 412:697–698
86. Deubel M, von Freymann G, Wegener M, Pereira S, Busch K, Soukoulis CM (2004) Direct laser writing of three-dimensional photonic-crystal templates for telecommunications. Nat Mater 3:444–448
87. Wang J, Xia H, Xu BB, Niu LG, Wu D, Chen QD, Sun HB (2009) Remote manipulation of micronanomachines containing magnetic nanoparticles. Opt Lett 34:581
88. Partridge JG, Field MR, Peng JL, Sadek AZ, Kalantar-zadeh K, Du Plessis J, McCulloch DG (2008) Nanostructured SnO_2 films prepared from evaporated Sn and their application as gas sensors. Nanotechnology 19:125504–125509
89. Guo CF, Zhang J, Miao J, Fan Y, Liu Q (2010) MTMO grayscale photomask. Opt Express 18 (3):2621–2631

Chapter 3
Laser Path-Guided Wrinkle Structures

3.1 Introduction

Wrinkle phenomenon exists widespreadly in landforms, skins of animals and plants, man-made materials, and so on. In our daily life, wrinkle is always regarded as a nuisance as a sign of aging or failure. For examples, the elders often have wrinkles in their faces; wrinkles often cause the damage of a workpiece; some of the major disasters happened in the early days of aviation were related to the wrinkles emerged in the wings, which were made of stiff layers capped on wood. In most cases, wrinkling occurs in a bilayer (or multilayer) system made of a stiff capping layer supported on a compliant material layer (or several compliant material layers). Compressive stress is essential for this process.

In 1998, N. Bowden et al. first proposed the fabrication of surface microstructures by wrinkling of a metal/polymer bilayer [1]. From then on, scientists began to explore how to utilize wrinkling with merits of low cost and simple process in microfabrication [2–24]. Many methods have been developed to control wrinkle patterns by introducing bas-reliefs into substrates [1, 9, 10], locally modifying mechanical properties of substrates [12–14], placing patterned elastomeric mold on a bilayer [15], stretching to produce anisotropic stress or strain [2, 5, 11, 17], placing nano-/microribbons and other shapes on a stretchable polymer [4], tuning the adhesive/slippery properties of the polymer/substrate interface [18, 22], and scanning the bilayer surface with a focused ion beam [23, 24]. Wrinkle patterns have also been used in many applications, including micro-/nanofluidics [5, 25], micro-optics [2, 3], smart adhesion [26], photovoltaics [27], particle alignment [5, 28–30], flexible electronics [4, 31], as well as detecting mechanical properties of thin films [7, 8, 32].

The quality of the wrinkle patterns, however, still needs to be further improved. Here the wrinkle quality refers to the ordering and controllability over wave configurations. To date, existing wrinkle patterns are not as perfect as the surface structures fabricated by conventional lithographic techniques, exemplified by the fact that defects in wrinkles are inevitable and configurations are quite limited. The imperfection of the wrinkle patterns should be related to the low controllability

Q. Liu et al., *Novel Optical Technologies for Nanofabrication*, Nanostructure Science and Technology, DOI 10.1007/978-3-642-40387-3_3,

of local stress contribution, which can hardly be well regulated on micron and submicron scale. However, the precise control over local stress on nanoscale or microscale is available by making patterns with fine features, and it might be a potential solution to make defect-free and arbitrary-shaped wrinkle patterns.

To date, there have been some works on locally patterning of surface by using energy beams. Huck and coworkers used ultraviolet (UV) light to photochemically modify the mechanical properties of the polymer layer, resulting in local control of stress contribution tailored by the patterns [14]. Moon et al. applied focused ion beam to scan polydimethylsiloxane (PDMS), leading to the formation of oriented wrinkles with the orientation and configuration related to scan speed and exposure dose [23, 24]. There were also other tries and these efforts finally got some specific wrinkle patterns. However, these approaches did not improve fundamentally the ordering of wrinkles because the feature size of their patterns is larger than the intrinsic wavelength of the wrinkles, λ_i.

A recently published paper authored by us reports that guiding paths (GPs) with a feature size much smaller than λ_i could effectively improve the quality, including the ordering and diversity of configurations of the wrinkle patterns [33]. A composite model was applied to explain the formation of the wrinkles, and a new concept, unit-wrinkle, was proposed as the basic unit of the wrinkle waves. In this chapter, we focus on laser direct writing of GPs for fabricating high-quality wrinkle patterns that have potential applications in micro-optics, particle alignment, etc.

3.2   Fundamentals of Wrinkling Based on a Bilayer or Multilayer System

H. G. Allen investigated wrinkle instability of a sandwiched structure [34]. He pointed out that wrinkling happens when the structure is under a critical compressive stress σ_c, which is expressed as [34]:

$$\sigma_c = \frac{E_m}{12}\left(\frac{l}{d}\theta\right)^2 + E_p\left(\frac{d}{l}\right)f(\theta) \qquad (3.1)$$

$$\theta = 2\pi\frac{d}{\lambda} \qquad (3.2)$$

and

$$f(\theta) = \frac{2}{\theta}\frac{(3-\nu_p)\sinh\theta\cosh\theta + (1+\nu_p)\theta}{(1+\nu_p)(3-\nu_p)^2\sinh^2\theta(1+\nu_p)^3\theta^2} \qquad (3.3)$$

where E_m is the Young's modulus of the capping layer, E_p is the Young's modulus of the compliant substrate (often a polymer), ν_p denotes the Poisson's ratio of the substrate, l and d are the thicknesses of the capping layer and the compliant substrate, respectively, and λ refers to the wavelength of emerged wrinkles.

For a metal/polymer/stiff substrate system with a finite thickness of the polymer layer, intrinsic wavelength of wrinkles can be expressed as [34, 35]:

$$\lambda_i = 2\pi l \left(\frac{Y}{1 + \sqrt{1 + 12YH^3}} \right)^{1/3} \tag{3.4}$$

where $H = l/d$, and Y is related to the modulus ratio of the two materials:

$$Y = \frac{E_m}{2(1 - \nu_m)^2 E_p} \tag{3.5}$$

here ν_m is the Poisson's ratio of the metal. From Eq. 3.4, wavelength increases with the increasing of the metal layer thickness, and it is linear to the thickness of the metal layer for a system with an infinitely thick polymer [6]. That is, in the case of $d \gg l$ in a metal/polymer bilayer system, wrinkle wavelength can be simplified as [6]:

$$\lambda_i = 2\pi l \left(\frac{E_m}{3(1 - \nu_m{}^2)} \frac{1 - \nu_s{}^2}{E_s} \right)^{\frac{1}{3}} \tag{3.6}$$

Therefore, wavelength of wrinkle can be easily tuned by changing the thickness of the metal layer. And accordingly, critical wrinkling stress σ_c and wave amplitude A can be described with the mechanical properties of different layers [1, 34]:

$$\sigma_c = \frac{1}{4} \left(\frac{3E_s}{1 - \nu_s{}^2} \left(\frac{E_m}{1 - \nu_m{}^2} \right)^{\frac{1}{2}} \right)^{\frac{2}{3}} \tag{3.7}$$

$$A = h \sqrt{\frac{\sigma}{\sigma_c} - 1} \tag{3.8}$$

The system applied to make wrinkle patterns in this chapter is a metal/PS bilayer supported on a glass slide. Thermal stress of this system stems from the difference in the linear thermal expansion coefficients of the metal and the glass substrate [36]:

$$\sigma = \frac{E_m(T - T_0)(\alpha_m - \alpha_s)}{1 - \nu_m} \tag{3.9}$$

where T and T_0 represent heating temperature and stress-free temperature, respectively; and α_m and α_s are the linear thermal expansion coefficients of the metal and the glass substrate, respectively. From Eq. 4.9, the thermal stress is independent of the polymer layer.

Equations 3.4, 3.5, 3.6, 3.7, 3.8, and 3.9 describe the relationships among wrinkle features (λ_i and A) and properties (film thickness, E, ν, α) of a bilayer or multilayer system. Therefore, we can design wrinkle wavelength and amplitude simply by tuning film thicknesses of the materials.

3.3 Well-Developed Methods for Fabricating Ordered Complexities via Surface Wrinkling

Early in the 1980s, scientists began to investigate wrinkling phenomenon with external load. For example, Tu et al. described radial wrinkles in a circular elastic plate stamped by a spherical punch [37]. However, only after N. Bowden et al.'s work people began to realize that wrinkling could be applied to make microstructures with periodic and sinusoidal surfaces. Before that, such sinusoidal microstructures were often fabricated by laser interference photolithography with a high cost. In contrast, wrinkling does not need expensive facilities, and hence, it is much cost-effective. N. Bowden and coworkers first made a PDMS thick layer on a glass substrate and then deposited in turn an adhering enhance layer (5 nm Ti) and a metallic layer (50 nm Au) to form a stiff layer [1], or used plasma oxidation to form a thin stiff silicate layer [10] on PDMS (see Fig. 3.1). They found disordered wrinkles generated spontaneously on the surface of the metal/PDMS bilayer. But if they make bas-reliefs in the PDMS thick layer before metal film deposition, oriented wrinkle patterns are formed perpendicularly to the relief edges after depositing a metal film (Fig. 3.2). The authors explained this phenomenon, and they attributed the anisotropic wrinkling as a result of nonuniform pre-wrinkling stress caused by the reliefs. For simplicity, pre-wrinkling stresses along x-direction and y-direction (σ_x and σ_y, and the x-direction is defined to be perpendicular to the step edge) near a step were analyzed, mathematically expressed as [1]:

$$\sigma_x = -\sigma_0 \left[1 - e^{-x/L} \right] \qquad (3.10)$$

$$\sigma_y = -\sigma_0 \left[1 - \nu_m e^{-x/L} \right] \qquad (3.11)$$

where σ_0 is equal biaxial compress thermal stress (stress when there is no pattern) and L is persistent length, expressed as:

$$L \approx 0.3 t_m \left(\frac{E_m}{E_p} \right) \qquad (3.12)$$

where t_m is the thickness of the metallic film. From Eqs. 4.10 and 4.11, stress perpendicular to the step edge (σ_x) is lower than σ_y, and wrinkles therefore grow perpendicular to the edges.

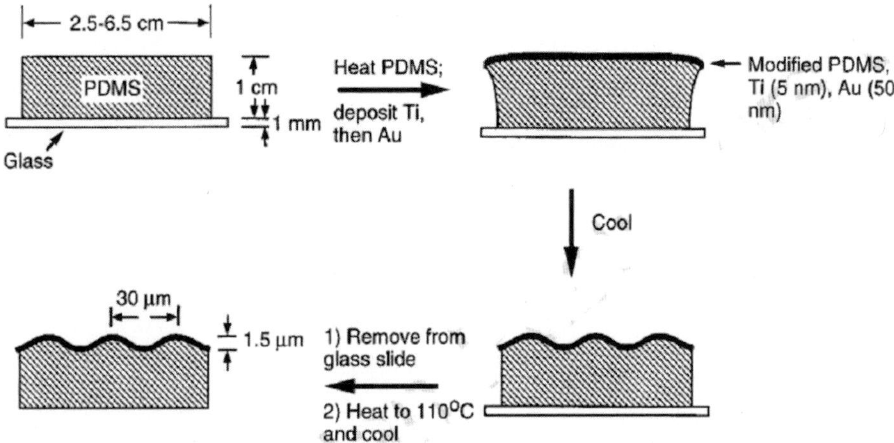

Fig. 3.1 Spontaneous formation of wrinkle patterns in a metal film supported on a polydimethylsiloxane thick layer (Reprinted by permission from Macmillan Publishers Ltd: Ref. [1], Copyright 1998)

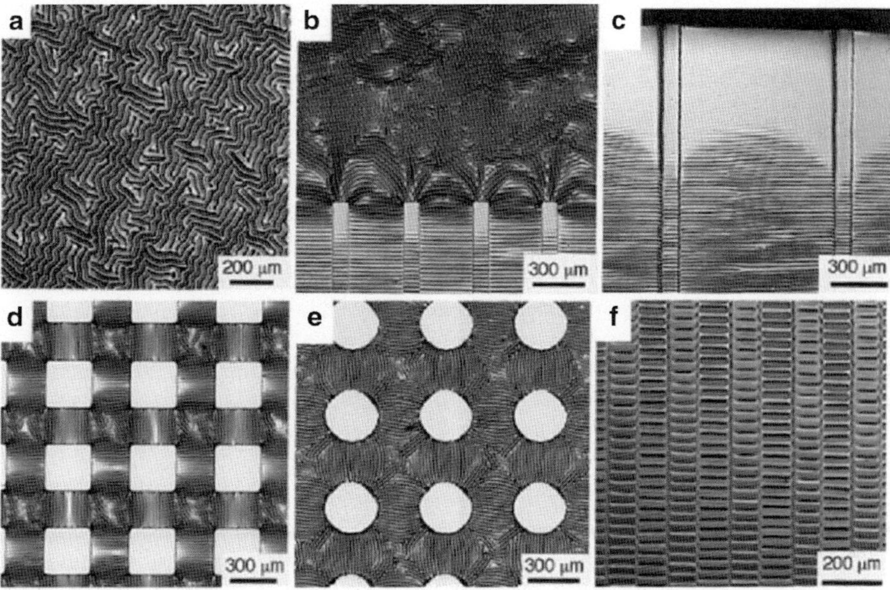

Fig. 3.2 (a) Disordered and (b–f) bas-relief-induced ordered structures in a metal/PDMS bilayer (Reprinted by permission from Macmillan Publishers Ltd: Ref. [1], copyright 1998)

Axial stretching is another common method to apply compressive load to the stiff/soft bilayer. Although stretching seems to directly cause tension, the Poisson's effect actually leads to compression in the perpendicular direction. Researchers seldom

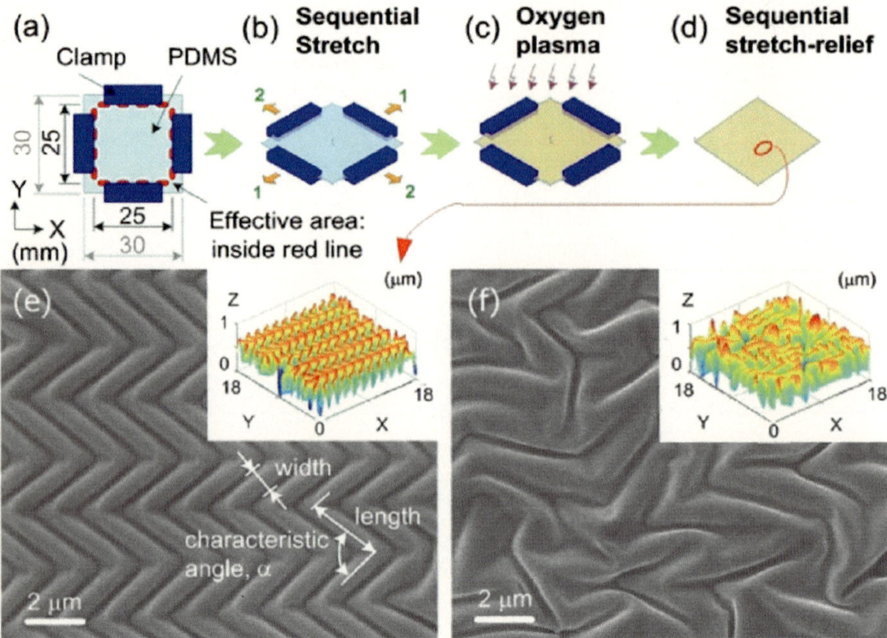

Fig. 3.3 Schematic fabrication process to generate wrinkle patterns (**a–d**). SEM and AFM images of surface patterns when stretching/releasing of PDMS film either sequentially (**e**) or simultaneously (**f**) (Reprinted with permission from Ref. [17]. Copyright 2007, American Institute of Physics)

adopt squeezing to generate compressive stress because of warping of the bilayer under compressive load. Pei-Chun Lin and coworkers developed a sequential and unequal biaxial stretching method to produce ordered herringbone structures [17]. By the sequential biaxial stretching and sequential releasing of a plasma-oxidized PDMS layer (or a PDMS layer capped with a metal film or flame treated [38]), wrinkle patterns transformed from one-dimensional ripples to two-dimensional herringbone structures. However, no ordered surface structures were found if without sequential stretching/releasing, as shown in Fig. 3.3. And if axial stress (only one direction) was applied, it would produce ripples in one dimensional only.

W. T. S. Huck et al. [14] developed a photochemical method for wrinkle complexities. This method locally changed stiffness and thermal expansion coefficient of a PDMS surface. After that a metallic thin film was deposited on the modified PDMS, it was found that wrinkles grew parallel to edges in the photochemically modified strips, while in the nonexposed area wrinkles grew perpendicularly to the edges, as shown in Fig. 3.4. Unlike bas-relief-induced wrinkle patterns, here, wrinkles in different areas are in different orientations. The mechanism was that the photochemically exposed areas became hardened, and this led to a strong nonuniformity of pre-buckling stress. Figure 3.5 shows schematically the stress along x-direction and y-direction. It is obvious that σ_y in hard areas (photochemically modified) is

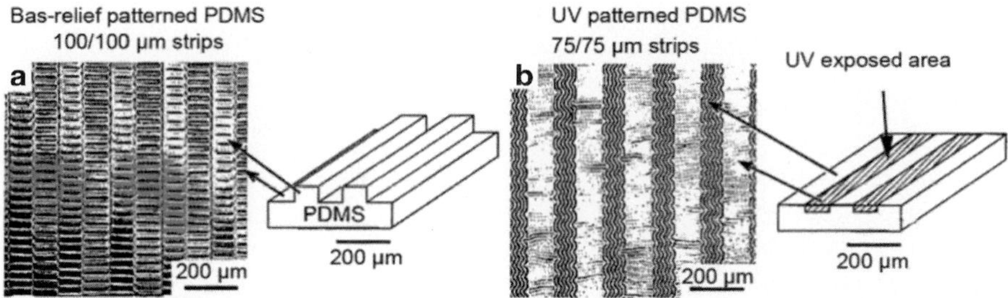

Fig. 3.4 Ordered wrinkle patterns directed by using (**a**) bas-reliefs and (**b**) selected-area hardening (Reprinted with the permission from Ref. [14]. Copyright 2000 American Chemical Society)

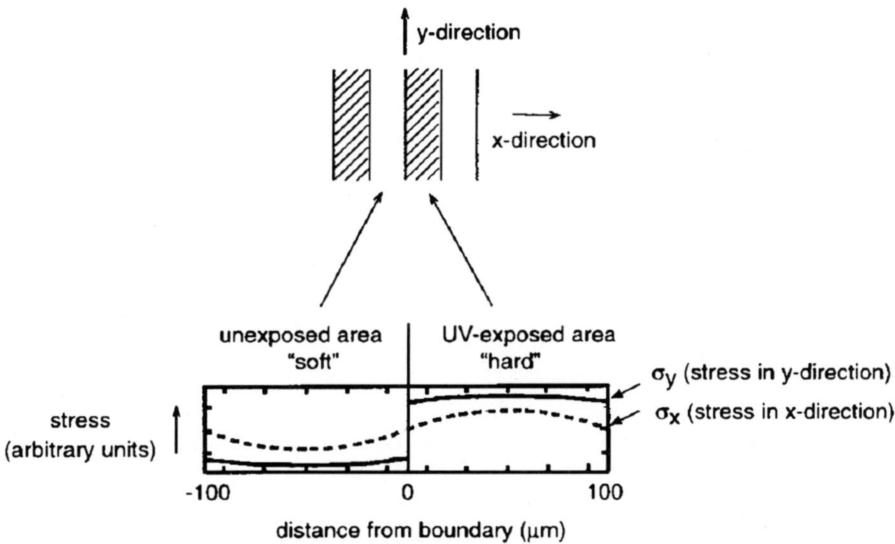

Fig. 3.5 Pre-buckling stress distribution in "soft" and "hard" areas (Reprinted with the permission from Ref. [14]. Copyright 2000 American Chemical Society)

larger than σ_x, while in the soft areas (nonexposed) the case is right opposite. The pre-buckling stress analysis could well explain the orientations of the wrinkles in different areas. One noteworthy thing is that the photochemically modified strips in Huck et al.'s work is at micron scale, typically several times of the wrinkle wavelength, and that might be a reason why their wrinkles are not highly ordered. We will show the modified area with a much smaller width could lead to very precisely control over the configuration of wrinkle patterns later.

H. Vandeparre et al. [18] reported a structure similar to that reported in Huck et al.'s paper published in Langmuir. They locally tuned the adhesion between the PS film and the substrate. After depositing a Ti film on PS followed by heating, wrinkles parallel to edges were formed in the slippery strips, and wrinkles

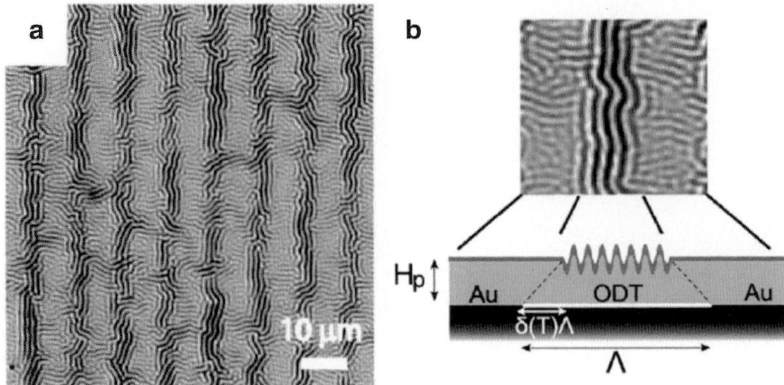

Fig. 3.6 (a) Wrinkle pattern assembled by selected-area-adjusting adhesion to the substrate. (b) Explanation of wrinkle alignment of a slippery interface (Reprinted with permission from Ref. [18]. Copyright (2007) by the American Physical Society)

perpendicular to edges were formed in the sticky stripes, as shown in Fig. 3.6a. A slight displacement of the PS film (Fig. 3.6b) on the slippery interface played an important role for the different wrinkle orientations, and wrinkle amplitude in slippery strips was higher. This work offers a simple method toward oriented wrinkles by chemically patterning substrate with different adhesion between the substrate and the polymer layer.

P. J. Yoo et al. developed elastomeric mold directed self-assembly of an aluminum/polystyrene (Al/PS) bilayer [15]. The PDMS mold was able to direct the configuration of wrinkle patterns (Fig. 3.7). If without a mold, the wrinkles would be randomly oriented when the bilayer was heated above the glass-transition temperature of the polymer. If using a mold which had a periodic pattern placing on the metallic film, periodic wrinkle pattern defined by the mold could appear. The difference between λ_i and the period of the mold determined the configuration of the surface structures. When the difference was not large, a sinusoidal surface profile could be obtained. This method can be used to make highly ordered surface structures, but it needs an expensive patterned mold with a pattern period close to λ_i of the bilayer.

The C. M. Stafford group reported diffusion controlled, self-organized symmetric wrinkle patterns [20]. In this work, the researchers first used UV exposure to make an oxidized and cross-linked layer on a PS film, after that the sample was placed in toluene vapor. Toluene went into PS from dot defects and swelling-induced wrinkle patterns were formed. Thin and thick surface layer led to the formation of spoke and target wrinkle patterns, respectively. Figure 3.8 well explains the two types of wrinkle patterns. The formation and growth kinetics of both spoke and target patterns originate from local defects in the cross-linked UVO-treated surface. Increasing UV oxidation time appears to limit the flux of solvent into the film at defect sites, resulting in a transition from Fickian to Case II diffusion behavior [39]. The differences in diffusion kinetics establish different stress states in the film, leading to spoke and target patterns, respectively. This work successfully made concentric circles by wrinkling for the first time.

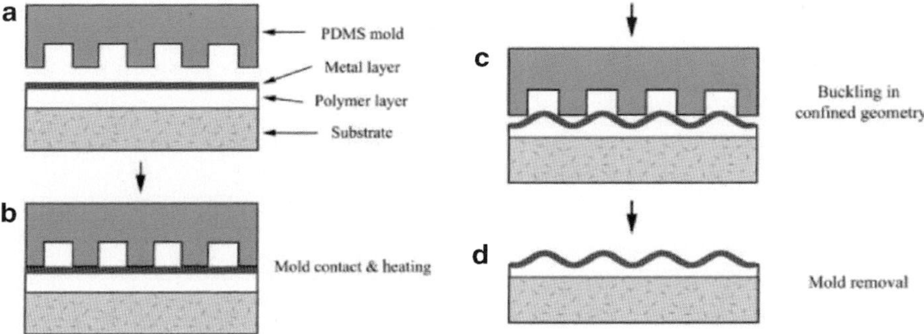

Fig. 3.7 (**a–d**) Schematic plan of wrinkle assembly by templating (Reproduced from Ref. [15] by permission of John Wiley & Sons Ltd)

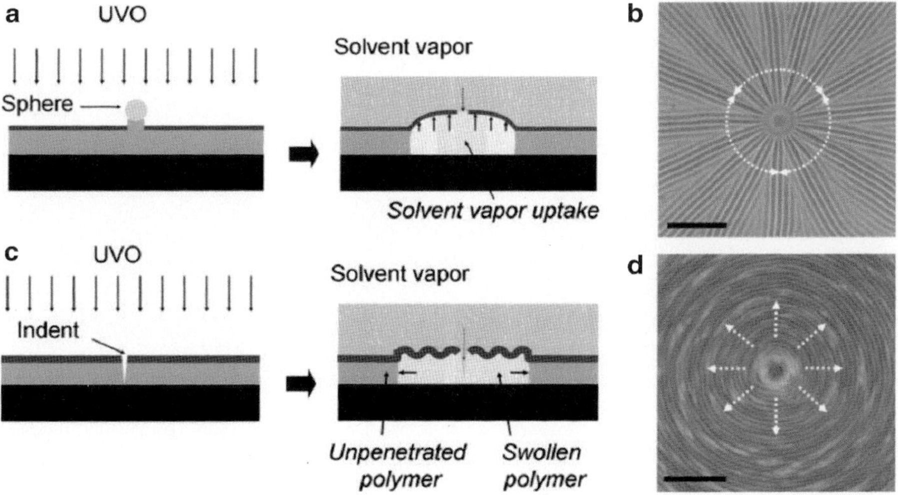

Fig. 3.8 The mechanism of spoke pattern (**a**) and its morphology (**b**); the mechanism of concentric pattern (**c**) and its morphology (**d**). The scale bars are 30 μm (Reproduced from Ref [20] by permission of John Wiley & Sons Ltd)

The methods reviewed above allow the fabrication of various wrinkle patterns. However, the wrinkle patterns in these works are very limited in quality. For example, these methods could not make arbitrary-shaped wrinkle patterns, and this might be ascribed that precise control of stress distribution at micro-/nanoscale is difficult. Therefore, scientists developed some lithographic techniques combining with wrinkling in order to obtain better wrinkle patterns or unique wrinkle patterns for a specific application.

Some lithographic methods combining with wrinkling were explored to make wrinkle patterns. M. W. Moon et al. reported a method with PDMS exposed under focused ion beam [23, 24]. By controlling the FIB intensity and area of exposure of

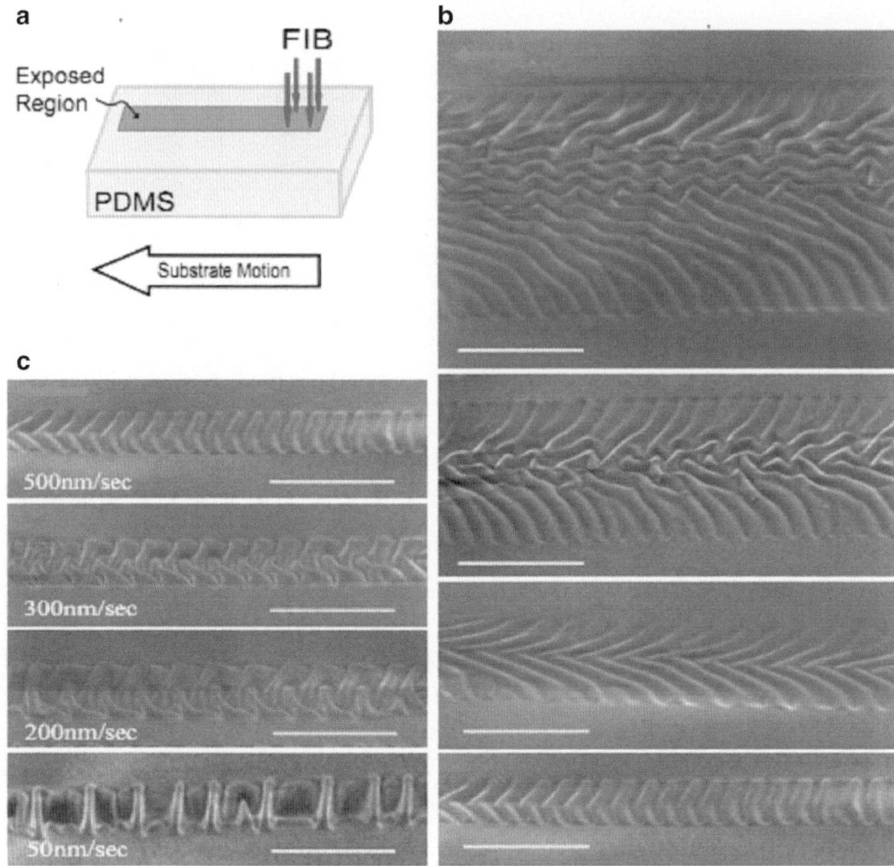

Fig. 3.9 (**a**) Schematic illustration of wrinkle generation by ion beam scan. (**b**) Morphologies of wrinkle strips with different width. (**c**) Dependence of morphology on scanning rate (Reproduced from Ref. [23] by permission of National Academy of Sciences, USA)

the PDMS, a variety of patterns with the wavelengths in the micrometer to sub-micrometer range could be created from simple one-dimensional wrinkles to peculiar and complex hierarchical nested wrinkles. It was found that PDMS under FIB exposure could form a thin amorphous-silica-like stiff layer, which expanded laterally to the ion beam scanning direction. Figure 3.9 shows the wrinkle patterns obtained by FIB exposure. However, these wrinkles were not high quality enough compared with the counterparts made by stretching or using a thermal method. But this method is valuable because it can make local wrinkle patterns. From Fig. 3.9b, the quality of wrinkle patterns is related to the spot size of the ion beam, and when the beam size is close to λ_i, wrinkles are more ordered. Besides, L. Guan et al. also found that nanodot-like wrinkles could emerge upon FIB exposure [40].

3.4   Laser Path-Guided Wrinkling on Metal Film

Laser direct writing (LDW) is a technique that enables the fabrication of micro-/nanostructures without a vacuum condition. In contrast to FIB and EBL, it is much cost-effective and capable of large-area fabrication, and hence suitable for practical applications. In the recent decades, LDW has been employed to make micro-optical elements, masks, gratings, nanoreliefs, photonic crystals, MEMS structures, etc. LDW based on two-photon absorption could even fabricate fine nanostructures with a feature size of ~100 nm [41], comparable to the resolution of EBL and FIB techniques. Moreover, some advanced laser direct writers are able to write a 4 in. wafer in tens of minutes. Therefore, LDW is a powerful tool for micro-/nanofabrication.

Here we show that LDW combining with wrinkling of a metal/PS bilayer supported on a Si wafer or a glass substrate is able to make high-quality wrinkle structures [33]. We have found that tuning the laser dose could lead to two different types of wrinkle assembly. Typically, the laser pulse is set to 200 ns, and laser spot size is ~350 nm in the laser direct writer equipped with a 533 nm laser and objective lens (0.95 NA), and the laser does is determined by laser power for a definite irradiation time. When laser power is high, the laser beam cuts the metal film. The function of these scanned lines is similar to the steps of bas-reliefs, and wrinkles align perpendicular to the lines. We call this type I laser-induced wrinkling (LW-I). When the laser power is significantly small, the laser beam causes a softening effect onto the metal film, and a novel wrinkling phenomenon happens: Wrinkles grow exactly along the laser-scanned paths after heating and we call it LW-II.

3.4.1   Experimental Results of LW-I

The laser beam could cut a metal film supported on a polymer thin film. Figure 3.10a shows an optical image of line structures written by LDW in a Sn (20 nm)/PS (120 nm) bilayer. Upon heating at 120 °C for 2 h, wrinkles emerge on the patterned bilayer with an orientation perpendicular to the lines, as shown in Fig. 3.10b. For these wrinkles, the physics is quite similar to N. Bowden et al.'s work, and hence their ordering is related to the pitch of the lines. The stresses can be expressed as [1]

$$\sigma_x = -\sigma_0 \left[1 - \frac{\cosh(x/L)}{\cosh(P/2L)} \right] \tag{3.13}$$

$$\sigma_y = -\sigma_0 \left[1 - \nu_m \frac{\cosh(y/L)}{\cosh(P/2L)} \right] \tag{3.14}$$

where L is persistent length, and P is the pitch between two lines.

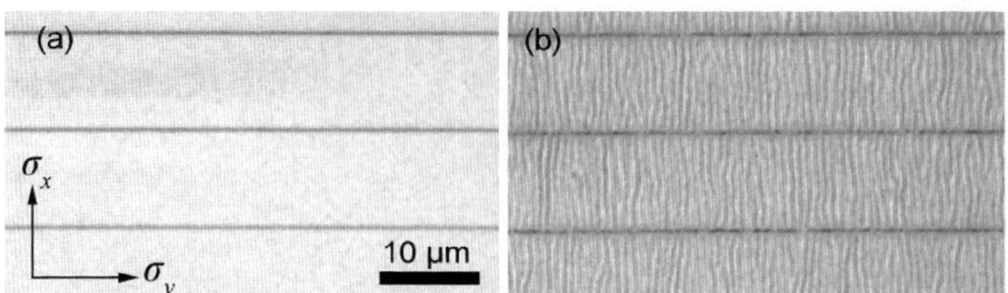

Fig. 3.10 (**a**) Line structures fabricated by LDW in the Sn capping layer of a bilayer. (**b**) Spontaneously formed ordered wrinkles directed by the lines

In contrast to bas-relief-induced wrinkle alignment, this method is able to make more complex patterns. Here we present a target pattern and a spoke pattern in Fig. 3.11, made by LW-I. To make the target pattern, we used a spoke pattern with a radius of 100 μm and an included angle of 11.25° between neighboring lines. The inset in Fig. 3.11a is a SEM image of the wrinkled surface, showing that the lines on Sn film scanned by laser are ablated, and wrinkles emerge perpendicular to the lines. Figure 3.11b clearly demonstrates that wrinkles form a target pattern. Figure 3.11c is a spoke pattern, made by writing a target pattern. The pattern made by laser writing and the resultant wrinkle pattern are reciprocal in phase; this is because wrinkles always grow perpendicular to the laser written lines.

Besides wrinkle alignment, LW-I can also be used to eliminate wrinkles. Wrinkles are often a nuisance in our daily life before they are harnessed for surface microfabrication. In fact, aligning wrinkles by using bas-reliefs or laser writing can also be regarded as eliminating wrinkles in just one direction while keeping wrinkles in the perpendicular direction. Simply by eliminating wrinkles in two perpendicular directions, it is possible to suppress all waves in the bilayer. Figure 3.12a is an interesting picture in which we could find disordered wrinkles, ordered wrinkles, and waveless zone, corresponding to the cases of equal biaxial compressive stress (non-released), axial stress, and biaxially released stress, respectively. We could find in squares there is no waves, as a result of stresses in both directions being released. And we also find that the length of waveless zone (l_1) is very close to the persistent length (l_2), which can be roughly explained by Eq. 3.13. Figure 3.12b shows some squares with different sizes, indicating that when the side length of the squares is smaller than 2 L, waves can be effectively suppressed. This is a closed graph effect (CGE). We could also see from Fig. 3.11c that there is no wave in the smallest circle, because of CGE.

The CGE might be useful where wrinkles are undesirable, e.g., in facial beauty. Wrinkles on human faces are a feature of body aging, and ladies always attempt to remove wrinkles from their faces. What they are doing is basically to change the mechanical properties of the skin, e.g., to make the epidermis softer, so that a higher critical wrinkle stress is needed, and emerged wrinkles are finer. Here LW-I is able to release stress, such that lines or wrinkles on face might be vanished or prevented if we are able to write invisible grids in derma.

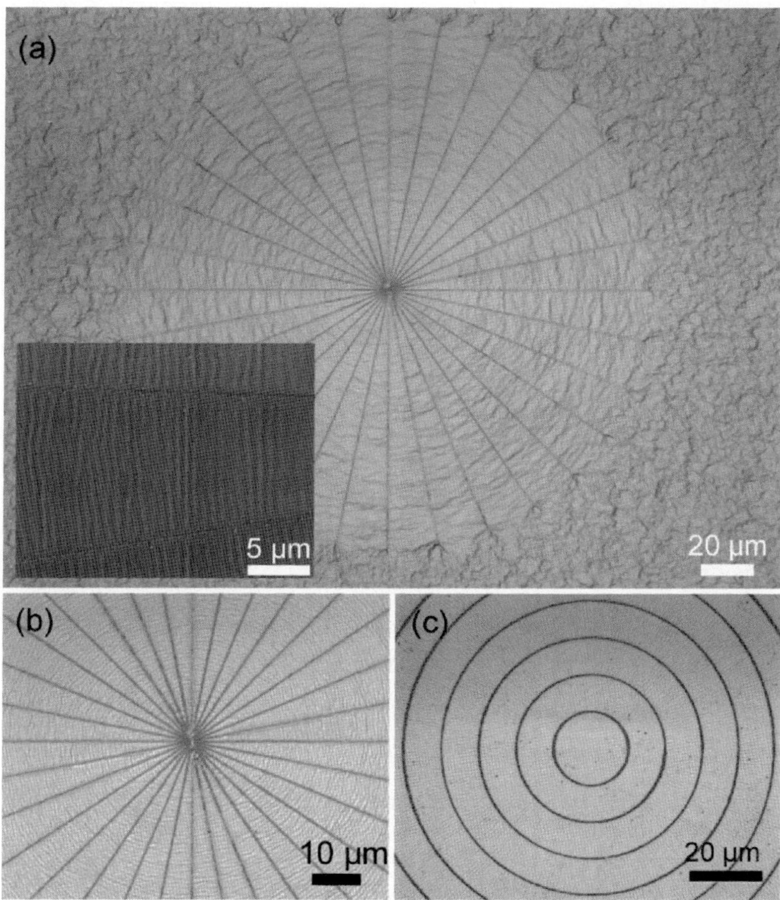

Fig. 3.11 (**a**) Curved wrinkles fabricated in a Sn/PS bilayer, *inset* is the corresponding SEM image. (**b**) Magnified image of (**a**). (**c**) Radially aligned wrinkles

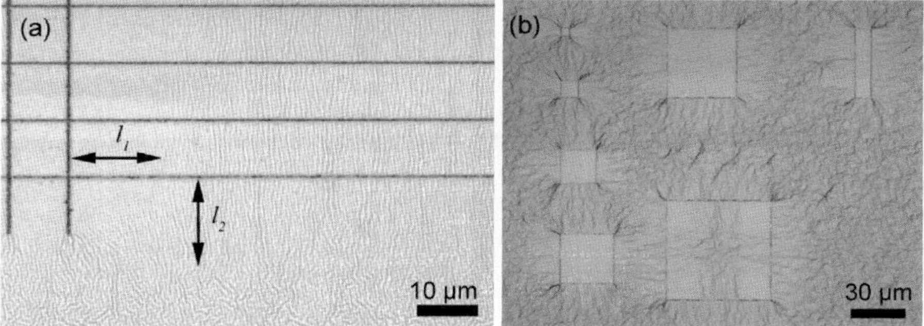

Fig. 3.12 (**a**) The suppression of wrinkle by scanning grid and paralleled lines on the Sn film. (**b**) Wrinkling can be suppressed in small closed graphics

3.4.2 LW-II for High-Quality Wrinkle Patterns

LW-I uses a high power laser beam to cut the metal film and form lines similar to the steps in N. Bowden et al.'s work, and it leads to wrinkle alignment or wrinkle suppression. However, when we use a lower laser power which does not cut the metal film, it will cause a softening effect. For a 20 nm thick Sn film supported on PS, we typically use a laser beam with a power of 0.3–1.0 mW and a pulse width of 200 ns to make "soft" lines. But for a 5–7 nm thick Au film, a higher power of 1–5 mW is required. We call the resultant lines guiding paths (GPs). The GPs can be lines, dots, curves, or arbitrary-shaped patterns, with a lower reflectivity compared to non-patterned film. The GPs are visible by optical microscopy, but only dark field image is clear. Figure 3.13 shows optical images of a dot array in a bright field and a dark field, for which the dots of the former is vague, owing to quite small change in reflectance for the GPs.

Unlike the lines in LW-I or the steps of bas-reliefs, wrinkles form always along the lines written by laser beam. Figure 3.14a, b illustrates the process of creating GPs, as well as guided wrinkle formation. Here the GPs serve as seeds for wrinkling, inducing the wrinkle crests forming exactly wherever laser writes. Figure 3.14c–j shows atomic force microscopy (AFM) topographic images of a set of wrinkle patterns guided by the GPs; the pitch of the GPs is designed to be around 2.1 μm to match λ_i. The wrinkle patterns include lines, circles, dots, and structures made up of lines and dots. Figure 3.14k shows the cross-section profile along the green line in Fig. 3.14i, indicating LW-II is able to make structures with a very homogenous surface.

These wrinkle patterns exhibit three unique features: (i) they are highly ordered and defect-free; (ii) the GPs define precisely the location and configuration of the wrinkles; (iii) the path-guided wrinkle patterns can be lines, curves, dots, and more complex shapes. These wrinkle patterns demonstrate that various high-quality surface microstructures can be fabricated in a full-dry process that might be widely used for high-throughput micro-/nanofabrication.

The wrinkle patterns made by LW-II are defect-free. In existing works, cracks and dislocation-like defects are unavailable, because perfect control of stress contribution on micron scale is very difficult. However, LW-II can change mechanical properties of the GPs, which are typically 300 nm in width (much smaller than λ_i). A similar method was reported by Huck et al. [14], where they photochemically modified the surface of an elastomer before depositing a metal thin film. As a result, the effective mechanical properties of the composite surface layer (the metal film and the modified elastomer surface) were patterned and wrinkles were aligned accordingly. The key difference of the present method is that we directly modify the gold thin film using patterns with feature sizes much smaller than λ_i to precisely control the location and shape of wrinkles.

Figure 3.15 is a set of optical images of dot-like or dot/line wrinkles. Panel (a) is a tetragonal dot array with a pitch close to λ_i, and corresponding FFT pattern reveals a tetragonal lattice. Panel (b) is a non-hexagonal dot array. And Fig. 3.15c is a

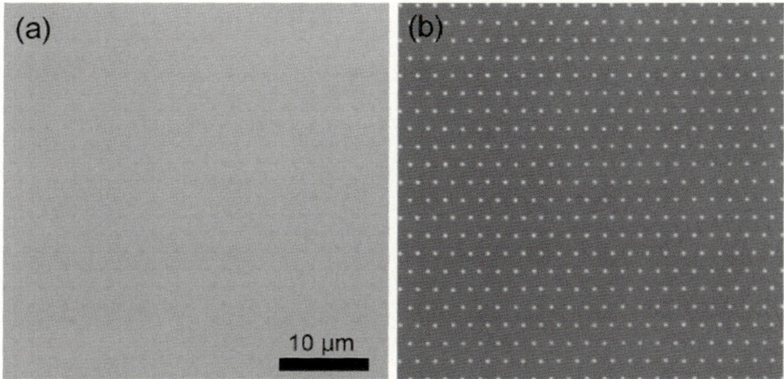

Fig. 3.13 Bright field and dark field front-lit images of a dot array fabricated in the metal layer

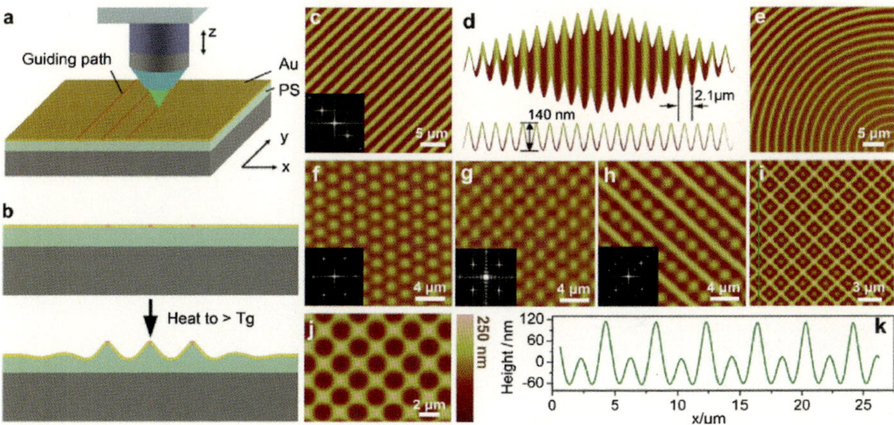

Fig. 3.14 Fabrication of wrinkle patterns. (**a**) Schematic illustration of laser scanning on Au films for making soft guide paths. (**b**) Controlled wrinkling along guiding paths. (**c–j**) AFM images of various microstructures: (**c**) Line–wrinkle array. (**d**) Tilt and side views of the wrinkles in (**c**), showing highly homogeneous and periodic surface structure. (**e**) Concentric circle array. (**f**, **g**) Hexagonal and tetragonal *dot*-wrinkle arrays, respectively. (**h**) A complex wrinkle pattern composed of *dots* and *lines*. Insets in (**c**) and (**e–h**) are the corresponding FFT patterns. (**i**) A wrinkle pattern composed of vertically aligned 0th order lines and 1st order dots. (**j**) An egg-crate structure. (**k**) Section analysis of the wrinkles along the *green line* in (**i**) (Reproduced from Ref. [33] by permission of John Wiley & Sons Ltd)

superlattice (similar to panel (a)), for which some dots (e.g., the two dots indicated by the black arrows) are not directed by GPs. Corresponding FT pattern displays clearly the difference from panel (b). Figure 3.15d, f shows dot/line composite surfaces with different pitches and/or configurations.

Figure 3.16 shows a set of curved wrinkles. Panels (a) and (b) are concentric circles; however, we can see that the latter has two rings in the FFT pattern,

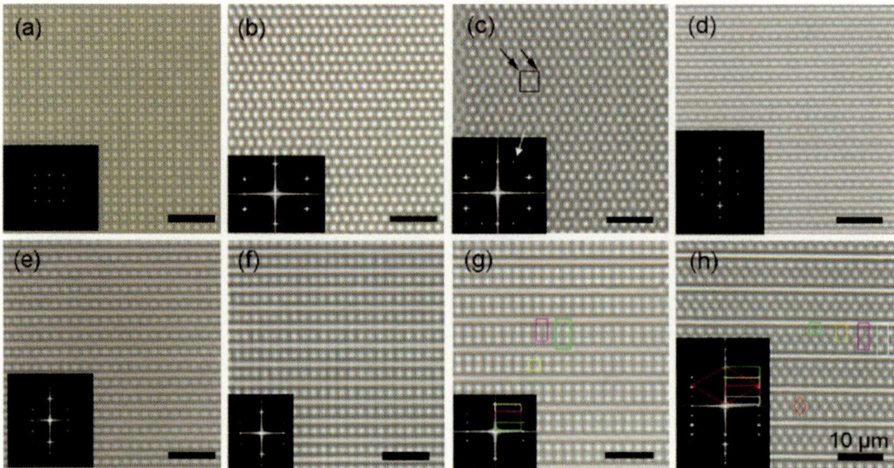

Fig. 3.15 Optical images of wrinkle patterns composed of dot (and line) wrinkles. All scale bars are 10 μm. *Insets* are corresponding FFT patterns

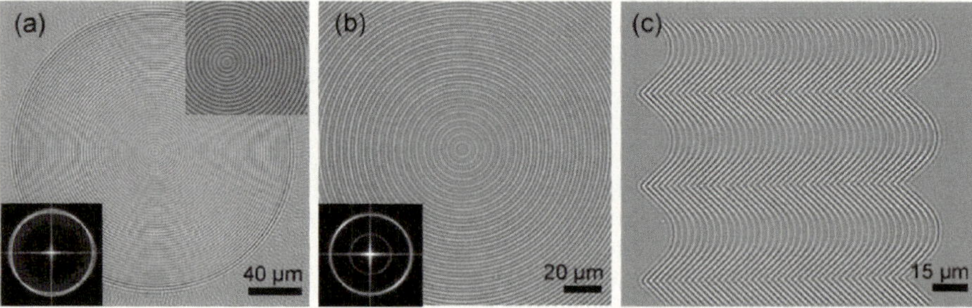

Fig. 3.16 Optical images of highly ordered and large-area wrinkle patterns composed of curve wrinkles. The *insets* in (**a**) and (**b**) are the corresponding FFT images. The wrinkle patterns in (**a**) and (**b**) are directed by guiding paths with a period of λ_i and $2\lambda_i$, respectively. (**c**) A configuration made up of curves

implying there are secondary structures with a period of 2λ. Figure 3.16c is a wrinkle pattern made of lines and curves. These images indicate that LW-II could make various curved wrinkle patterns.

The optical images, however, could not reveal accurately the surface profiles of the wrinkle patterns. Here Fig. 3.17 shows a set of AFM images with height information of the patterns. Panels (a–c) are tetragonal dot array with pitches of λ_i, 1.4 λ_i, and $2\lambda_i$, respectively. The dots, which are actually cones, locate exactly in positions where laser writes; they deviate to a sinusoidal shape because the tip is softer than other zones, leading to a larger deformation. Obviously, the dot pitch significantly affects surface profiles, and for panels (b) and (c), sinusoidal waves are found to emerge between the cones. Figure 3.17d is made up of dots and lines.

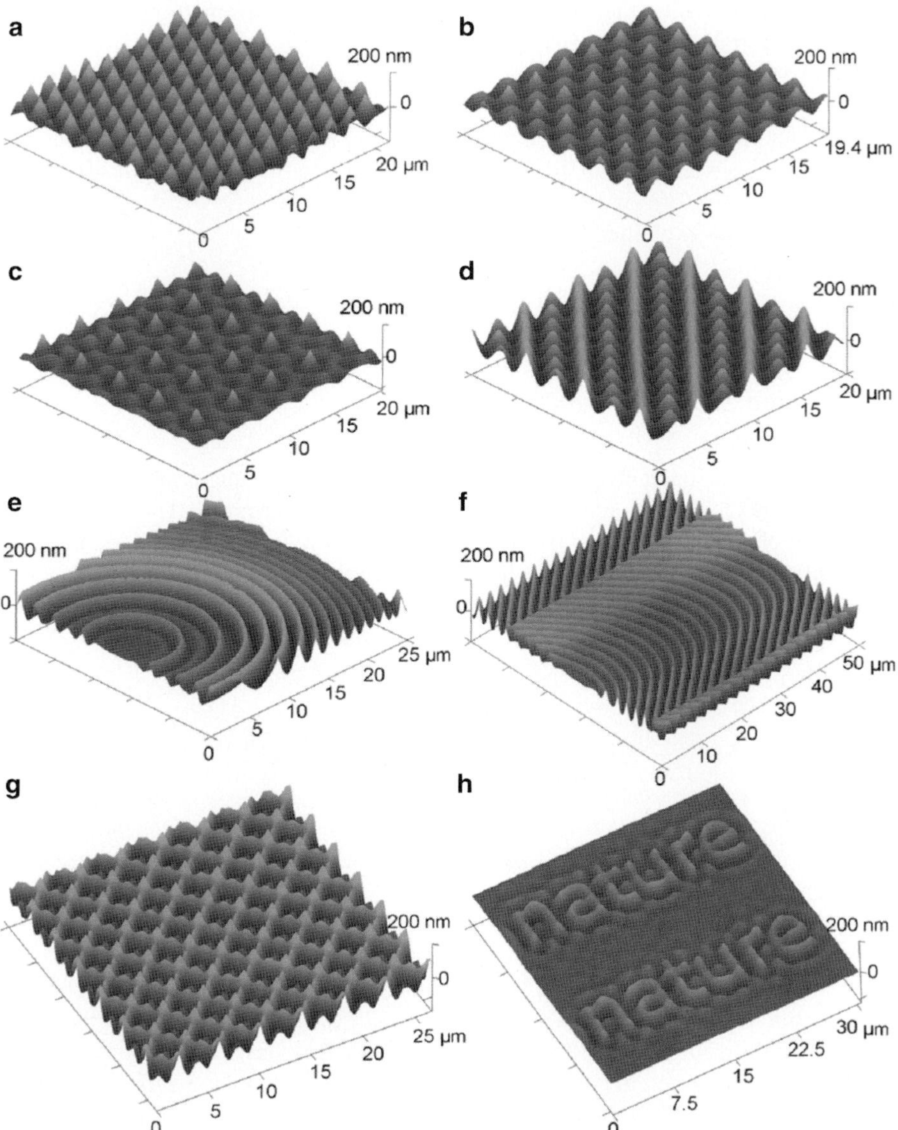

Fig. 3.17 AFM 3D view of a group of configurations fabricated in a Au/PS bilayer via LDW-assisted assembly II

Figure 3.17e is composed of aperiodic circles, which is quite different from the fact that spontaneously formed wrinkles often have an intrinsic wavelength. Figure 3.17f is made up of linear and curved structures. Panel (g) shows a pattern directed by a grid with a pitch of ~ 1.4 λ_i. We could find that the height along a line is periodically changed, and there are dots between neighboring lines. Panel (h) depicts a wrinkle pattern writing the word "nature," indicating that this method enables us to fabricate arbitrary-shaped wrinkle patterns.

Compared with conventional photolithography, electron beam lithography, and focused ion beam milling, this technique is simple and cost-effective, especially suitable for fabricating large-area wavy-surfaced structures. However, it does not mean that the wrinkles are limited to wavy-surfaced structures. Besides, why wrinkles can also form in area without GPs, as shown in Fig. 3.17b, g? All these questions will be explained hereinafter.

3.4.3 Quantitative Design of Wrinkle Patterns: Unit-Wrinkle and Superposition Effect

In existing works, scientists focused their interest on the alignment of wrinkles. However, recently C. F. Guo et al. proposed a new concept, unit-wrinkle, as the basic unit of wrinkle patterns. By and large, a basic unit is helpful for human beings to understand their objects, e.g., unit cell for crystals and monomer for polymers. A unit-wrinkle takes the shape of a laterally damped wave, with the maximum amplitude at the location of GP. As the basic unit of the LDW-guided wrinkles, the unit-wrinkle can be described mathematically as:

$$h(x) = A_0 \cdot \cos\left(\frac{2\pi x}{\lambda_i}\right) \cdot \exp\left(-\left|\frac{x}{l_c}\right|\right) \tag{3.15}$$

where A_0 is the maximum height at the center of GP ($x = 0$) and l_c is the effective damping length that can be determined experimentally (which will be discussed in details below).

The profile of the unit-wrinkle is attributed to modification of the elastic modulus of the metal film as a result of laser irradiation. The SEM image shows clearly lower densification in the laser-exposed area, as shown in Fig. 3.18. Consequently, the elastic modulus of the metal film is expected to be lower in the laser-exposed area [42]. Numerical simulations of wrinkling based on a composite film model were performed to confirm this hypothesis. In particular, a softening parameter S is used to reflect the effect of laser exposure, with the Young's modulus $E_S = SE_m$ ($0 < S < 1$) for the laser-exposed metal film, while E_m is the Young's modulus of the unexposed film. Owing to that wrinkle wavelength, λ_i, is related to the Young's modulus of the metal film (Eq. 3.6), the parameter S can be estimated by comparing the wrinkle wavelength in a large area exposed to laser with λ_i in the unexposed area. By taking $S = 0.4$, the simulated wrinkle profile agrees reasonably well with the experimental data. In Fig. 3.19, the measured profile, the fitted profile based on Eq. 3.15, and the simulated profile based on a softened GP agree quite well with each other.

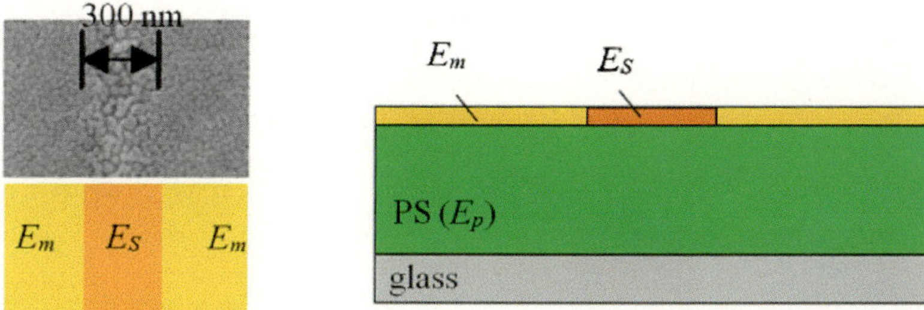

Fig. 3.18 SEM image of the film surface shows morphological difference between exposed and unexposed areas by laser. Apparently, the exposed area has a lower density and hence lower Young's modulus. The *right part* is a composite film model, in which the elastic modulus of the metal film is assumed to be lower in the laser-exposed area ($E_S < E_m$)

Fig. 3.19 Surface profile of a unit-wrinkle, comparing the experimental result with numerical simulation based on a composite film model and the approximation by an exponentially damped wave function in Eq. 3.15. *Inset* is corresponding AFM image of the unit-wrinkle

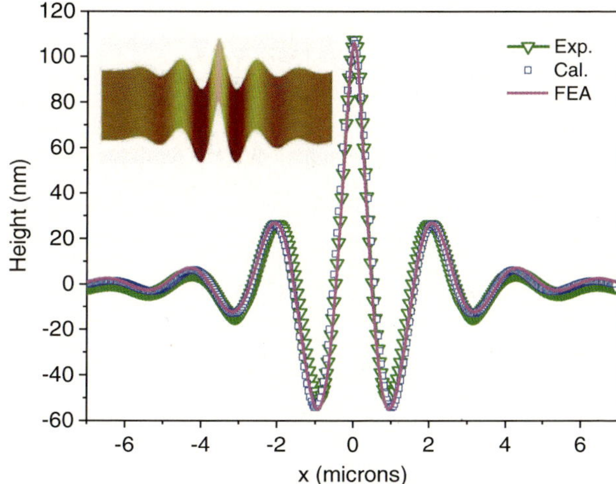

In contrast to the previous work by Huck et al. [14], the use of nanoscale feature size of GPs is essential for creating unit-wrinkles and hence precise control of the wrinkle patterns. In Huck et al.'s work, the feature size of their strip is much larger than λ_i, and multiple wrinkle crests appear within the wide strip. Consequently, the wrinkle pattern cannot be fully controlled unless the feature size of the strips is sufficiently small. The LDW technique offers a similar case. If the GP are much wider than λ_i, then multi-crests are formed, as shown in the AFM image of Fig. 3.20a. Moreover, it is noted that the pre-wrinkling stress distributions in the metal film depend on the feature size (Fig. 3.20b). When the feature size of a GP is significantly smaller than λ_i, the stress outside the GP is nearly unaffected while the stress in the laser-exposed area becomes highly anisotropic, showing a small size effect.

A question still exists: according to the conventional theory of wrinkling [43, 44], a homogeneous film with a lower Young's modulus would require a higher

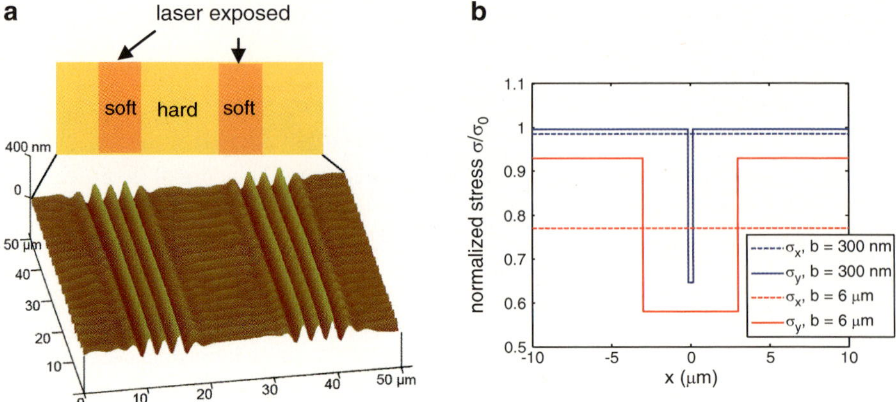

Fig. 3.20 (**a**) When the feature size of the laser-exposed pattern is several times the intrinsic wrinkle wavelength, wrinkles are aligned parallel to the pattern boundaries in the exposed area but perpendicular to the boundaries in the unexposed areas, similar to the wrinkle patterns made by Huck et al. (**b**) Finite element analysis of pre-wrinkling stress distributions in the metal film for a narrow and a wide GPs (GP width b = 300 nm and 6 µm, respectively). We assume that the Young's modulus in the GPs is 0.5 E_m, where E_m is the Young's modulus outside the GPs. The stresses are normalized by the reference stress for the case of a homogeneous film ($\sigma_x = \sigma_y = \sigma_0$). Since the PS underlayer is very soft at the elevated temperature ($T > T_g$), the stress along the x-direction (σ_x) is nearly uniform in the film, while the stress in the y-direction (σ_y) is much lower inside the GPs. For the narrow GP (b = 300 nm), σ_x is very close to σ_y outside the GP. For the wide GP (b = 6 µm), σ_x is much lower than σ_y outside the GP. The relative magnitudes of the two stresses are consistent with the observed wrinkle alignments inside and outside the GPs in panel (**a**)

critical strain but a lower critical stress for the onset of wrinkling. That means there would not be a crest formed at the GPs. Actually, upon heating, in contrast to a large-area laser-exposed zone, unexposed zone does wrinkle first and have larger wrinkle amplitude, as shown in Fig. 3.21. Moreover, from the corresponding FFT pattern, the laser-exposed region has a smaller wavelength, implying the smaller Young's modulus. Therefore, the laser exposure does cause a softening effect of the metal film. Let us come back to the question, why does it first wrinkle at the GPs? In fact, for the composite film (especially when the GPs are quite narrow), the critical condition for wrinkling in general cannot be predicted by the conventional theory. A similar phenomenon has been noticed recently for an elastic film on a compliant substrate with preexisting interfacial delamination [45].

The unit-wrinkle makes it possible to understand the formation of surface waves and thereby enables the design and fabrication of more complex wrinkle patterns. A simple way to investigate the interaction of unit-wrinkles is to measure surface profile of two parallel unit-wrinkles with different distances. Figure 3.22a shows the result simply by summing the profiles of two unit-wrinkles with a distance of 3.6 µm, and it is quite well consistent with the experimental profile of two unit-wrinkles. Moreover, a simulated profile based on finite elemental analysis (FEA) of

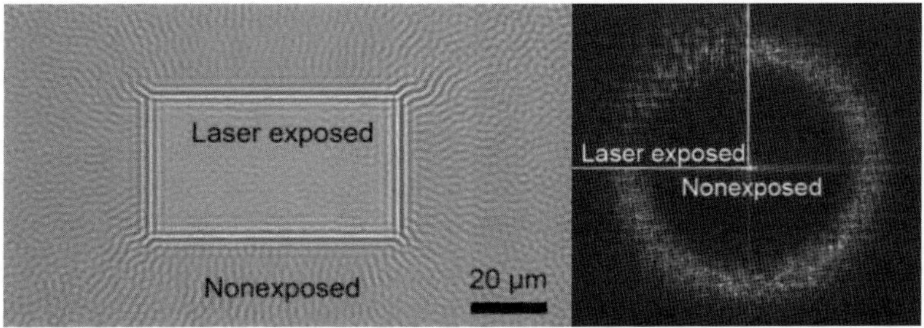

Fig. 3.21 Wrinkling of a gold film with a large *rectangular area* exposed to laser. This image and corresponding FFT pattern show that the wrinkle wavelength and amplitude in the laser-exposed area are both smaller than that in the unexposed area

a composite film with two GPs at a distance of 3.6 µm also perfectly fits those two profiles as shown in Fig. 3.22b. The interaction may be treated in a similar manner as interaction of waves so that the resultant wrinkle pattern can be predicted by superposition of the two unit-wrinkles:

$$H(x) = k[h(x) + h(x - d)] \tag{3.16-1}$$

where d is the distance between two GPs, $h(x)$ and $h(x{-}d)$ are the height profiles of the unit-wrinkles as given by Eq. 3.15, and k is a dimensionless parameter. The superposition effect is even valid in two-dimensional case,

$$H(x, y) = k \sum_{i=1}^{n} h_i(x, y) \tag{3.16-2}$$

where $h_i(x,y)$ represents the profile of one unit-wrinkle that depends on the location and orientation of the corresponding GP. Figure 3.22c shows an AFM image of two perpendicular unit-wrinkles, for which the wrinkle profile calculated by Eq. 3.16-2 matches the experimental profile remarkably well.

The parameter k, however, is not always equal to 1. Actually, k is related to the interval of the GPs. In Fig. 3.23a, k is a calculated value as a function of the pitch of GPs, d. The parameter k could be larger or smaller than 1, depending on the pitch of GPs, and it oscillates between peaks at $(2n + 1)\lambda_i/2$ and valleys at $n\lambda_i$, where n is a natural number. Experimental result of k shown in Fig. 3.23b agrees reasonably well with the calculated result, but there is a difference compared to calculated k. And k finally approaches to 1 as the pitch d gets large. The parameter k is significant for quantitative design of wrinkle structures. Figure 3.24 shows a set of wrinkle patterns with two parallel GPs in different pitches. By introducing the parameter k, the experimental surface profiles measured by AFM fit perfectly well with the designed profiles.

Equation 3.16 offers the possibility to quantitatively design and fabricate wrinkle patterns. Based on this equation, we can also explain why wrinkle patterns with

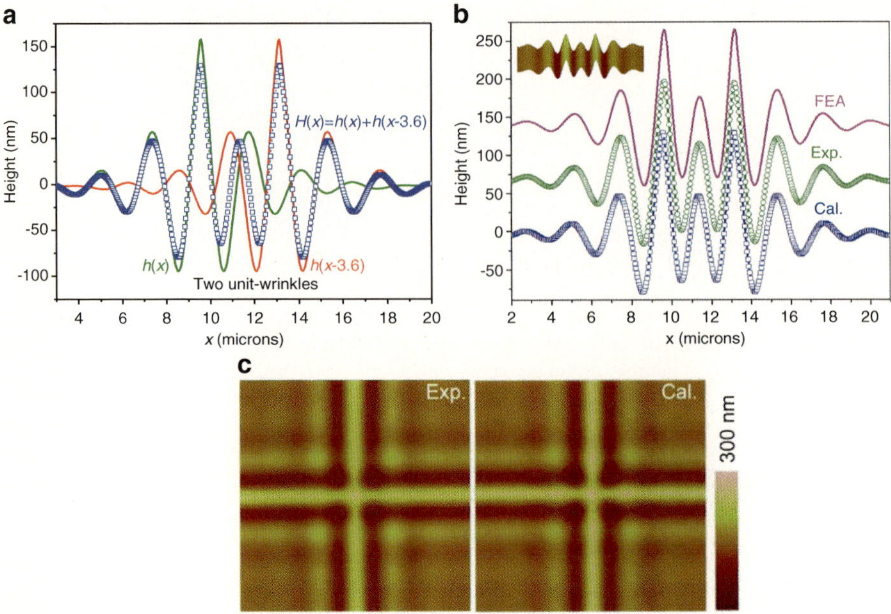

Fig. 3.22 (**a**) Calculated profile (by superposition) of two parallel unit-wrinkles separated by a pitch distance of 3.6 μm. (**b**) Comparison among the experimental, the numerical simulation, and the calculated wrinkle profiles with two parallel GPs (pitch = 3.6 μm). *Inset* is the corresponding AFM image. (**c**) AFM image and calculated topography of the wrinkle pattern with two perpendicular GPs (Reproduced from Ref. [33] by permission of John Wiley & Sons Ltd)

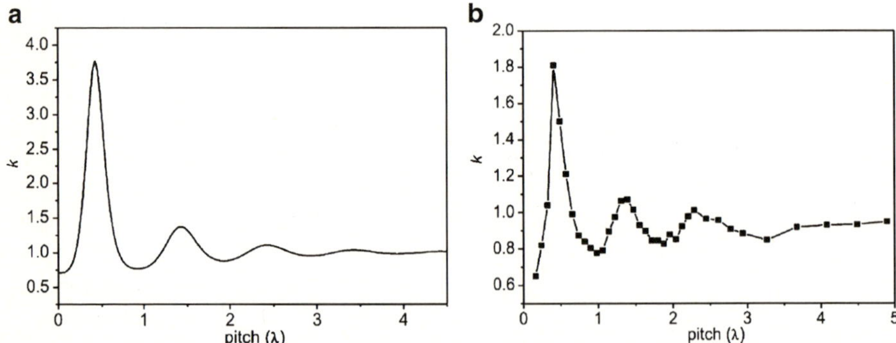

Fig. 3.23 The dimensionless parameter k was determined as a function of the pitch (d) by comparing experimental data of two parallel unit-wrinkles with the superposition prediction in Eq. 3.16-1. For relatively large pitches ($d > 4\ \lambda_i$), k is approximately 1, suggesting a linear superposition. For relatively small pitches, however, k varies with d and oscillates around 1

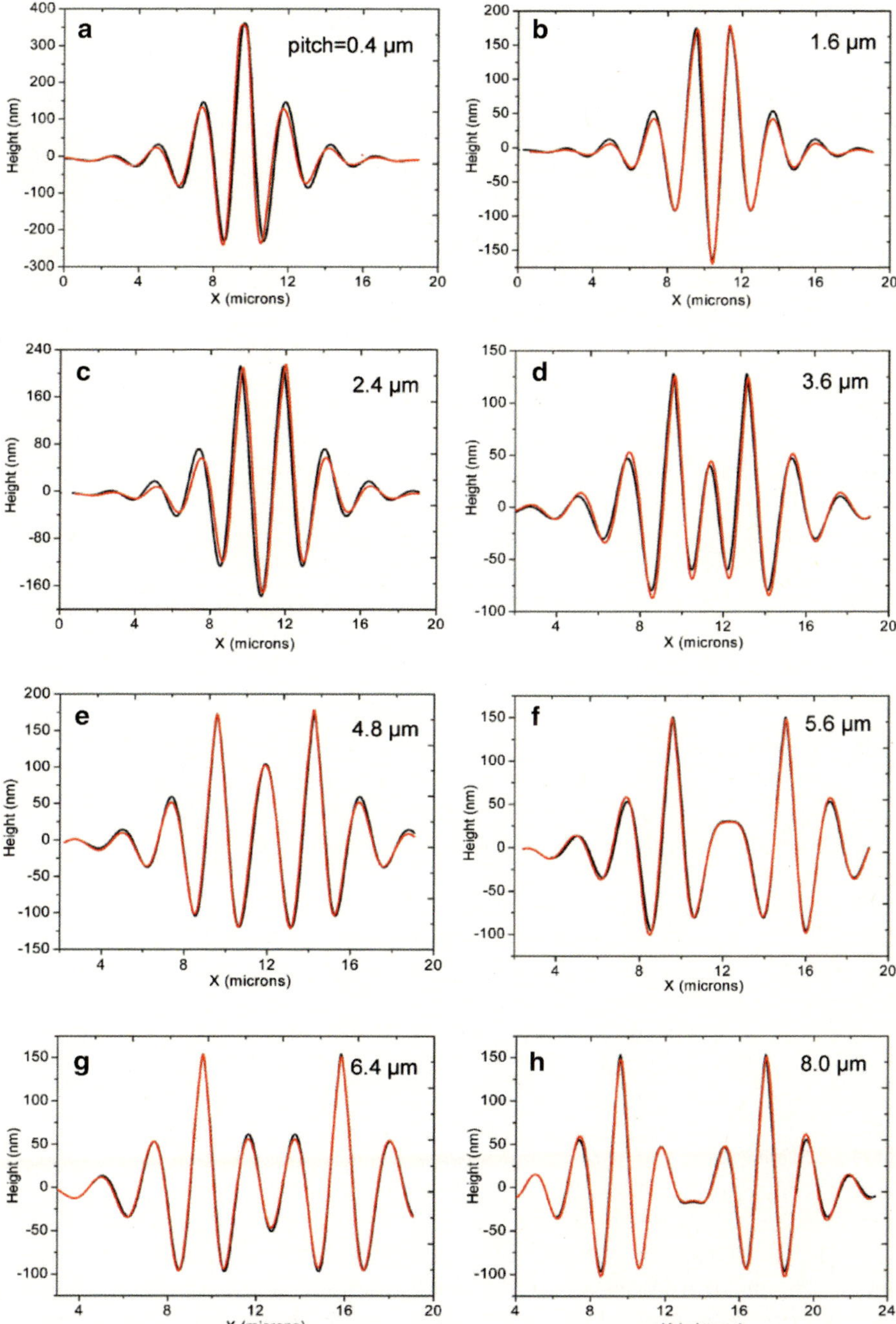

Fig. 3.24 Comparison of calculated (*dark*) surface profiles by superposition to experimental (*red*) profiles with two unit-wrinkles at different pitches, showing that two set of curves can almost perfectly overlap. Note that *k* is considered

Fig. 3.25 Trapezoidal and inverted trapezoidal structures cannot be obtained by other existing methods

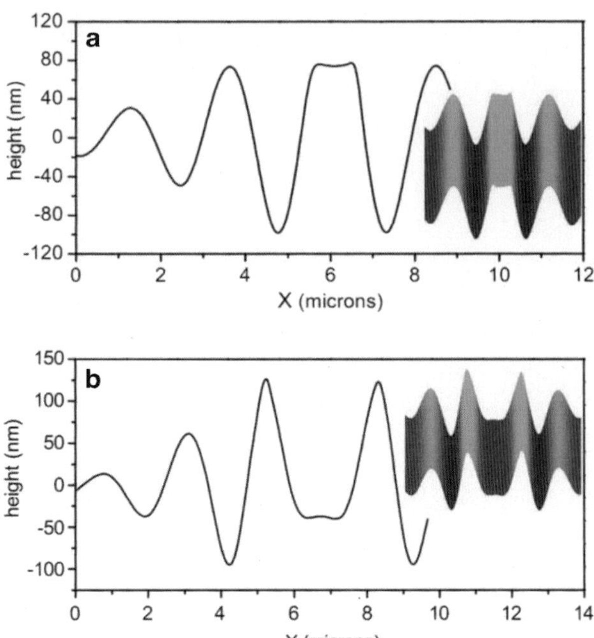

a trapezoidal or inverted trapezoidal surface can be obtained, as shown in Fig. 3.25. Therefore, GP-directed wrinkling could produce diverse topographies, including sinusoidal, flat, or other surfaces, only by simply changing the pitch of GPs.

It has been mentioned that a unit-wrinkle has a shape of evanescently damped profile, with 0th, $\pm$1st, $\pm$2nd... orders. The 0th order centers at the GPs with the largest height, while the $\pm$1st and $\pm$2nd orders with smaller heights are two sides of 0th order. Obviously, the suborders cannot be neglected because their heights are comparable to that of the 0th order. For example, in Fig. 3.17b, c, and g, there are structures made of superposed 1st orders.

The superposed patterns at the edge of line–wrinkle patterns with different pitches are interesting. The misfit between intrinsic wavelength λ_i and line pitch d leads to various patterns, relying on the value of misfit, which is defined as:

$$f = \frac{|d - \lambda_i|}{\lambda_i} \tag{3.17}$$

Figure 3.26 shows one set of optical micrographs of line structures with different misfits. When f approaches to 0, a hexagonal dot lattice is formed outside the lines. As f increases close to 0.414, a tetragonal lattice is formed. The results are obviously in accordance with the wave superposition effect. Upon this, it is possible to design complex surface patterns by varying the pitches of GPs.

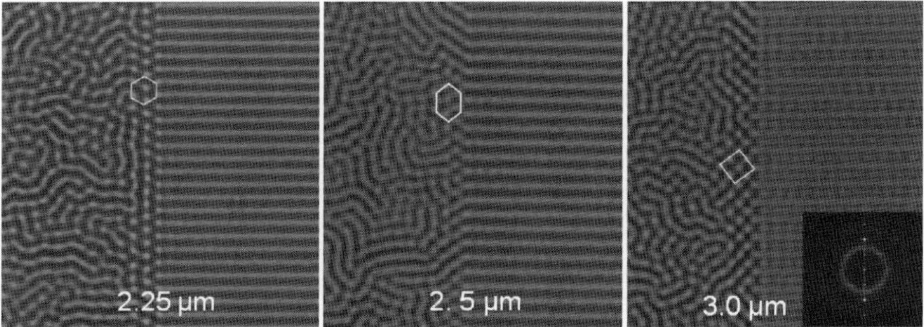

Fig. 3.26 Wrinkle patterns at the edge of GP-guided lines. The lines are directed by GPs with different pitches of 2.25, 2.50, and 3.00 µm, respectively ($\lambda_i = 2.19$ µm), and corresponding $f = 2.7$ %, 14.2 %, and 37.0 %, respectively

3.4.4   Versatile Controllability of Wrinkles

3.4.4.1   Wavelength Control

The wavelength of wrinkle depends on the structure and mechanical properties of the bilayer. It is often an intrinsic value for a given system. However, GP-directed wrinkling could mandatorily tailor the wavelength of wrinkles by changing the pitch of GPs. For the Au/PS bilayer system with an intrinsic wrinkle wavelength of $\lambda_i = 2.1$ µm, the LDW-guided wrinkle wavelengths ranging from ~0.6 to ~2.8 µm can be obtained (Fig. 3.27). The range of the tunable wrinkle wavelength is related to the intrinsic wrinkle wavelength λ_i of the bilayer system, and it can be calculated with the superposition principle. Figure 3.27a, b is experimental and calculated profiles with GP pitches of 0.6, 1.8, and 2.8 µm, respectively, showing that experimental and calculated profiles are consistent with each other. By changing pitches from 0.59 to 3.15 µm, we could find that the peak splits only in the pitch range from ~0.6 to ~2.8 µm. When the pitch is larger than 2.8 µm, a suborder peak emerges between the 0th orders, and we consider the wrinkle wavelength halves.

It is obvious the adjustment of wavelength is also limited by the feature size of GPs. By using GPs with smaller feature sizes (which might be realized with a scanning near-field microscope or AFM), wrinkle patterns with wavelengths less than 300 nm might be achieved, and accordingly, bilayer with a smaller λ_i should be employed. The tunability of the wrinkle wavelength is very useful for making surface structures with different periods such as Fresnel lenses (Fig. 3.28), which have been proven to be potentially useful in micro-optics.

3.4.4.2   Height Control

In addition, the wrinkle height can be controlled by adjusting the laser exposure dose in selected areas. Generally speaking, wrinkle height related to compressive

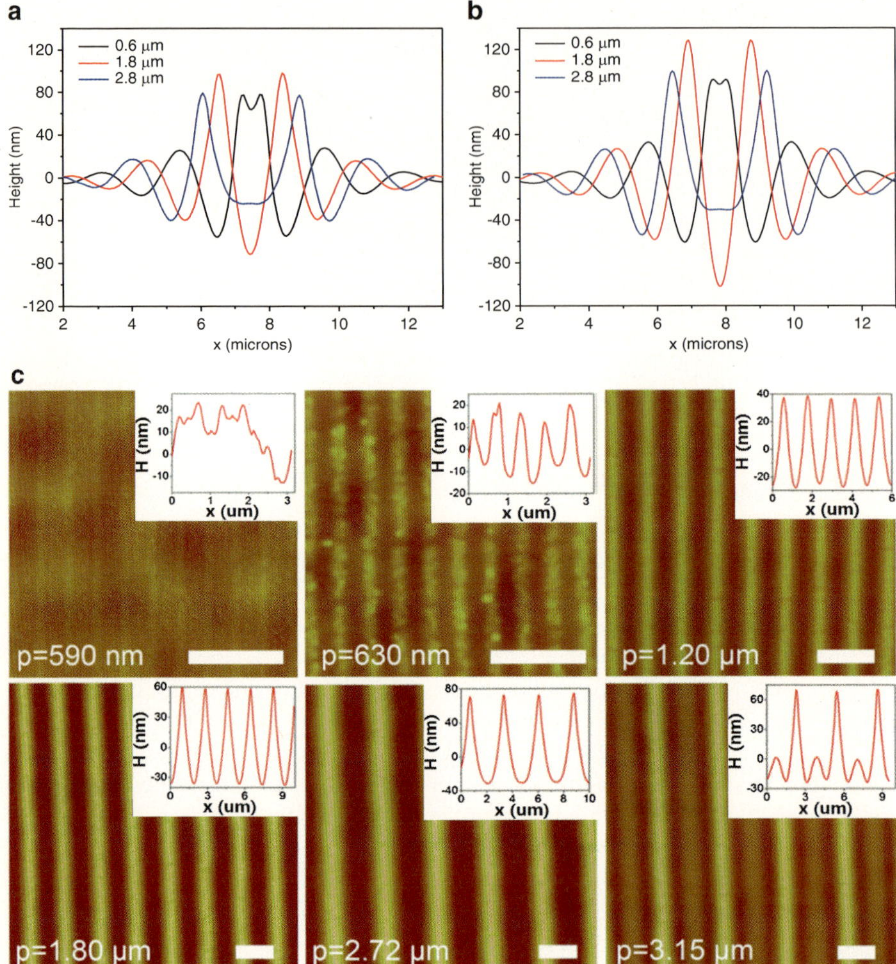

Fig. 3.27 Pitch control of wrinkles. (**a**) Experimental surface profiles of wrinkle patterns with two parallel GPs that are with pitches of 0.6, 1.8, and 2.8 μm, respectively. (**b**) Calculated surface profiles by superposition of two unit-wrinkles with pitches of 0.6, 1.8, and 2.8 μm, in good agreement with the experimental results in panel (**a**). (**c**) AFM images of parallel wrinkles formed by using multiple GPs with pitches (*p*) of 590 nm, 630 nm, 1.20 μm, 1.80 μm, 2.72 μm, and 3.15 μm. Scale bars are 2 μm. *Insets* are corresponding profiles by cross-sectional analysis

stress and mechanical properties of materials can be adjusted by heating the bilayer at different temperatures and/or for different durations or by changing the mechanical properties of the bilayer. Local control of wrinkle height, however, should be too hard to be realized by locally changing stress distribution at the micron scale, and this has been identified by many previous works. With the capability of selected-area patterning, we could locally modify the elastic modulus of the metallic film and thus control the wrinkle height selectively. As shown in

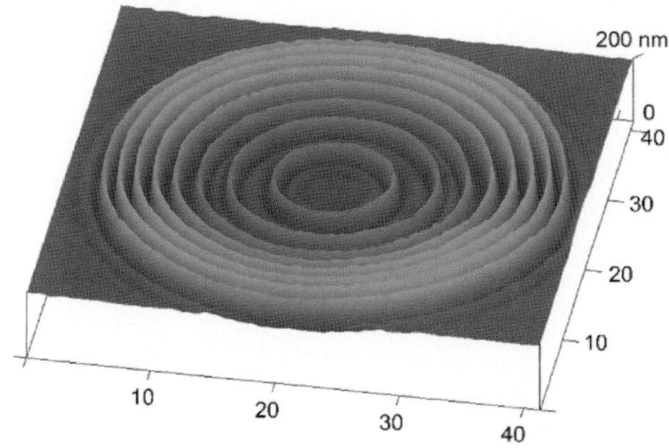

Fig. 3.28 A Fresnel lens obtained by GP-directed wrinkling (Reproduced from Ref. [33] by permission of John Wiley & Sons Ltd)

Fig. 3.29a, the height of the unit-wrinkle increases with the laser power within a certain power range (1.0–2.8 mW). This can be understood as a result of decreasing elastic modulus in the laser-exposed area due to increasing laser power, as shown in Fig. 3.30, with numerical simulations. The laser exposure onto the metal film causes a softening effect, leading to a decreased Young's modulus $E_S = SE_m$ $(0 < S < 1)$ of the metal film in the exposed area. Higher laser power leads to smaller S, and corresponding unit-wrinkle has a higher 0th order. However, no further softening effect is observed as the laser power increases to higher than 2.4 mW, indicating a saturation exists in the softening effect.

One thing worth noting is that the laser exposure on metal film may also cause a hardening effect for the skin of the underlying PS film. Thus unit-wrinkle can still emerge in areas where laser writes even if the Au film exposed becomes apparently discontinuous, and this will be further discussed elsewhere.

Interestingly, by varying the laser power along one GP, the height of the unit-wrinkle can be controlled continuously along the path (e.g., from 0 to 148 nm as shown in Fig. 3.29b). Furthermore, we note that the unit-wrinkle can have a much larger height-to-wavelength aspect ratio than that of typical wrinkle patterns $(A/\lambda \sim 0.1)$. By heating the bilayer to a relatively high temperature and holding for a longer time (~4 h), a higher wrinkle amplitude can be achieved at the 0th order while the wrinkle wavelength does not change significantly, as predicted by the theory of viscoelastic wrinkling [43]. A unit-wrinkle with a height of 586 nm and a width of ~1,700 nm is demonstrated in Fig. 3.31, corresponding to an aspect ratio of 0.34. Although previous works with high aspect ratios by sequential deposition have been reported elsewhere [21], the wrinkle patterns are not well ordered.

3.4.4.3   Local Patterning and Pattern Alignment

The GPs buckle earlier upon heating, implying smaller critical wrinkling stress (σ_c) at the GPs. Thus, it is possible to generate wrinkles only in the patterned areas by

Fig. 3.29 Height control of wrinkles. (**a**) Wrinkle height versus laser power, showing that wrinkle height increases with the increase of laser power. The *inset* defines the wrinkle amplitudes (0th, 1st, and 2nd orders). (**b**) Tilt view of an AFM image showing the wrinkle pattern with the height varying continuously along the guiding path due to varying laser power (Reproduced from Ref. [33] by permission of John Wiley & Sons Ltd)

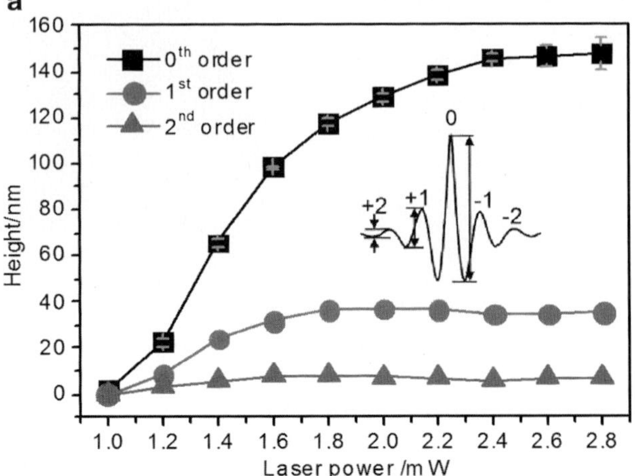

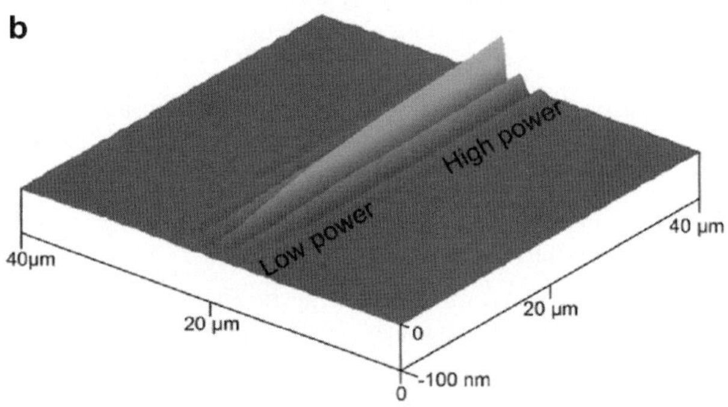

Fig. 3.30 Comparison between simulated (*solid line*) and experimental (*open circles*) profiles of unit-wrinkles with different softening parameters/laser powers. The 0th order wrinkle amplitude increases with the decrease of Young's modulus in the softer region, due to increasing laser power

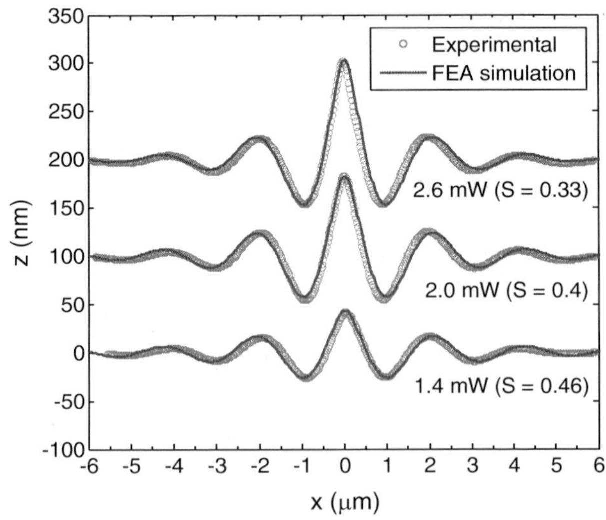

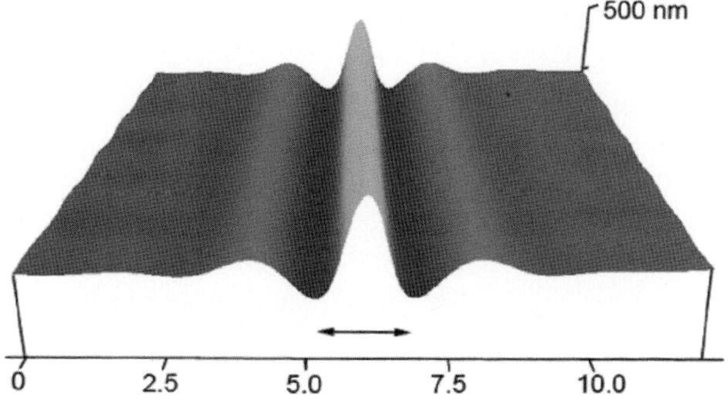

Fig. 3.31   A unit-wrinkle with an aspect ratio of 0.34 for the 0th order. The height is 586 nm and the width is 1,700 nm. Such unit-wrinkles can be obtained by heating the bilayer with a GP at 140 °C for several hours

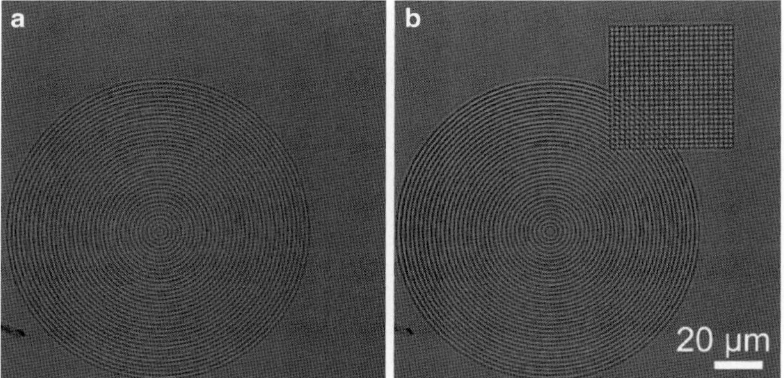

Fig. 3.32   (**a**) Locally patterned structure. (**b**) New wrinkle patterns can be generated by rewriting and reheating. It provides the possibility to alter original wrinkle structures

precisely controlling the pre-buckling stress slightly higher than the σ_c of patterned areas. This is quite valuable because we often just need the desired wrinkles and hope to prevent the generation of disordered wrinkles in non-patterned areas. Figure 3.32a shows an optical micrograph of a GP-guided concentric ring structure, outside of which there is no wrinkle at all. It is interesting that non-wrinkled or even wrinkled areas can be rewritten, followed by reheating to generate new wrinkle patterns. This is similar to the alignment process in conventional photolithography. Figure 3.32b shows that after rewriting and reheating, a new tetragonal pattern forms in the sample with a small part overlapping with the original target pattern. The local patterning and alignment of wrinkles offers a route to alter original wrinkle patterns.

3.5   Applications of Path-Guided Wrinkle Structures

Path-guided wrinkling enables the fabrication of the high-quality surface structures via a simple process. And these structures have potential applications in many areas, including optics, MEMs, microfluidics, photovoltaics, particle alignment, and mechanical property test. Here we show some applications of wrinkle structures in optics and particle alignment.

3.5.1   Fresnel Lenses

Wrinkle structures have been used in optical applications, such as one-dimensional sinusoidal phase gratings based on line–wrinkle structures, a wrinkle-based complex eye structure developed by E. Chan [3]. These works convince that wrinkled surface is useful in some optical applications.

Since path-guided wrinkling is able to make wrinkle patterns in high quality, and we can exactly control the amplitude and wavelength of wrinkles, it is possible to obtain optical devices with higher performance. For example, Fig. 3.33a shows a Fresnel lens made up of wrinkles with tapered pitches. This Fresnel lens presents good imaging effect as shown in Fig. 3.33b. The resolution of the image (the letter "A") could reach up to ~600 nm by using a green light source (546 nm). Figure 3.33d shows the imaging effect of a 3×3 lens array. Figure 3.33c demonstrates the good focusing effect of a 2×2 Fresnel lens array irradiated by the light beam with the spot size of ~1.5 μm. All the imaging and focusing experiments were performed under an Olympus microscope (Olympus BX-51) with a method similar to that reported by E. Chan et al. [3]. The focal length of the lenses was measured to be ~42 μm, matching perfectly well with the designed focal length according to:

$$f_N = \frac{\rho_N^{\,2}}{N\lambda_l} \tag{3.18}$$

where f_N is the focal length, ρ_N is the radius of the wave zone plate, N is the amount of circles, and λ_l is the wavelength of the light source. Note that this is a design for amplitude type Fresnel lens instead of phase type, and it is still unknown how it works in our Fresnel lens, since the wavy-surfaced lens also leads to phase change, which is actually dominant.

In comparison, concentric ring structure with a fixed pitch does not show imaging and focusing effects. Figure 3.34a demonstrates a concentric ring structure, and panel (b) is the corresponding focusing experiments without intensely focused spot. This indicates that the focusing effect of the Fresnel lenses stems from a periodic design rather that the circles themselves.

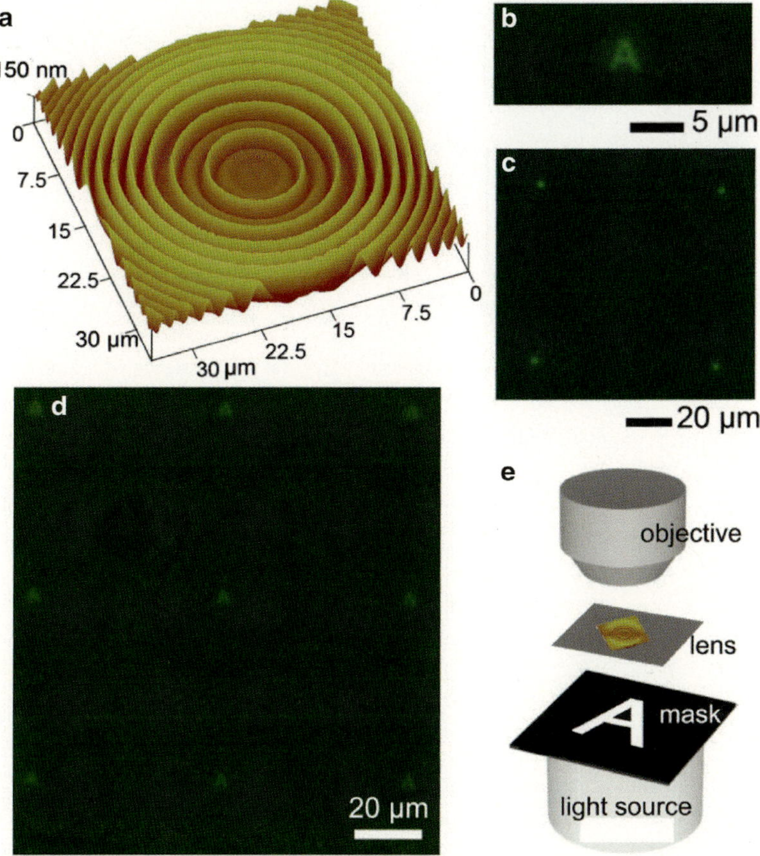

Fig. 3.33 (**a**) AFM image of a Fresnel lens made up of wrinkles with tapered pitches. (**b**) Imaging effect (showing a letter "A") of a Fresnel lens. (**c**) Focusing effect of a 2 × 2 lens array. (**d**) Imaging effect of a 3 × 3 lens array. (**e**) Schematic illustration of the focusing and imaging experiments

3.5.2   Talbot Effect

The Talbot effect refers to an optical phenomenon that periodic object can have a series of self-images at regular distances behind the object plane under illumination of a plan wave or a spherical wave. The repeated image is called Talbot image, and the regular distance is called Talbot distance. Also there are self-images with a phase shift of π, at a series of half Talbot distances. A wrinkle grating with a pitch

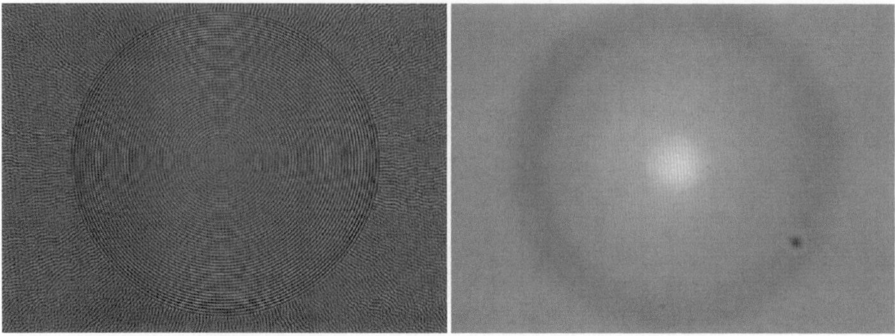

Fig. 3.34 A concentric ring array and its focusing effect (at the z distance with a *brightest spot*)

(λ) of 1.73 µm presents a Talbot distance Z_T of 10.7 µm at 546 nm, quite close to the calculated value of 10.9 µm, according to:

$$Z_T = \frac{2\lambda^2}{\lambda_l} \tag{3.19}$$

and from

$$n \cdot Z_T = \frac{N\lambda}{2\lambda_l} \tag{3.20}$$

many Talbot images behind such 200 µm $\times$ 200 µm ($N\lambda \times N\lambda$) wrinkle grating should be observed in theory, however, the number (n_{exp}) of Talbot images observed in experiment is ~10, which depends on diffraction efficiency of the wrinkle grating. The Talbot experiment indicates that the wrinkle patterns have strict periodicity and are in good order.

The Talbot experiment can also be applied to detect defects in wrinkle patterns. For example, the wrinkle-grating image and the Talbot image are quite different at a defect (indicated by the white arrow and might be caused by a small particle), as shown in Fig. 3.35.

3.5.3 Beam Splitter

The two-dimensional sinusoidal phase grating (2DSPG) is very useful in optics, especially when used as beam splitter. Beam splitter is an important optical device in medical science, military, and optical communications. The 2DSPGs are often made by means of laser interference lithography, which is relatively expensive. Patterned wrinkling could produce one-dimensional sinusoidal gratings, as reported by C. Harrison et al. [2]. However, there is still no report on fabricating 2DSPG by

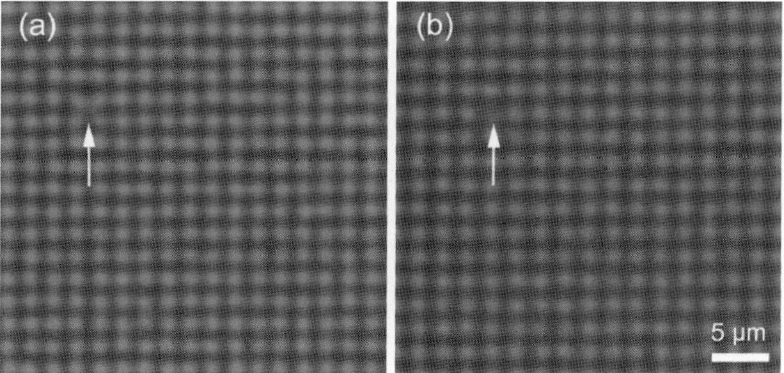

Fig. 3.35 (**a**) Wrinkle grating image. (**b**) Talbot image of wrinkle grating at Z_T

surface wrinkling. Although Lin et al. developed a sequential and unequal biaxial stretching method to produce ordered herringbone structures made up of wrinkles in two directions [17], this method, however, is still incapable of making 2DSPGs.

The fabrication of tetragonal 2DSPG can be simply realized by writing a tetragonal grid and then heating. The mesh size of the tetragonal grid should be close to λ_i. A 2DSPG is composed of two sets of perpendicularly woven wrinkles by superposition. Figure 3.36 is a simple setup for the beam splitting experiment. The laser beam coming from left hits the 2DSPG, and the diffractive spectrum is recorded on a screen placed behind the grating. Figure 3.37 shows a schematic illustration together with an experimental diffractive pattern. The first-order diffractive beams have a diffractive angle:

$$\theta = \arcsin\left(\frac{\lambda_l}{2\lambda}\right) \tag{3.21}$$

The diffractive efficiency of the 2DSPG, however, is not high, because there is a strong reflection from Au film on the PS film. Moreover, the profile of the 2DSPG should be optimized for higher diffraction efficiency: the crest-to-trough height is about 200 nm in our case (which does not cause enough phase change), this is too small compared with the desired height of ~1,000 nm, implying that we have to use much thicker PS film.

3.5.4 Template

The wrinkle pattern can be a template for the alignment of nanoparticles, microspheres, etc [28–30]. However, previous wrinkle templates were not perfect, and thus, the configuration of nanoparticles or microspheres was not so good. By GP-guided wrinkle patterns, we employed the wrinkle structures as the templates to

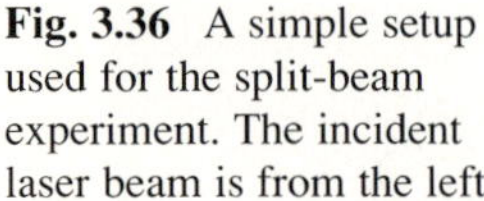

Fig. 3.36 A simple setup used for the split-beam experiment. The incident laser beam is from the left

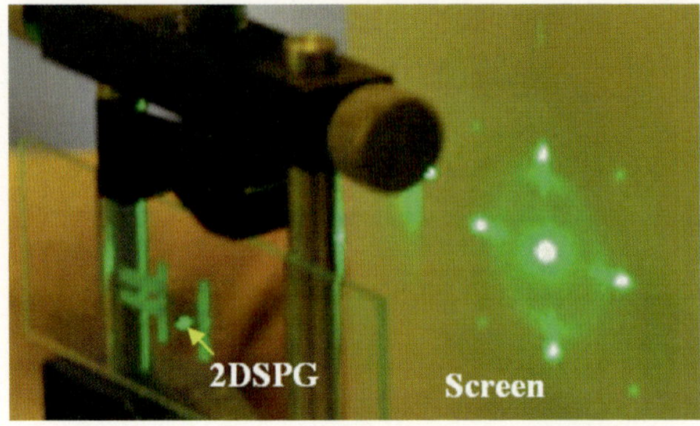

Fig. 3.37 Schematic illustration and experimental diffractive pattern obtained by the wrinkle-based 2DSPG

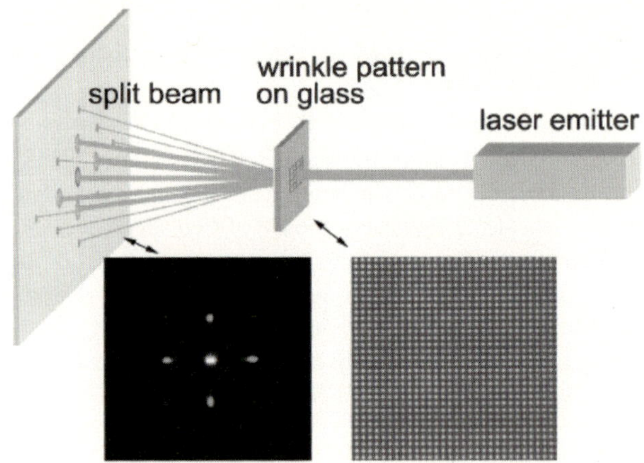

make regular lattice of PS spheres. Figure 3.38a, b show a tetragonal wrinkle pattern (actually it is a 2DSPG), which is very similar to an egg-crate structure (Fig. 3.38c). This egg-crate structure could regulate PS spheres which have a diameter quite close to the wavelength of the wrinkles. In a tetragonal lattice shown in Fig. 3.38d, PS spheres fall exactly on the troughs. PS spheres can also be regulated to other shapes, including circles, lines, or two-dimensional structures, as displayed in Fig. 3.38e, f.

3.6   Conclusion and Perspective

This chapter mainly focuses on path-guided wrinkling, which enables the fabrication of highly ordered, defect-free, and arbitrary-shaped surface structures. The surface profile of the wrinkle structures can be designed quantitatively, and the

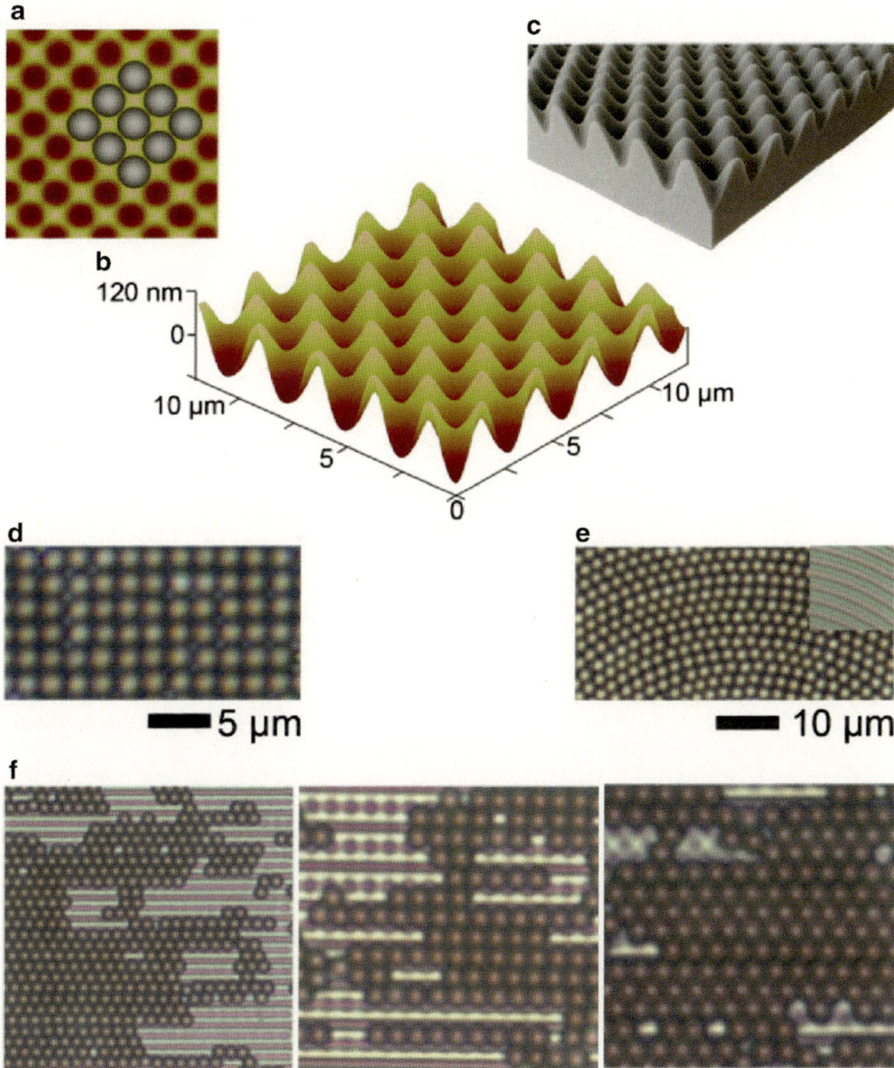

Fig. 3.38 (**a**) An egg-crate structure used as the template for aligning microspheres. (**b**) Tilt view of the egg-crate structure. (**c**) A real egg-crate. (**d**) Tetragonal PS microsphere lattice directed by an egg-crate wrinkle structure. (**e**) PS microspheres aligned in a concentric pattern. The *inset* shows the optical image of the template made of concentric circular wrinkles. (**f**) Other PS sphere structures regulated by different wrinkle patterns. Note that the last one is a distorted hexagonal structure (Reproduced from Ref. [33] by permission of John Wiley & Sons Ltd)

formed surface patterns have many potential applications in micro-/nanofluidics, solar cells, MEMs, and so on, besides those mentioned in this chapter. It should be noted that the unit-wrinkle could be a brand-new and powerful probe for the test of mechanical properties of thin films because it has three parameters, A_0, λ_i, and the critical damping length l_c, among which l_c is very sensitive to the modulus and thickness of the metal film and has not been used for measuring mechanical properties of thin films before.

In the future, there will be more efforts to improve the guided-path wrinkle technique for higher ordering and more desired configuration. Some new wrinkle modes such as "fold" [46] will be paid more and more attention by scientific researchers, and corresponding applications will be extended to more areas.

References

1. Bowden N, Brittain S, Evans AG, Hutchinson JW, Whitesides GW (1998) Spontaneous formation of ordered structures in thin films of metals supported on an elastomeric polymer. Nature 393:146–149
2. Harrison C, Stafford CM, Zhang W, Karim A (2004) Sinusoidal phase grating created by a tunably buckled surface. Appl Phys Lett 85:4016–4018
3. Chan EP, Crosby AJ (2006) Fabricating microlens arrays by surface wrinkling. Adv Mater 18:3238–3242
4. Sun Y, Choi WM, Jiang H, Huang Y, Rogers JA (2006) Controlled buckling of semiconductor nanoribbons for stretchable electronics. Nat Nanotechnol 1:201–207
5. Efimenko K, Rackaitis M, Manias E, Vaziri A, Mahadevan L, Genzer J (2005) Nested self-similar wrinkling patterns in skins. Nat Mater 4:293–297
6. Stafford CM, Harrison C, Beers KL, Karim A, Amis EJ, VanLandingham MR, Kim H-C, Volksen W, Miller RD, Simonyi EE (2004) A buckling-based metrology for measuring the elastic moduli of polymeric thin films. Nat Mater 3:545–550
7. Stafford CM, Vogt BD, Harrison C, Julthongpiput D, Huang R (2006) Elastic moduli of ultrathin amorphous polymer films. Macromolecules 39:5095–5099
8. Chung JY, Chastek TQ, Fasolka MJ, Ro HW, Stafford CM (2009) Quantifying residual stress in nanoscale thin polymer films via surface wrinkling. ACS Nano 3:844–852
9. Kwon SJ, Yoo PJ, Lee HH (2004) Wave interactions in buckling: self-organization of a metal surface on a structured polymer layer. Appl Phys Lett 84:4487–4489
10. Bowden N, Huck WTS, Paul K, Whitesides GW (1999) The controlled formation of ordered, sinusoidal structures by plasma oxidation of an elastomeric polymer. Appl Phys Lett 75:2557–2559
11. Ohzono T, Shimomura M (2004) Ordering of microwrinkle patterns by compressive strain. Phys Rev B 69:132202–132206
12. Ohzono T, Matsushita SI, Shimomura M (2005) Coupling of wrinkle patterns to microsphere-array lithographic patterns. Soft Matter 1:227–230
13. Muller-Wiegand M, Georgiev G, Oesterschulze E, Fuhrmann T, Salbeck J (2002) Spinodal patterning in organic–inorganic hybrid layer systems. Appl Phys Lett 81:4940–4942
14. Huck WTS, Bowden N, Onck P, Pardoen T, Hutchinson JW, Whitesides GM (2000) Ordering of spontaneously formed buckles on planar surfaces. Langmuir 16:3497–3501
15. Yoo PJ, Suh KY, Park SY, Lee HH (2002) Physical self-assembly of microstructures by anisotropic buckling. Adv Matter 14:1383–1387

16. Jiang C, Singamaneni S, Merrick E, Tsukruk VV (2006) Complex buckling instability patterns of nanomembranes with encapsulated gold nanoparticle arrays. Nano Lett 6:2254–2259
17. Lin P-C, Yang S (2007) Spontaneous formation of one-dimensional ripples in transit to highly ordered two-dimensional herringbone structures through sequential and unequal biaxial mechanical stretching. Appl Phys Lett 90:241903–241905
18. Vandeparre H, Léopoldès J, Poulard C, Desprez S, Derue G, Gay C, Damman P (2007) Slippery or sticky boundary conditions: control of wrinkling in metal-capped thin polymer films by selective adhesion to substrates. Phys Rev Lett 99:188302–188305
19. Vandeparre H, Damman P (2008) Wrinkling of stimuloresponsive surfaces: mechanical instability coupled to diffusion. Phys Rev Lett 101:124301–124304
20. Chung JY, Nolte AJ, Stafford CM (2009) Diffusion-controlled, self-organized growth of symmetric wrinkling patterns. Adv Mater 21:1358–1362
21. Ahmed SF, Rho GH, Lee KR, Vaziri A, Moon M-W (2010) High aspect ratio wrinkles on a soft polymer. Soft Matt 6:5709–5714
22. Garnier GM, Croll AB, Davis CS, Alfred J (2010) Crosby, contact-line mechanics for pattern control. Soft Matt 6:5789–5794
23. Moon M-W, Lee SH, Sun J-Y, Oh KH, Vaziri A, Hutchinson JW (2007) Wrinkled hard skins on polymers created by focused ion beam. Proc Natl Acad Sci 104:1130–1133
24. Moon M-W, Lee SH, Sun J-Y, Oh KH, Vaziria A, Hutchinson JW (2007) Controlled formation of nanoscale wrinkling patterns on polymers using focused ion beam. Scripta Mater 57:747–750
25. Chung S, Lee JH, Moon M-W, Han J, Kamm RD (2008) Non-lithographic wrinkle nanochannels for protein preconcentration. Adv Mater 20:3011–3016
26. Chan EP, Smith EJ, Hayward RC, Crosby AJ (2008) Surface wrinkles for smart adhesion. Adv Mater 20:711–716
27. Kim JB, Kim P, Pegard NC, Oh SJ, Kagan C, Fleischer JW, Stone HA, Loo Y-L (2012) Wrinkles and deep folds as photonic structures in photovoltaics. Nat Photonics 6:327–332
28. Hyun DC, Moon GD, Cho EC, Jeong U (2009) Repeated transfer of colloidal patterns by using reversible buckling process. Adv Funct Mater 19:2155–2162
29. Lu H, Mohwalda H, Fery A (2007) A lithography-free method for directed colloidal crystal assembly based on wrinkling. Soft Matter 3:1530–1536
30. Schweikart A, Fortini A, Wittemann A, Schmidt M, Fery A (2010) Nanoparticle assembly by confinement in wrinkles: experiment and simulations. Soft Matter 6:5860–5863
31. Rogers JA, Someya T, Huang Y (2010) Materials and mechanics for stretchable electronics. Science 327:1603–1607
32. Cerda E, Chandar KR, Mahadevan L (2002) Thin films: wrinkling of an elastic sheet under tension. Nature 419:579–580
33. Guo CF, Nayyar V, Zhang Z, Chen Y, Miao J, Huang R, Liu Q (2012) Path-guided wrinkling of nanoscale metal films. Adv Mater 24:3010–3014
34. Allen HG (1969) Analysis and design of structural sandwich panels. Pergamon, New York
35. Yoo PJ, Lee HH (2005) Morphological diagram for metal/polymer bilayer wrinkling: influence of thermomechanical properties of polymer layer. Macromolecules 38:2820–2831
36. Okayasu T, Zhang HL, Bucknall DG, Briggs GAD (2004) Spontaneous formation of ordered lateral patterns in polymer thin-film structures. Adv Funct Mater 14:1080–1088
37. Tu TX, Stronge WJ (1985) Wrinkling of a circular elastic plate stamped by a spherical punch. Int J Solids Struct 21:995–1003
38. Bullough M, Cui Y (2012) A library of large-scale surface patterns induced by flame on elastomers. Soft Matt 8:3304–3307
39. McDonald PJ, Godward J, Sackin R, Sear RP (2001) Macromolecules 34:1048
40. Guan L, Peng K, Yang Y, Qiu X, Wang C (2009) The nanofabrication of polydimethylsiloxane using a focused ion beam. Nanotechnology 20:145301–145305
41. Kawata S, Sun H-B, Tanaka T, Takada K (2001) Finer features for functional microdevices. Nature 412:697–698

42. Emery RD, Povirk GL (2003) Surface flux limited diffusion of solvent into polymer. Tensile behavior of free-standing gold film. Part I. Coarse-grained films. Acta Mater 51:2067–2078
43. Huang R (2005) Kinetic wrinkling of an elastic film on a viscoelastic substrate. J Mech Phys Solids 53:63–89
44. Huang ZY, Hong W, Suo Z (2005) Nonlinear analyses of wrinkles in a film bonded to a compliant substrate. J Mech Phys Solids 53:2101–2118
45. Mei H, Landis CM, Huang R (2011) Concomitant wrinkling and buckle-delamination of elastic thin films on compliant substrates. Mech Mater 43:627–642
46. Kim JB, Kim P, Pegard NC, Oh SJ, Kagan CR, Fleischer JW, Stone HA, Loo Y-L (2012) Wrinkles and deep folds as photonic structures in photovoltaics. Nat Photonics 6:325–332

Chapter 4
Laser Micro-/Nanofabrication and Applications Based on Multiphoton Process

4.1 Introduction

Optical lithography is the engine that has powered the semiconductor revolution. The fabrication of electronic circuits on chips relies on the patterning of surfaces by optical lithography, which is used to control where different components – metal wires, semiconductor gates, and oxide insulators – form. The demand for increasingly powerful integrated circuits has spurred remarkable progress in optical lithographic techniques in the past decades [1]. In optical lithography, a light-sensitive film, called photoresist, is exposed in selected areas by using a patterned mask. The light triggers chemical reactions that change the film's solubility. Solvents are then used to remove the exposed or unexposed areas, so that only selected areas on the chip undergo the next processing step. For example, after selective removal of photoresist, protected parts of a semiconductor layer become separated gate regions, whereas exposed regions are open for doping or deposition of electrodes. Feature sizes as small as 32 nm can now be achieved in device fabrication, beating the diffraction limit set by the wavelength of the deep-ultraviolet (DUV) light used for exposure (193 nm) through clever optical tricks [2]. However, the light sources and the masks that create the patterns are costly; even higher costs can be anticipated for the shorter wavelengths needed for even smaller feature sizes.

On the other hand, the materials processing with lasers take advantage of virtually all of characteristics of laser light. The high energy density and directionality achieved with lasers permit strongly localized heat or photo treatment of materials. The developed laser processing techniques, in particular, laser direct writing technique, provide a cheap, fast, and designable tool for fabricating microstructures, which have been widely used in microelectronic industry. However, with the increasing demands for the spatial resolution at nanometric scale, the conventional laser processing technique is difficult to obtain a spatial resolution of better than 100 nm due to the optical diffraction limit.

Q. Liu et al., *Novel Optical Technologies for Nanofabrication*, Nanostructure
Science and Technology, DOI 10.1007/978-3-642-40387-3_4,
© Springer-Verlag Berlin Heidelberg 2014

As a high-power ultrashort laser, femtosecond laser has gained much attention due to its unique physical property. Because the pulse durations of these lasers are shorter than the typical material relaxation time, the physics and technology of the laser–matter interactions are generally determined by the laser irradiation. Under the intense intensity of femtosecond laser pulse, various nonlinear optical effects between laser and matter can occur. Among them, one of the nonlinear optical effects, multiphoton absorption (MPA), is easily induced by femtosecond laser pulse, which provides an opportunity to achieve the spatial resolution smaller than optical diffraction limit with laser direct writing technique. In the past decade, multiphoton-induced polymerization (MPP) and multiphoton photoreduction (MPR) have been developed as two promising techniques for micro-/nanofabrication by using femtosecond laser [3, 4]. MPP and MPR techniques have been successfully used in fabricating three-dimensional (3D) micro-/nanostructures due to its capability of 3D micro-/nanofabrication and high spatial resolution at nanometric scale. The applications based on MPP and MPR techniques have been developed with various functional materials, such as luminescent polymers, metallic materials, and polymer nanocomposites.

In this chapter, we first simply describe the basic principle of the nonlinear optical effect, MPA, and related materials that provide the explanation on the reason that multiphoton nanofabrication can overcome the optical diffraction limit and achieve spatial resolution at nanometric scale. Then, we discuss the materials, typical optical setup, and the fabrication modes used in multiphoton nanofabrication. After figuring out the latest progress on the improvement of spatial resolution with femtosecond laser direct writing technique based on multiphoton lithography, we highlight the 3D micro-/nanostructures fabrication with functional materials by MPP. Finally, the applications of multiphoton micro-/nanofabrication in micro-/nanodevices and micro-/nanoelectromechanical systems (MEMS/NEMS) are presented, which have the potential to make major leaps in a broad range of applications in the near future.

4.2   Principle and Molecular Design for Multiphoton Absorption

4.2.1   Nonlinear Optics and Multiphoton Absorption

The great advantage of ultrafast laser pulse is that all of their energy is crammed into a very short time, so they have very high peak power and intensity. A typical ultrashort pulse from a Ti–sapphire laser oscillator has a paltry nanojoule of energy, but it is crammed into 100 fs, so its peak power is 10,000 W. It can be focused on a spot with the diameter of micrometer or less, yielding an intensity more than 10^{12} W cm^{-2}. And it is easy to amplify such pulse by a factor of 10^6. By this means, it is easy to achieve high-intensity effects which cannot be obtained with an ordinary light source.

The fundamental equation of optics is the wave equation:

$$\frac{\partial^2 E}{\partial z^2} - \frac{1}{c^2}\frac{\partial^2 E}{\partial t^2} = \mu_0 \frac{\partial^2 P}{\partial t^2} \qquad (4.1)$$

where μ_0 is the magnetic permeability of free space and c is the speed of light in vacuum. E is the real electric field, and P is the real induced polarization. The induced polarization contains the effects of light on the medium, including linear optical effects, such as absorption and refraction and also nonlinear optical effects.

At low intensity, the induced polarization is proportional to the electric field:

$$P = \varepsilon_0 \chi^{(1)} E \qquad (4.2)$$

where ε_0 is the electric permittivity of free space and the linear susceptibility, $\chi^{(1)}$, describes the linear optical effects. Under a high intensity of electric field E, the induced polarization is described by the following equation:

$$P = \varepsilon_0 \left[\chi^{(1)} E + \chi^{(2)} E^2 + \chi^{(3)} E^3 + \chi^{(4)} E^4 + \ldots \right] \qquad (4.3)$$

where $\chi^{(2)}$ and $\chi^{(3)}$ are the second-order and the third-order susceptibilities, respectively.

Medium can interact with optical fields in two possible ways, one is through dissipative processes and the other one is through parametric processes. In parametric processes, there is an energy–momentum exchange between different modes of optical field, but there is no energy exchange between the optical field and the medium. However, in dissipative processes, the energy is exchanged between the optical field and the medium through absorption and emission.

For the linear optical effects, the optical response of a medium at an optical frequency ε can be expressed equivalently by the complex refractive index n_c as:

$$n^2{}_c(\omega) = \varepsilon(\omega) = 1 + 4\pi\chi^{(1)}\omega \qquad (4.4)$$

The complex refractive index can be represented as the sum of the real and the imaginary parts as:

$$n_c = n + ik \qquad (4.5)$$

where the real part n corresponds to the dispersion of refractive index, linear parametric process, and the imaginary part k corresponds to linear absorption, linear dissipative process.

In the nonlinear optical effects, the even-order susceptibilities like $\chi^{(2)}$ and $\chi^{(4)}$, do not make a contribution to the dissipative processes except when the external field is a DC field. Therefore, nonlinear dissipative processes are related to multi-photon absorption. The imaginary part of nonlinear susceptibility corresponds to energy transfer from the optical field to the medium, and this energy transfer rate can be expressed as [5]:

$$\frac{dW}{dt} = \langle E \cdot P \rangle \tag{4.6}$$

where E and P are the electric field and the polarization, respectively. The brackets in this equation mean a time average over several cycles of the field. If we only consider nonlinear dissipative processes in the expression for P, the lowest-order nonlinear absorption will be described by the imaginary part of $\chi^{(3)}$, which corresponds to two-photon absorption (TPA), Stokes Raman gain, or anti-Stokes Raman attenuation, and similarly, the imaginary part of $\chi^{(5)}$ relates to three-photon absorption.

The processes of multiphoton absorption were first predicted theoretically by Maria Goeppert-Mayer [6]. Because of the high photon intensities required, even two-photon absorption was not demonstrated until the advent of the laser [7]. Figure 4.1 illustrates the schematic energy level diagram of the simplest case of multiphoton absorption [8]. Two-photon absorption involves the simultaneous absorption of two photons. The photons can have the same energy, E_1 (degenerate case, $E_f = 2E_1$), or different energies, E_1 and E_2 (nondegenerate case, $E_f = E_1 + E_2$). In the case of nondegenerate two-photon absorption, a medium simultaneously absorbs two photons with energies, E_1 and E_2, when the energy $E_1 + E_2$ is in resonance with one of the electronic states. As a result of the absorption process, the two absorbed photons are lost from the excitation beams, correspondingly reducing their intensity, and the medium is brought to an excited state (state f in Fig. 4.1), at E_1+E_2 above the ground state (state g). After excitation, the system relaxes quickly to state r, the lowest vibronic level of the lowest-energy excited state, by internal conversion or vibrational relaxation (dashed arrow). The system finally returns to the ground state by radiative or nonradiative pathways (bold dashed arrow). This process can be regarded as an initial interaction of a photon of energy E_1 with the molecule, which is thus left in a temporary virtual state of energy E_1 above the ground state [9, 10]. This is not a real state (eigenstate) of the molecule, and it exists only for a short time interval, τ_v. If a photon of energy E_2 interacts with the molecule during τ_v, it can be excited to state f. The order of magnitude for τ_v, which can be estimated from the uncertainty principle, is 10^{-15} to 10^{-16} s for photon energies in the visible and near-IR ranges [9, 11]. The qualifier "simultaneous" for TPA is used to indicate that the two photons interact with the molecule within the time τ_v and that no real states act as an intermediate state in this process.

For the monochromatic waves with frequency ω, Eq. 4.6 can be written as:

$$\frac{dW}{dt} = \frac{1}{2}\omega\mathrm{Im}(E \cdot P) \tag{4.7}$$

Fig. 4.1 Schematic energy level diagram showing the excitation of a medium from the ground state, g, to an excited state, f, located at energy E_f above the g state by the absorption of two photons (*vertical solid arrows*) (Reproduced from Ref. [8] by permission from Optical Society of America)

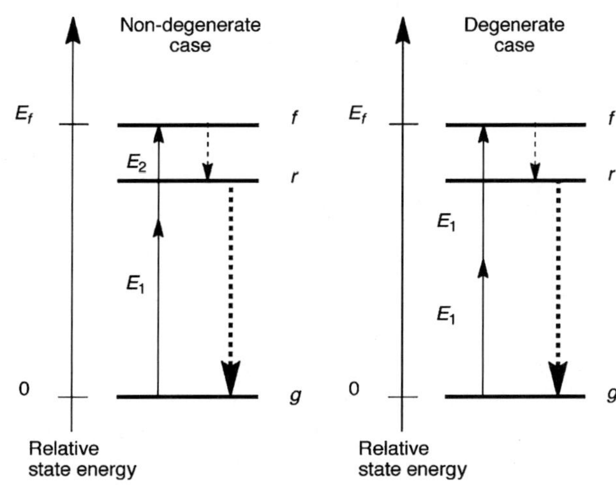

For a degenerate two-photon absorption process, Eq. 4.7 can be transformed into the following expression [12]:

$$\frac{dW}{dt} = \frac{8\pi^2\omega}{n^2c^2}I^2\mathrm{Im}\left(\chi^{(3)}\right) \tag{4.8}$$

where I is the intensity of light and is defined as $I = EE^*nc/8\pi$. From this equation, it should be noted that the rate of energy absorption in this nonlinear absorption process is not linearly but quadratically dependent on the light intensity. Instead of using the extinction coefficient for a linear absorption process, two-photon absorption is described in terms of two-photon absorption cross section, σ_2, as:

$$\frac{dn_{photon}}{dt} = \sigma_2NF^2 \tag{4.9}$$

where dn_{photon}/dt is the number of photons absorbed per unit time, N is the number of absorbing molecules per unit volume, and $F = I/hv$ is the photon flux of light source. Because $dW/dt = (dn_{photon}/dt)hv$, one can obtain the following, commonly used theoretical expression for σ_2 [12]:

$$\sigma_2 = \frac{8\pi^2hv^2}{n^2c^2N}\mathrm{Im}\left(\chi^{(3)}\right) \tag{4.10}$$

Molecular two-photon absorption coefficient β (cm/GW) and σ_2 can be expressed as follows:

$$\beta = \sigma_2N_0 = \sigma_2N_Ad_0 \times 10^{-3} \tag{4.11}$$

N_0 is the molecule density (cm^{-3}), N_A is Avogadro's number, and d_0 is the concentration in moles. The unit for σ_2 is named Goeppert-Mayer (GM) and is defined as 1 GM $= 10^{-50}$ cm^4/photon $\bullet$ molecule. It is widely used as a scale to compare two-photon absorption activity of a material. Currently, several methods can be used to measure σ_2, including nonlinear transmission, two-photon-induced fluorescence emission, and Z-scan measurement.

4.2.2 Molecular Design for Multiphoton Absorption

In order to realize the full potential of the multiphoton technology, improvement is needed in design and synthesis of highly active organic molecules. Since 1990s, the research on design and synthesis of new molecules with enhanced two-photon absorption cross section σ_2 has been widely carried out. Numerous molecules have been designed and synthesized for investigating the relationship between molecular structures and σ_2, in order to establish well-defined structure–property relationship with systematically varied molecular structural factors and precisely reproducible characterization of two-photon properties. The detailed knowledge of molecular design parameters of new molecules with enhanced σ_2 has been achieved.

In principle, molecules with π-conjugated system always exhibit excellent TPA activity. The efforts of molecular design for enhancing TPA activity have been focused on the enhancement of intramolecular electronic transfer by modifying electronic structures of molecules through expanding the length of π-conjugated system and using strong electronic donative and attractive groups at the ends of the π-conjugated system [13]. Main strategy of molecular design for achieving large TPA activity can be summarized as three types according to molecular symmetry as shown in Fig. 4.2.

These can be classified into centrosymmetric (A-π-A, D-π-D) and noncentrosymmetric (D-π-A) molecules. The centrosymmetric molecules are the most commonly studied molecules. As shown in Fig. 4.3, the extended D-A-π-A-D, D-A-D-A-D, and D-D-A-D-D have been proposed to increase TPA activity through constructing structures with improved intramolecular charge transfer (ICT) properties. Marder's group extended the π-conjugation and introduced cyano (CN) moiety as electron accepting elements on the π-centers or π-bridges and alkoxy moiety as donating elements of the phenylene-vinylene derivatives [14]. SJ-1 is a D-A-π-A-D molecule with CN group on the π-center phenyl ring; SJ-2 is a D-A-D-A-D molecule with cyano groups on π-bridge rings and alkoxy on the π-center. The larger σ_2 of SJ-2 originates from stronger electron affinity caused by more CN moieties and addition of donating character. They also reported that bis (diarylaminostyryl) chromophores with electron-rich heterocycle show large TPA property [1, 15]. Compared to SJ-1 and SJ-2 having acceptors on the π-bridge, SJ-3, a D-D-A-D-D type molecule, having CN on the center ring, exhibited largest σ_2 due to more efficient centrosymmetric charge redistribution than D-π-D structure.

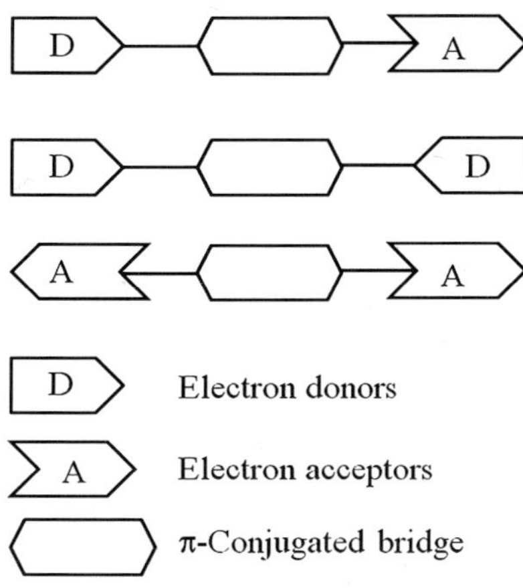

Fig. 4.2 Schematic principle of molecular design for enhancing TPA activity with different molecular symmetries

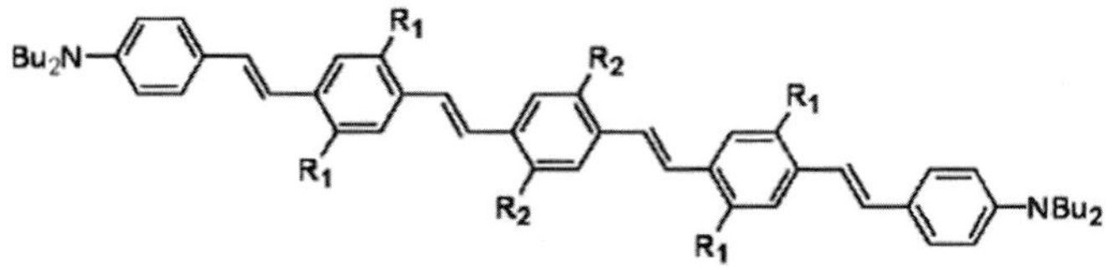

TPA dyes	R1	R2	$\lambda^{(2)}_{max}$ (nm)	σ_2 (GM)
SJ-1	CN	H	970	3600
SJ-2	CN	$OC_{12}H_{25}$	960	4400
SJ-3	OCH_3	CN	790	5300

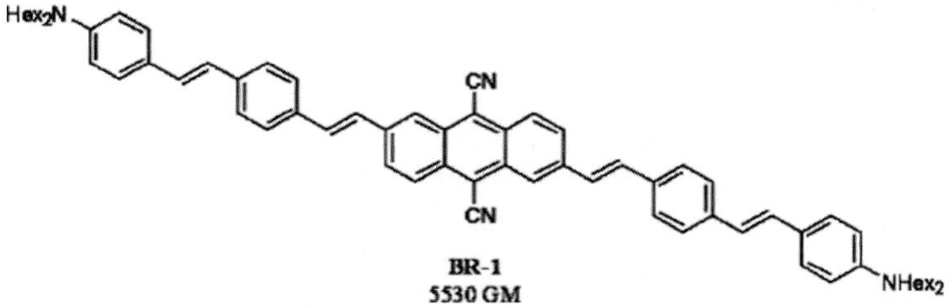

Fig. 4.3 Chemical structures and TPA activities of D-A-π-A-D, D-A-D-A-D and D-D-A-D-D type of chromophores (Reprinted from Ref. [1], Copyright 2008, with permission from Elsevier)

We have designed and successfully synthesized a series of novel 3,6-bis (ethynyl)carbazole derivatives (Fig. 4.4, 4a–d) in a facile way through a Sonogashira coupling reaction. Those 3,6-bis(ethynyl)carbazole derivatives worked as A-π-D-π-A V-shaped two-photon polymerization (TPP) initiators with high sensitivity combining large σ_2 and facilitated radical formation [16, 17]. It was found that 4c and 4d exhibited large σ_2 from 780 to 820 nm and low fluorescence quantum yields. Two-photon polymerization experiments verified that the low

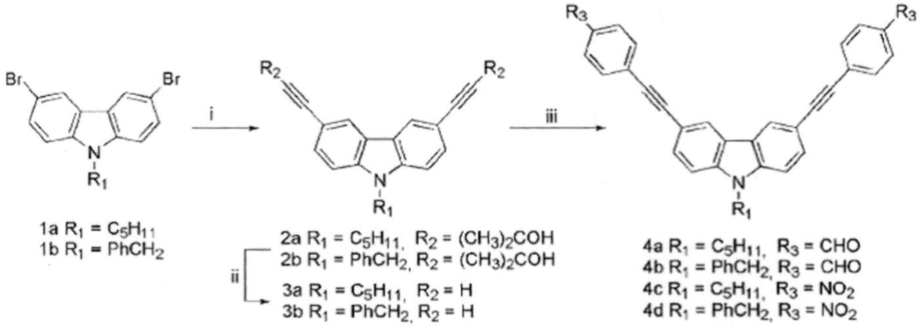

1a R_1 = C_5H_{11}
1b R_1 = $PhCH_2$

2a R_1 = C_5H_{11}, R_2 = $(CH_3)_2COH$
2b R_1 = $PhCH_2$, R_2 = $(CH_3)_2COH$
3a R_1 = C_5H_{11}, R_2 = H
3b R_1 = $PhCH_2$, R_2 = H

4a R_1 = C_5H_{11}, R_3 = CHO
4b R_1 = $PhCH_2$, R_3 = CHO
4c R_1 = C_5H_{11}, R_3 = NO_2
4d R_1 = $PhCH_2$, R_3 = NO_2

Fig. 4.4 Synthesis route of carbazole derivatives used as multiphoton polymerization initiator (Reproduced from Ref. [16] by permission from the Royal Society of Chemistry)

quantum yield, large σ_2, and benzyl group of 4d contributed to the high efficiency in acrylate resins, which exhibited a much lower threshold power of 0.8 mW at the concentration of 0.18 mol% than that of 6.37 mW for benzil. The corresponding threshold of laser exposure intensity for TPP was 3.0×10^7 mJ cm^{-2}. The lowest loading concentration of 4d was up to 0.012 mol% with a threshold power of 3.2 mW. The result indicated its extremely high sensitivity as a TPP initiator. The concept of combining radical stabilization and large σ_2 within a molecular structure is significant for developing highly efficient two-photon polymerization initiators.

Compared to the centrosymmetric molecules, noncentrosymmetric molecules are another kind of typical molecules with larger nonlinear optical properties. Although this kind of molecules has been widely used as second-order nonlinear optical materials, they also exhibited large TPA activity. C_{2v} symmetric compounds as well as their one-dimensional (1D) carbazole-based hemicyanines, in which methyl pyridinium, methyl indolium, and methyl benzothiazolium were used as acceptor group, were synthesized by Knoevenagel condensation (Fig. 4.5) [18, 19]. One-photon absorption, fluorescence, and two-photon fluorescence spectra have been investigated. The results indicated that the different ionic acceptors affected their one-photon and two-photon properties. Among them, 2D methyl pyridinium carbazole derivatives exhibited low quantum yields and large σ_2 of 1,600 GM. The synthesized compounds have been used as photoinitiator of TPP, and 3D microstructure has been successfully fabricated by TPP lithography.

Zhan et al. has reported various kinds of low-bandgap donor–acceptor conjugated copolymers. As shown in Fig. 4.6, a new low-bandgap π-conjugated D-A copolymer of squaraine and pyridopyrazine (P1) and a new small molecule squaraine–pyridopyrazine model compound (P2) were synthesized and compared [20]. P1 and P2 exhibited strong NIR absorption and low bandgap (1–1.3 eV). P2 in solution showed an intense and sharp absorption peak at 764 nm, while P1 in solution exhibited a red-shifted and broad absorption peak at 808 nm. The HOMO and LUMO levels of P1 were estimated to be 5.02 and 4.15 eV, while those of P2 were estimated to be 5.27 and 3.22 eV, respectively. Polymer P1 exhibited strong TPA at telecommunication wavelengths with σ_2 per repeat unit

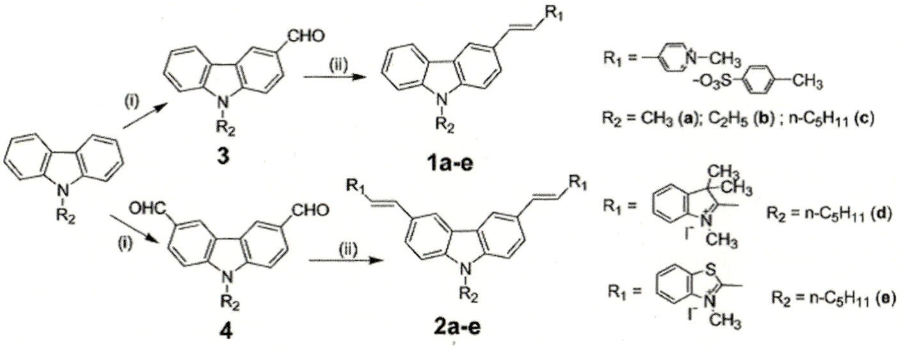

Fig. 4.5 Synthesis route of carbazole derivatives *1a–e* and *2a–e*. Reagents and conditions: (*i*) POCl$_3$/DMF, ClCH$_2$CH$_2$Cl, reflux; (*ii*) salts, piperidine (catalytic amount), ethanol, reflux (Reproduced from Ref. [18] by permission from the Royal Society of Chemistry (RSC) on behalf of the Centre National de la Recherche Scientifique (CNRS) and the RSC)

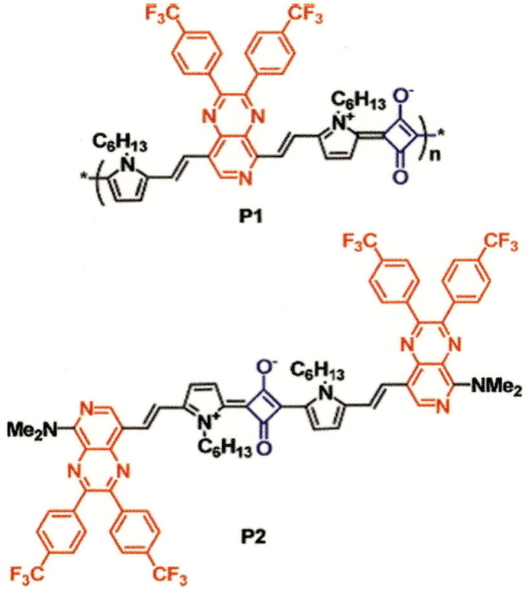

Fig. 4.6 Chemical structures of a series of low-bandgap conjugated copolymer P1 and P2 (Reprinted with the permission from Ref. [20]. Copyright 2011 American Chemical Society)

as high as 2,300 GM, which was 3–5 times that of the small molecule P2. The higher HOMO, the lower LUMO levels, lower bandgap, red-shifted absorption, and stronger two-photon absorption of P1 were attributed to higher degree of conjugation and delocalization of π-electrons in the polymer.

Another kind of low-bandgap, conjugated donor–acceptor copolymers of porphyrin with 2,3-bis(4-trifluoromethylphenyl)pyrido [3,4-b]pyrazine (P1) and perylene diimide (P2) were synthesized through Sonogashira coupling polymerization and compared with porphyrin-dithienothiophene D-D copolymer (P3) as

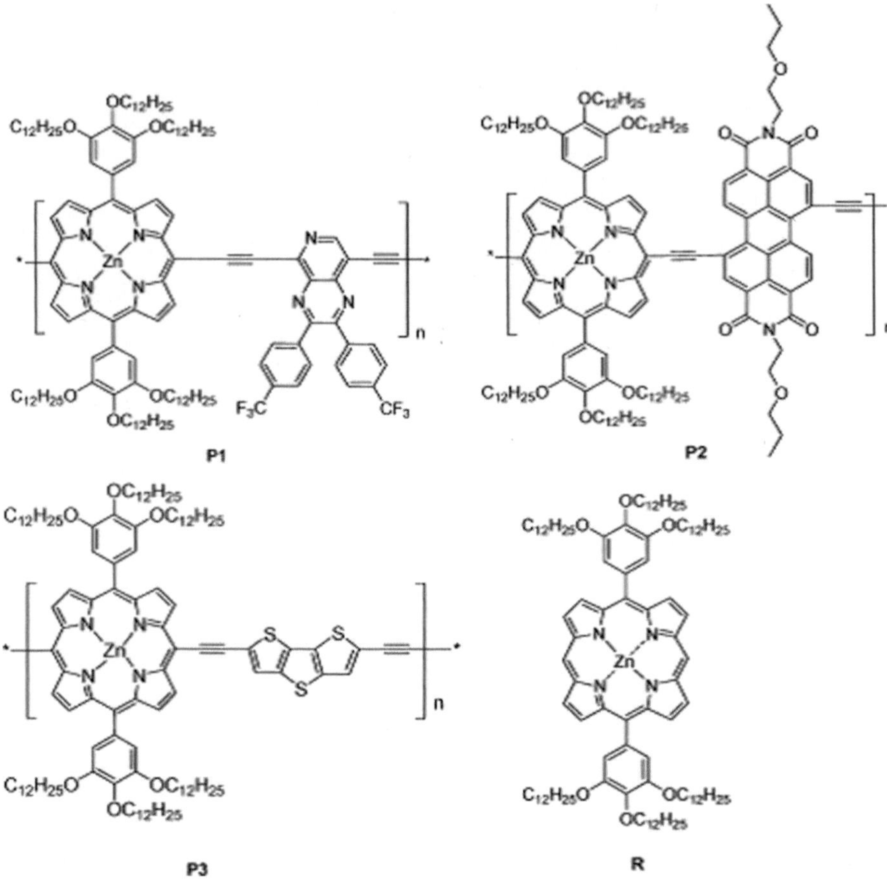

Fig. 4.7 Chemical structures of a series of low-bandgap conjugated copolymer with strong two-photon absorption (Reprinted with the permission from Ref. [21]. Copyright 2010 American Chemical Society)

shown in Fig. 4.7 [21]. All of these polymers possessed good thermal stability with decomposition temperatures over 300 °C. Polymers P1 and P2 in films exhibited strong absorption in near-IR (820–950 nm) with optical bandgaps as low as 1.15 eV; their Q-bands red shift 60–190 nm were comparable to that of P3, while the Soret bands were similar. The HOMO (-5.3 to -5.4 eV) and LUMO (-3.6 to -4.0 eV) of the D-A polymers were lower than that of the D-D polymer. Two-photon absorption properties of the polymers were investigated using the femtosecond z-scan method. The D-A polymer P2 exhibited σ_2 over 7,000 GM/ repeat unit at telecommunication wavelengths (1,320 and 1,520 nm), which was larger than that of P1 and P3 owing to the strong, rigid, and coplanar perylene diimide acceptor and strong D-A intramolecular charge transfer.

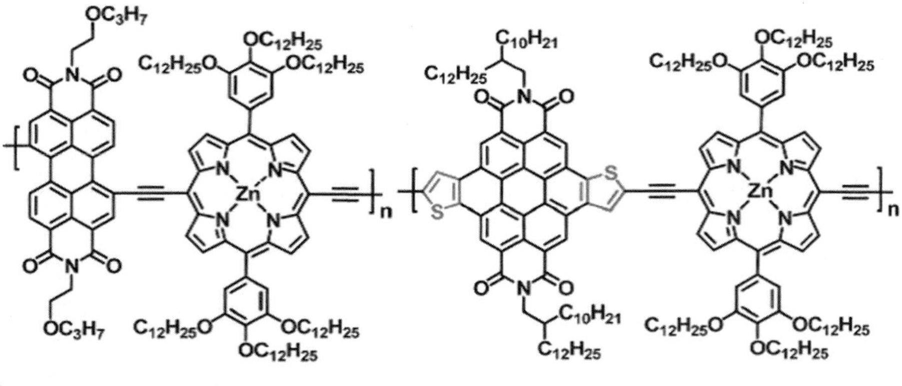

Fig. 4.8 Chemical structures of a series of low-bandgap conjugated copolymer with strong two-photon absorption (Reprinted with the permission from Ref. [22]. Copyright 2012 American Chemical Society)

Recently, low-bandgap D-A conjugated copolymer poly(DTCDI-POR) of planar acceptor dithienocoronene diimide (DTCDI) and strong donor porphyrin (POR) has been successfully synthesized through Sonogashira coupling polymerization (Fig. 4.8) [22]. Poly(DTCDI−POR) exhibited good thermal stability (decomposition temperature of 323 °C), strong absorption (molar extinction coefficient per repeat unit is 1.05×10^5 L mol^{-1} cm^{-1} at 468 nm in $CHCl_3$ solution) in visible and near-infrared region (300–900 nm), low bandgap (1.44 eV), and strong TPA at telecommunication wavelengths with σ_2 per repeat unit as high as 7,809 GM at 1,520 nm.

Subsequently, a new family of dendrimers with a naphthalene-core flanked on both sides with triphenylamine branching was successfully synthesized and presented an increasing σ_2 from 959 to 9,575 GM with the generation number from 1 to 3 (Fig. 4.9) [23]. Wang et al. found that these dendrimers can efficiently initiate the polymerization of acrylate resins to achieve regular diamond structures and display higher TPP efficiency and sensitivity with the generation number. The overall TPP processes involved in TPA process, such as the intramolecular charge transfer and intramolecular energy transfer as well as intermolecular electron transfer between initiator and monomer, are illustrated in Fig. 4.9. Steady-state fluorescence and time-resolved decay dynamics revealed that the light energy was absorbed by peripheral triphenylamine unit and then transferred to generation 1, the energy funnel. Although strong interaction between dendritic initiator and monomer has been observed based on fluorescence quenching measurements, no intermolecular energy transfer but electron transfer was confirmed by the cyclic voltammograms and HOMO–LUMO measurements. That is, the dendritic initiator first produces the excited state via TPA process, then transfers an electron to an acrylate monomer, and finally induces the latter to polymerize. Moreover, A. Rebane and coworkers studied a series of dendrimers with an amino-branching

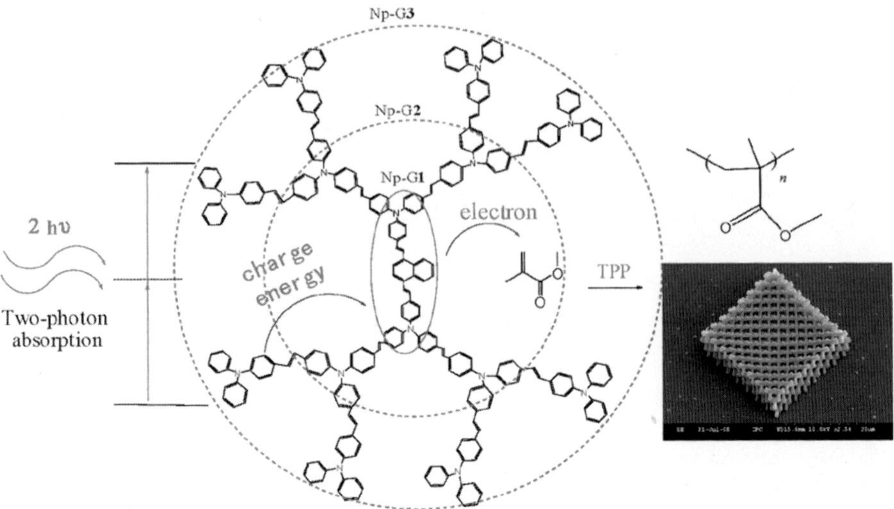

Fig. 4.9 A new family of dendrimers with naphthalene core and triphenylamine branching as a TPA initiator. The overall TPP approach involved in TPA, intramolecular charge transfer, and intramolecular energy transfer as well as intermolecular electron transfer (Reprinted with the permission from Ref. [23]. Copyright 2011 American Chemical Society)

unit and discovered that the compact size of branched structures was more effective in enhancing TPA activity than linear oligomeric or polymeric molecules. One of the series, TP-6 exhibited large TPA activity, 2,700 GM at 680 nm [24].

4.3 Multiphoton Micro-/Nanofabrication

4.3.1 Methods of Multiphoton Micro-/Nanofabrication

4.3.1.1 Multiphoton Polymerization

The multiphoton polymerization process can be generally described in Fig. 4.10 [1, 25]. During the multiphoton micro-/nanofabrication process, the photosensitizer in the photopolymerizable resin is excited by absorbing two or more photons simultaneously and then emits fluorescence in the UV–vis region. Consequently, the fluorescence will be absorbed by photoinitiators with good chemical reactivity and give rise to radicals. Then, the radicals react with monomers or oligomers, resulting in the production of the monomer radicals, which would expand in a chain reaction until two radicals meet. Finally, the low molecular weight monomers are polymerized into cross-linked, high molecular weight materials.

$$\text{Initiation: } S \xrightarrow{h\upsilon, h\upsilon} S^* \cdots \xrightarrow{I} I^* \rightarrow R^{\cdot}$$

$$\text{Propagation: } R^{\cdot} + M \rightarrow RM^{\cdot} \rightarrow RMM \cdots \rightarrow RM_n^{\cdot}$$

$$\text{Termination: } RM_n^{\cdot} + RM_m^{\cdot} \rightarrow RM_{n+m}R$$

Fig. 4.10 The schematic illustration of the multiphoton polymerization process. *S* stands for the photosensitizer, *I* denotes the photoinitiator, *R* is the radical, and *M* is the monomer. S^* and I^* are the excited states of the photosensitizer and photoinitiator after absorbing the photon energy, respectively

4.3.1.2   Multiphoton Fabrication of Negative-Tone Photoresists

The photopolymerizable resin can be broadly classified into negative and positive photoresists. Negative photoresist is the more widely used one, in which the portion of the photoresist that is exposed to light becomes insoluble to the photoresist developer. The unexposed portion of the photoresist is dissolved by the photoresist developer. In contrast, the positive resist behaviors in the opposite way, in which the portion of the photoresist that is exposed to light becomes soluble to the photoresist developer. The portion of the photoresist that is unexposed remains insoluble to the photoresist developer.

By using the negative photoresist, various micro-/nanostructures can be created by using the multiphoton lithography. During the multiphoton fabrication process, the femtosecond laser is tightly focused into the photoresist by a high numerical aperture (NA) objective lens. The photoresist will be polymerized to achieve the predesigned pattern by using a computer-controlled program. After fabrication, the sample with both of the exposed and unexposed resist will be developed with a certain organic solvent by removing the unpolymerized low molecular weight photoresist, and thus the cross-linked, high molecular weight material will be remained.

The typical commercial negative photoresists such as SCR500 and SU-8 have been widely used in research. SCR500 is composed of urethane acrylate oligomers (molecular weights 480 and 1,200) and two types of photoinitiators (Irgacure 369 and Irgacure 184). SU-8, as a kind of cross-linkable epoxide negative photoresist, has eight epoxy groups per monomer and contains a triaryl sulfonium salt photoacid generator (PAG) [26]. The chemical structure of SU-8 molecule is shown in Fig. 4.11. SU-8 is used extensively in the conventional photolithography of MEMS because it has the ability to be cast in films with a high thickness that can yield structures with high aspect ratios. Note that one of the disadvantages is that the polymerized material would absorb some of the organic solvent molecules that used as developer during the development, leading to the swelling of the micro-/nanostructures. Due to this reason, the closely spaced geometries would come into contact together, resulting in the defect of the pattern or the stuck structure. At a certain degree, positive-tone resists can help minimize the shrinkage and distortion typically encountered in the photoprocessing of acrylates and other negative-tone resists.

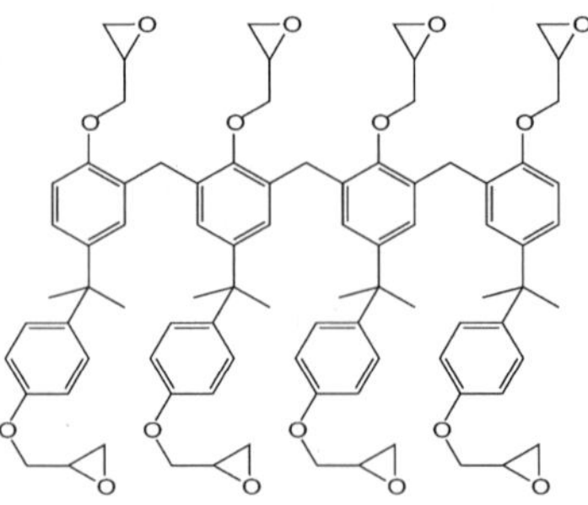

Fig. 4.11 The chemical structure of SU-8 molecule

4.3.1.3   Multiphoton Lithography of Positive-Tone Photoresists

With the development and the demand of the semiconductor industry, the positive photoresist becomes popular. During the multiphoton lithography process, the femtosecond laser is also tightly focused into the photoresist by an objective lens with high NA. Unlike the negative photoresist, the laser-exposed part will be removed after the photoacid protolysis of the copolymers in positive-tone photoresists. Thus, the unexposed resist would remain on the substrate. There are many kinds of commercial positive photoresists, such as Hoechst AZ4620, Hoechst AZ 4562, Shipley 1400-17, Shipley 1400-27, Shipley 1400-37, and Shipley microposit developer.

Marder, Perry, and coworkers demonstrated a solid-state, acid-sensitive, positive-tone, chemically amplified resist, designed for 3D microfabrication by two-photon laser excitation using the initiator BSB-S2 (Fig. 4.12) as the PAG in a random copolymer consisting of tetrahydropyranyl methacrylate, methylmethacrylate (MMA), and methacrylic acid units. The tetrahydropyranyl ester groups are converted into carboxylic acids after the photoacid protolysis in the presence of acid, causing them to be soluble in an aqueous basic developer. The MMA groups provide strength and optical clarity. The carboxylic acid groups generated after exposure and development provide a chemically active site for surface functionalization [27]. In our group, we have designed and synthesized a kind of T-Boc protected calix resorcinarene derivatives and built the positive-tone molecular glass resist [28]. The chemical structure of the derivative is shown in Fig. 4.13. The result gives us an evidence that molecular glass resist could improve the line edge roughness of the micro-/nanostructure that achieved by multiphoton lithography.

4.3.1.4   Optical Setup for Multiphoton Fabrication

The experimental setup of a typical micro-/nanofabrication system is schematically shown in Fig. 4.14. Experimentally, multiphoton micro-/nanofabrication is carried out through an optical setup with a mode-locked femtosecond laser, which produces pulses with a center wavelength of 780 nm, a pulse width of 80 fs, and a repetition rate of 80 MHz, respectively. A neutral density filter and a mechanical shutter are

Fig. 4.12 The chemical structure of BSB-S2

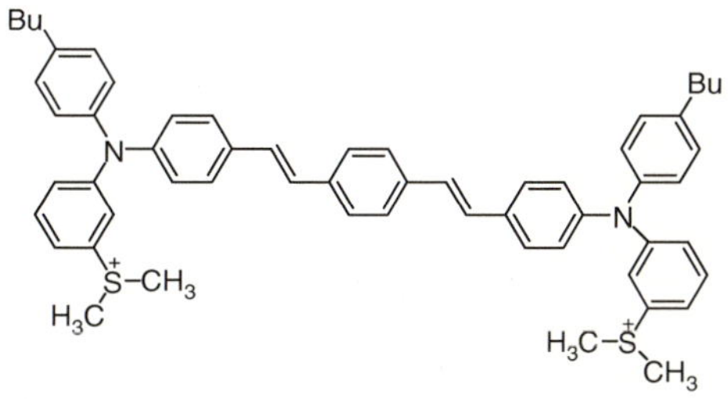

Fig. 4.13 The chemical structure of the T-Boc protected calix resorcinarene derivative

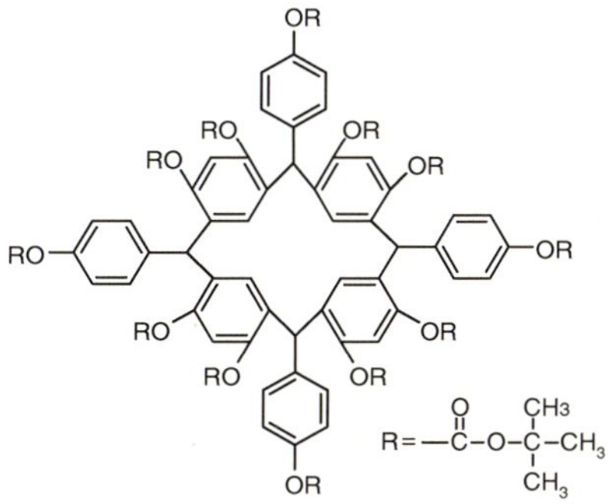

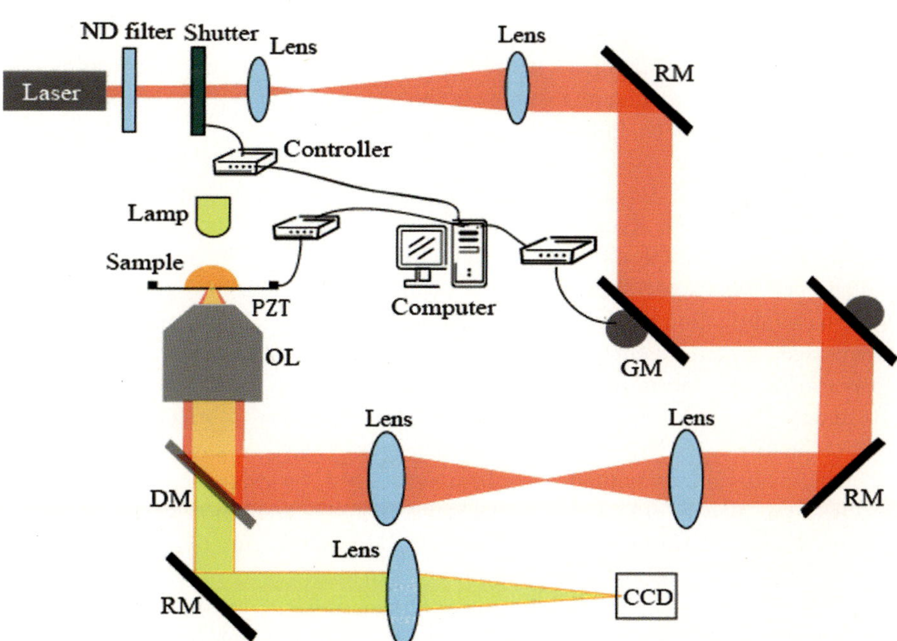

Fig. 4.14 Schematic illustration of a typical experimental setup for multiphoton fabrication. ND filter represents neutral density filter, *RM* stands for reflective mirror, *GM* is the galvano-mirror, *DM* means the dichroic mirror, *OL* denotes the objective lens, *PZT* is high-precision nanopositioning piezostage, and *CCD* represents charge-coupled device camera

used for controlling the incident laser power and exposure time. The average output power can be tuned from several milliwatts to more than 1 W. The beam is generally expanded in order to overfill the back aperture of the objective lens. A microscope is usually adopted to hold, position, and view the sample. In the experimental setup shown here, the laser beam passes through the dichroic mirror of the inverted microscope and is tightly focused into the sample by a high NA objective lens. A computer is used to control the scanning operation of the galvano-mirrors according to the preprogrammed pattern. Transmitted light and reflected illuminated light are used to view the sample with a CCD camera and a video screen by tracing the positioning of the sample and making it possible to monitor the micro-/nanofabrication in real time. The sample rests on a piezostage that can be moved in three dimensions relative to the focus of the laser beam. As an alternation, a pair of galvano-mirrors can be employed to move the focus spot relative to the sample. After the preprogrammed pattern is completely fabricated, the corresponding posttreatment will be employed.

4.3.1.5   Scanning Modes for Multiphoton Polymerization

In general, two basic types of scanning modes have been used in research related to the spatial resolution of MPP nanofabrication: pinpoint scanning mode (PSM) and continuing scanning mode (CSM) [29]. In MPP nanofabrication with PSM, the laser spot is typically stationary with respect to the sample during laser irradiation, which leads to the creation of a single photopolymerized voxel. The scaling laws of single voxels based on PSM have been investigated by Kawata and coworkers [30–33]. In their reports, a lateral spatial resolution (LSR) of 100 nm was obtained on the surface of a substrate [32], and an LSR of 60 nm was achieved with suspended lines [33]. Recently, our group has successfully achieved an LSR of less than 100 nm on the substrate surface and a sub-15 nm LSR in suspended lines, which corresponds to 1/50th of the fabrication wavelength of 780 nm, by using MPP nanofabrication under CSM [34]. In the CSM mode, the galvano-mirrors and piezostage worked together to realize the fast fabrication process due to the rapid response of galvano-mirrors. The fabrication speed in the CSM mode can be 10 times improved compared to the PSM mode for fabricating the same microstructures.

Additionally, other methods have been developed, for example, many triangular patches, the size of which generally stands for the resolution of the designed computer-aided design (CAD) modeling data, lead to 3D stereolithography (STL) data. The 3D STL data is sliced vertically into many layers to generate the 2D scanning paths data of a focused beam. As shown in Fig. 4.15, the intersection line between a triangular patch and a slicing plane at given z-coordinates is calculated using the two intersection points. The spatial relationship between a triangular patch and a slicing plane can be summarized as (a) a patch above the plane, (b) a patch below the plane, (c) a patch on the plane, and (d) a patch intersecting with a slicing plane. In case a triangular patch is located on a slicing plane, it can be

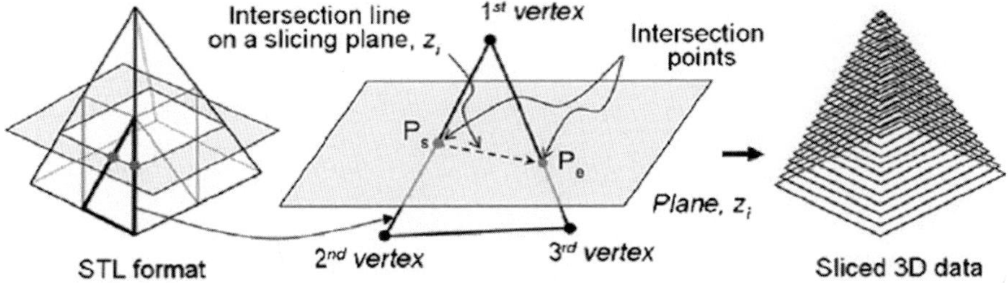

Fig. 4.15 Schematic illustration of the slicing procedure for the generation of 2D scanning data. Each triangular patch in a STL file consists of three line segments; hence, the intersection line between a triangular patch and a slicing plane is obtained by calculating two intersection points of P_s and P_e (Reprinted from Ref. [1], Copyright 2008, with permission from Elsevier)

subdivided into two cases: one vertex of a patch on the plane and two vertexes of a patch on the plane [35, 36].

Direct fabricating of 2D nanoscale patterns is not suitable to realize mass production, but it is of increasing interest in a wide variety of applications for the rapid fabrication of prototype patterns, or master patterns. Recently, a nanoreplication printing (nRP) process has been developed for producing complicated 2D patterns via direct laser writing. In this nRP process, a voxel matrix scanning (VMS) technique is initially proposed to fabricate 2D patterns. For the creation of the voxel matrix, a two-tone (black and white) bitmap figure is employed. The image is transformed into a voxel matrix consisting of two codes: "1" for a black color pixel and "0" for a white color pixel. Figure 4.16a illustrates the VMS method using TPP [35, 36].

The positions of the voxels on the outer contour are specified by the voxel matrix. Therefore, the fabricated pattern is generally larger than the designed shape, with undesirable shape-error terms, which arise from the intrinsic shortcomings of the VMS method, as shown in Fig. 4.16b. To overcome these problems, a contour offset algorithm (COA) was developed to improve the precision of the nRP process [35]. The offset for modifications in the COA can be readily calculated using the relations among the desired pattern size, initial design size, and voxel size provided as $R_{offset} = (L \times r)/l$, where R_{offset} is the offset, L is the size of the pattern in the CAD system, r is the voxel radius, and l is the size of the fabricated pattern. The outer contour of the pattern is reconstructed by employing a CAD tool while considering the offset; a new voxel matrix is then generated from the reconstructed bitmap figure of the 2D pattern. Using the nRP process with the COA, precise 2D patterns can be created readily as shown in Fig. 4.17, and this precise patterning is applicable in various applications.

Furthermore, multiple beams multiphoton processing has been developed as a simple and practicable method for fabricating integrated MEMS/NEMS with parallel laser microfabrication by designing configurations of multiple beams. Assembled 2D and 3D microstructures have been fabricated by using multiple beam multiphoton processing. There are two ways to achieve a controllable configuration

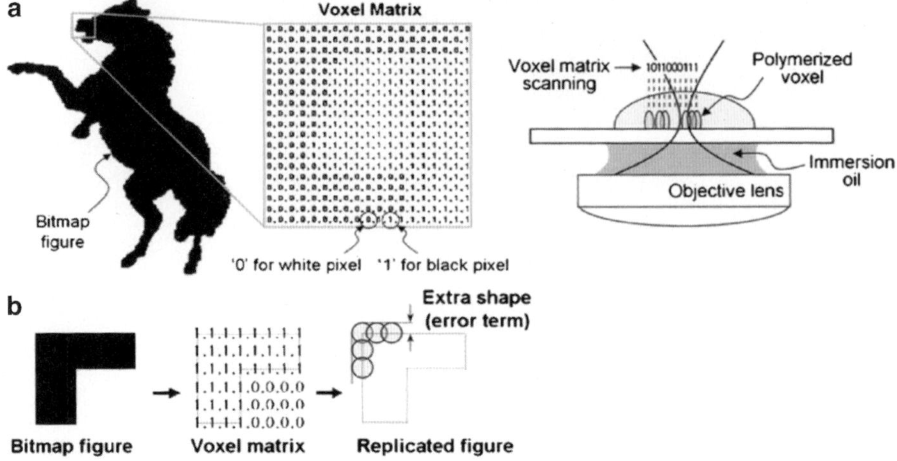

Fig. 4.16 (**a**) The bitmap file of a horse is transformed into a voxel matrix form. The *rectangular box* is a part of the full voxel matrix that consists of two kinds of components, "1" for laser beam on and "0" for laser beam off. The schematic diagram on the *right* side shows the voxel matrix scanning (VMS) procedures; a laser beam is scanning along the row of a voxel matrix, and a "1" represents the laser beam being on and a "0" the laser beam being off. The center of a voxel is located in the coordinate of each "1" of the voxel matrix. (**b**) Schematic view of the generation of extra shape due to extra shape term appears as much as voxel radius in VMS (Reprinted from Ref. [1], Copyright 2008, with permission from Elsevier)

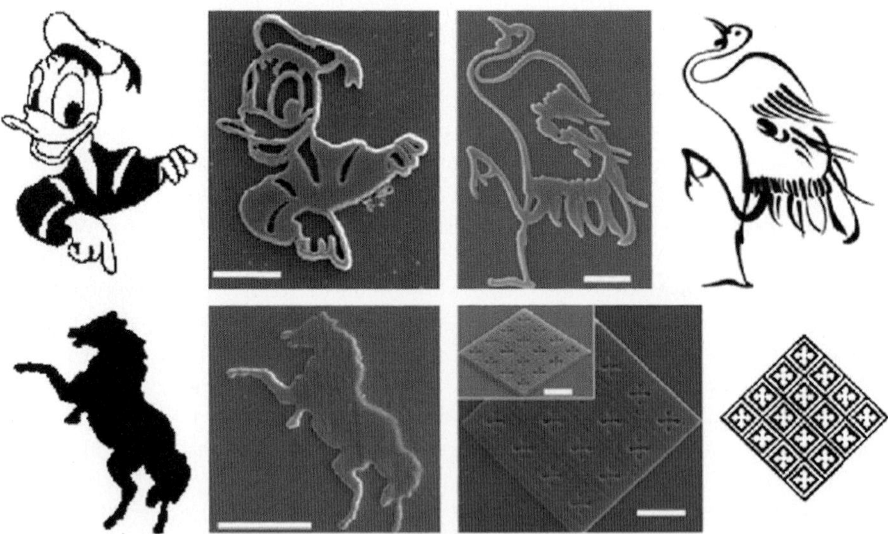

Fig. 4.17 Various patterned examples via the nRP process; a Donald character, horse, and Korean traditional patterns in a counterclockwise sequence. The *left-end* and *right-end* bitmap figures are the originally designed data, and they are replicated as microscale patterns, shown in the middle (All of the scale bars are 5 μm in length) (Reprinted from Ref. [1], Copyright 2008, with permission from Elsevier)

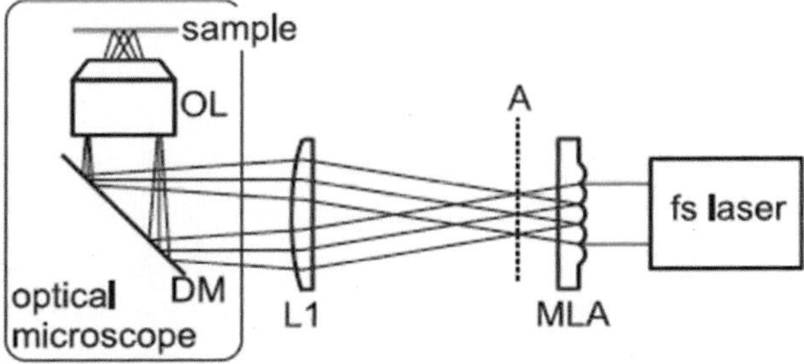

Fig. 4.18 Schematic illustration of the optical setup. *MLA* microlens array, *L1* lens, *DM* dichroic mirror, *OL* objective lens. A is the plane where multiple focuses are generated (Reprinted from Ref. [37], with kind permission from Springer Science+Business Media)

of multiple beams. One way is to use microlens array (MLA) to produce multiple spots for parallel fabrication, which has been reported by Matsuo and Kawata [37, 38]. The related optics is schematically shown in Fig. 4.18. 2D and 3D multiple structures such as microletter set (Fig. 4.19a), self-standing microspring array (Fig. 4.19b), and other periodic patterns are demonstrated. More than 200 spots have been fabricated simultaneously by optimizing the exposure condition for the photopolymerizable resin, which is a two-order increase of yield efficiency. Since the laser beam power is divided into several hundred beams to produce multiple spots simultaneously, the laser pulse energy is amplified by a regenerative amplifier (Titan 527DP, Quantronics) in order to keep the nonlinear photochemical reaction occurring at the identical excitation level.

The other approach to realize parallel processing of assembled structures is to assemble and control the configuration and geometry of multiple beams by changing the parameters of lenses set as well as aperture mask (AM) [39]. An optical system is schematically shown in Fig. 4.20. A diffractive beam splitter (DBS, G1029A, MEMS Optical, Inc.) is used to split a single input beam into nine beams, where the central beam is in zero-order diffraction and eight beams distributed surrounding the central beam with radii of R are generated by first-order diffraction. These multiple beams are made into parallel by a lens L1 and are selected by an AM to obtain the desired configuration of multiple beams. We then gather these multiple beams by a lens L2 and scan them in the horizontal dimension by a pair of galvano-mirrors. The multiple beams are collected by a pair of lenses L3 and L4, and finally the multiple laser spots are tightly focused into the photocurable resin by an objective lens. The sample is mounted on a piezostage to move the laser focus spots in the axial direction. Various assembled 2D and 3D microstructures, including microgears and photonic crystals (PhCs), have been demonstrated with high fabrication efficiency and nanometer scale resolution (Fig. 4.21). By using this protocol, we can realize the handling and assembling of several parts into a designed MEMS/NEMS. The proposed nanofabrication method provides the capability for realizing assembled microstructures and thus could be expected to play an important role in fabricating micromachines and microdevices.

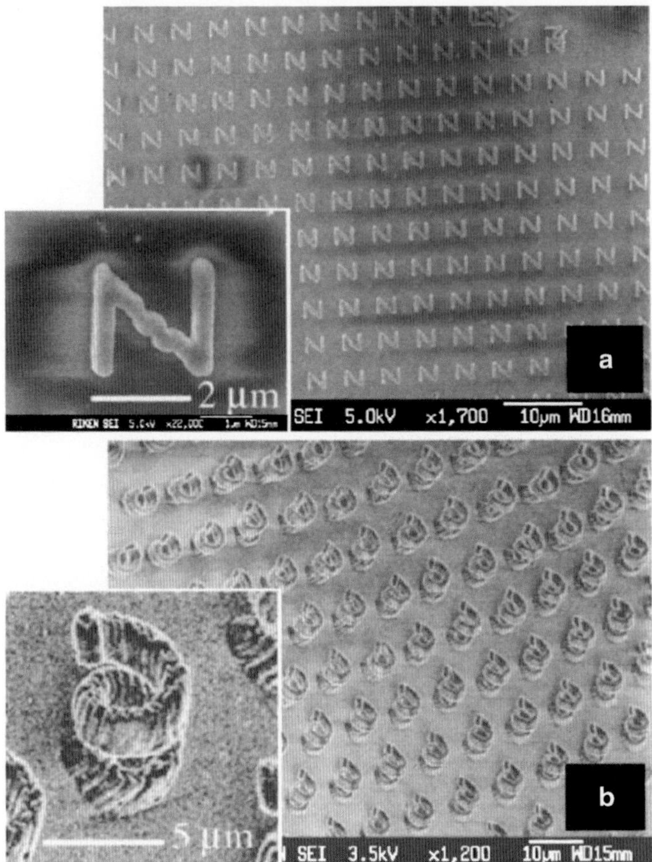

Fig. 4.19 Examples of two- and three-dimensional fabrication with the parallel processing technique: (**a**) the microletter array of "N" was fabricated with 28 exposure points in two dimension, in which totally 227 structures were fabricated using the relay lens of 150 mm focal length; (**b**) the example of 3D fabrication, where the self-standing microspring array was fabricated with the relay lens of 80 mm focal length. The *insets* give a better view of the individual structures (Reprinted with permission from Ref. [38]. Copyright 2005, American Institute of Physics)

4.3.2   The Optical Diffraction Limit

In principle, the resolution beyond the optical diffraction limit can be achieved in the multiphoton micro-/nanofabrication process due to the nonlinear optical effect. For linear exposure such as lithography, materials respond to light excitation to the first-order effect, while for two- and multiphoton absorption, the response is restricted to two and high orders. The square light intensity distribution is spatially narrower than that of linear one, resulting in a reduction of light–matter interactive volume, and therefore, the improvement of fabrication resolution. The resolution depends on the

Fig. 4.20 Schematic optical setup of multiple beam multiphoton processing approach for laser nanofabrication of assembled 3D microstructures (Reprinted with permission from Ref. [39]. Copyright 2007, American Institute of Physics)

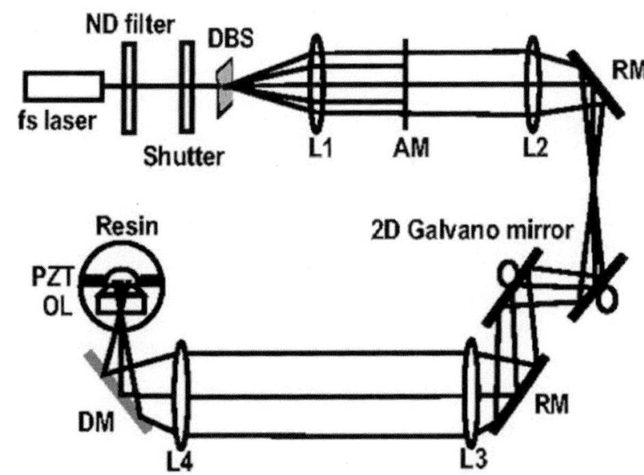

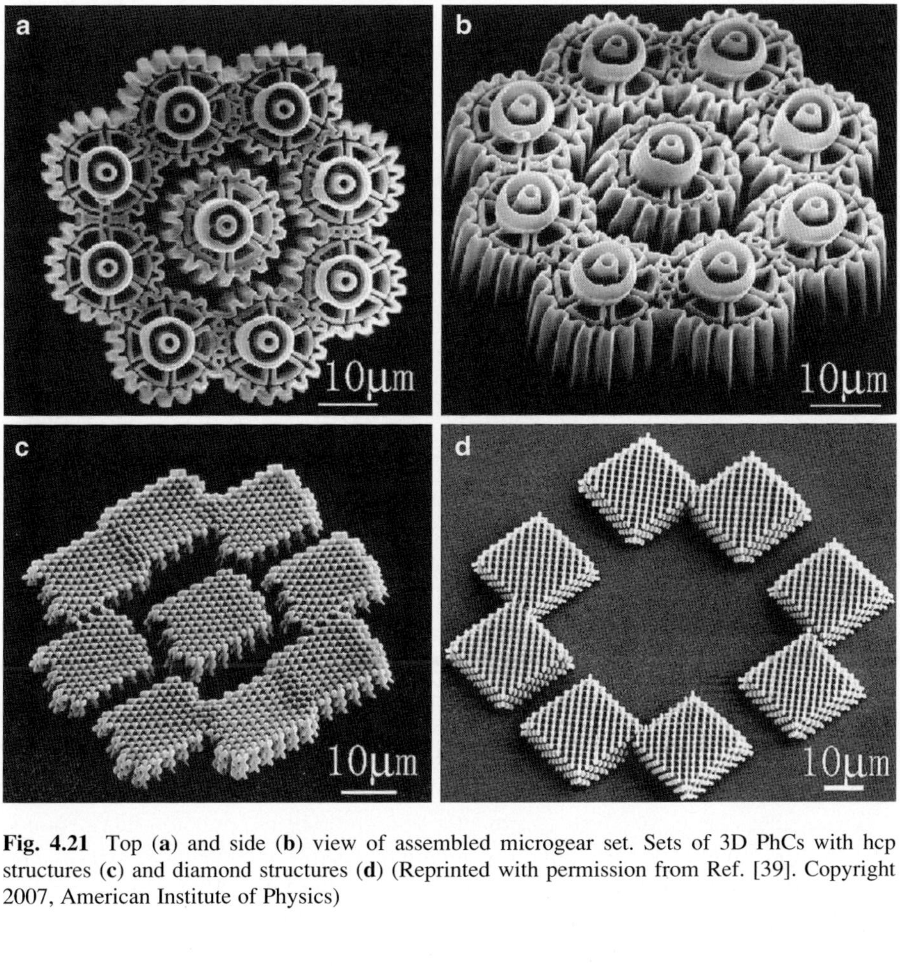

Fig. 4.21 Top (**a**) and side (**b**) view of assembled microgear set. Sets of 3D PhCs with hcp structures (**c**) and diamond structures (**d**) (Reprinted with permission from Ref. [39]. Copyright 2007, American Institute of Physics)

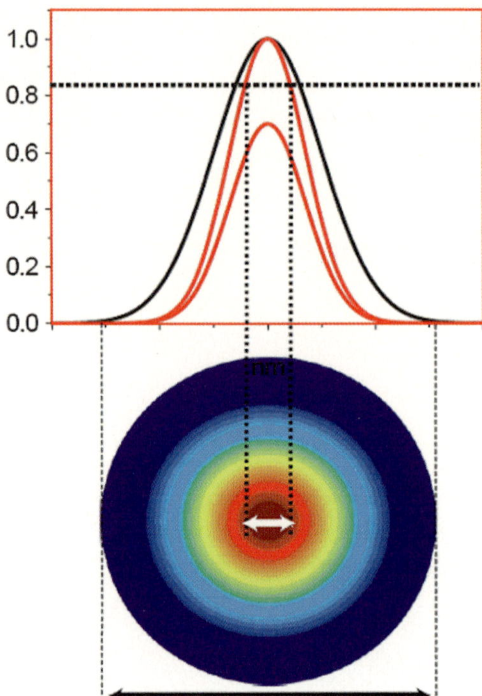

Fig. 4.22 Schematic illustration of breaking the optical diffraction limit in multiphoton micro-/nanofabrication

parameters such as the wavelength of the femtosecond laser, the numerical aperture of objective lens, the laser intensity, the scanning speed, the exposure time, as well as the sensitivity of the photopolymerizable resin. As illustrated in Fig. 4.22, the excitation area is highly confined in a micro area due to the nonlinear phenomenon. Therefore, the intrinsic 3D resolution is able to be achieved.

Multiphoton micro-/nanofabrication techniques have been established as powerful and widely used tools in fabricating micro-/nanometer-scale structures. In particular, their unique capability for intrinsic 3D processing provides the unmatched opportunity for developing arbitrary 3D micro-/nanostructures. Various micromachines and microdevices such as microgears and 3D photonic crystals fabricated by MPP have been demonstrated in MEMS/NEMS systems [3]. Meanwhile, efforts toward improving the spatial fabrication resolution of multiphoton nanofabrication have continued to overcome the optical diffraction limitations of the light source and achieve resolution at nanometric scale. Kawata et al. showed the improvement of the spatial resolution to 120 nm performed by multiphoton micro-/nanofabrication with a femtosecond pulse laser of 780 nm, which is almost one-sixth of the laser wavelength [40, 41].

Physically, one approach to improve the resolution of multiphoton fabrication is based on the nonlinear optical effect. As mentioned above, the matter will absorb two or more photons simultaneously during the multiphoton fabrication process, leading to the high-order effect. By employing near-infrared femtosecond laser, two- or three-

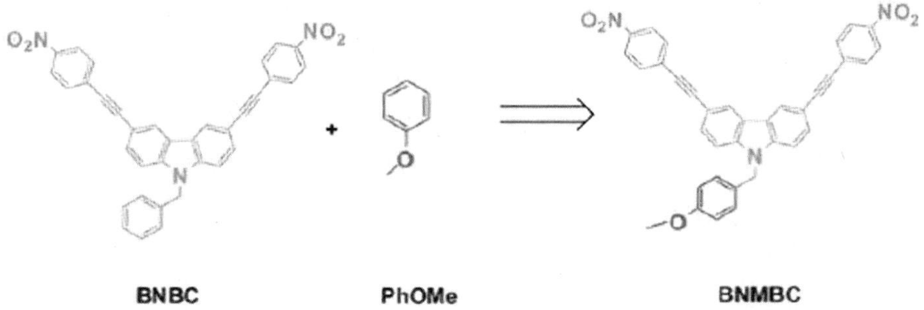

Fig. 4.23 The chemical structure of photoinitiators (Reproduced from Ref. [43] by permission from the Royal Society of Chemistry)

order effect can be created. The square light intensity distribution is spatially narrower than that of linear case [41], resulting in the reduction of light–matter interactive volume, and thus the improvement of the fabrication resolution.

Aside from the optical nonlinearity, another method is to utilize the chemical reaction and the molecular design to achieve high fabrication resolution of tens of nanometers, which is much smaller than the optical diffraction limit. Take two-photon-induced polymerization of resins, for example, photogenerated radicals are spatially distributed obeying the square law of the light intensity function. The rate of photopolymerization of monomers is therefore expected to follow the same distribution. However, due to the existence of the radical quenchers like dissolved oxygen molecules, radicals survive and initiate polymerization only at the region with higher light intensity, and thus the low threshold than that defined by optical diffraction limit can be achieved. For the particular case of photopolymerization of resins, if the written feature is sufficiently small such as suspended thin fibers, the diameter can be further reduced [34]. This effect is called materials nonlinearity. Recently, Harke and Petrozza et al. proposed the polymerization inhibition by triplet state absorption for nanometer-scale lithography, which represents the key for achieving direct laser writing lithography with sub-diffraction resolution photophysically [42].

Considering the effect of optical nonlinearity, the chemical and material effects, much research has been carried out to achieve high spatial resolution. By designing and synthesizing the two-photon-induced photoinitiator with radical quenching moiety, 3,6-bis[2-(4-nitrophenyl)-ethynyl]-9-(4-methoxybenzyl)-carbazole (BNMBC), possessing high initiating efficiency in two-photon-induced polymerization and capability for fabricating 3D structures at high resolution has been successfully demonstrated [43]. The chemical structures are shown in Fig. 4.23. The high resolution in TPP using BNMBC was achieved compared to the reported photoinitiator 3,6-bis[2-(4-nitrophenyl)-ethynyl]-9-benzylcarbazole (BNBC). BNMBC possesses a large σ_2 of 2,367 GM by z-scan measurement. We have investigated the photopolymerization properties of resins R1–R3, in which BNBC, BNBC/phenyl methyl ether (PhOMe), and BNMBC were used as photoinitiators, respectively. TPP experiments demonstrated that BNMBC exhibited high TPP initiating efficiency and an effective

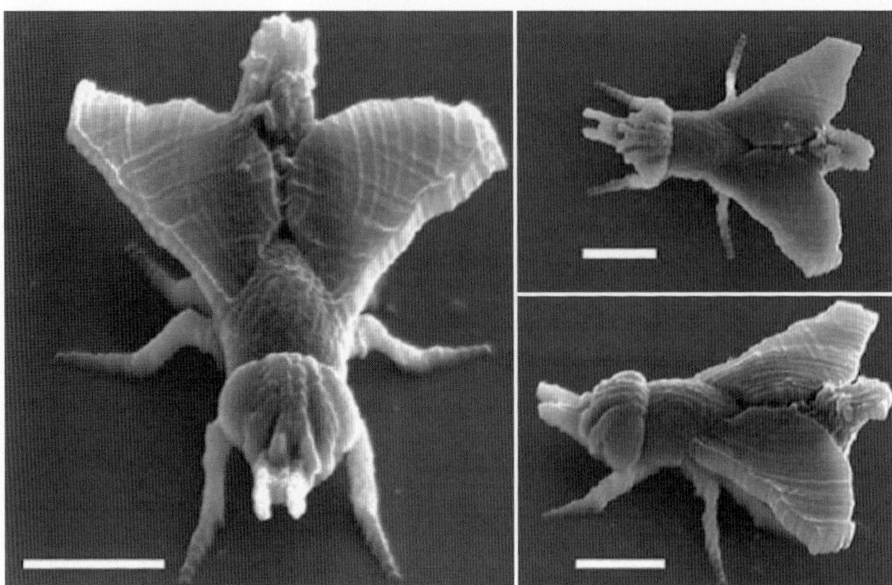

Fig. 4.24 SEM images of a fly fabricated with a laser power of 7.70 mW and a scanning speed of 66 μm s^{-1} by using the photoresist R6. Scale bars are 5 μm (Reproduced from Ref. [43] by permission of the Royal Society of Chemistry)

radical quenching effect. The volumes of polymer fibers fabricated by the TPP of the photoresist using BNMBC as the photoinitiator were decreased to 20–30 % of those synthesized using BNBC as a photoinitiator. The introduction of a radical quenching group to the photoinitiator resulted in the effective confining capability of radical diffusion compared to that with the same molar ratio of radical quencher. This concept of using photoinitiator with radical quenching group could result in the fabrication of precise structures.

Furthermore, we have demonstrated the fabrication of arbitrary complex 3D structures using the photoresist with our designed photoinitiator BNMBC. Figure 4.24 shows SEM images of the sample of a fly fabricated using R6 with a laser power of 7.70 mW and a scanning speed of 66 μm s^{-1}. The fly is of 6.47 μm in height, 20.90 μm in length, and 17.61 μm in width, which is fully supported by six legs with the diameter of about 650 nm. It indicated that the radical quenching effect of BNMBC did not decrease the cured polymer strength for a photoresist with high concentration of cross-linker, which could form high cross-linking 3D polymer networks. The fine striates on the wings of the fly were clearly reproduced at a scale smaller than 100 nm by TPP on the photoresist containing our novel photoinitiator, BNMBC. The high resolution was contributed by the confining radical diffusion of BNMBC in TPP.

By using the same photoinitiator, high resolution and the complex 3D structure with massive overhangs such as the crossed columns of the bird's nest can be observed clearly [44]. Thus, the photoresist consisting of the photoinitiator with

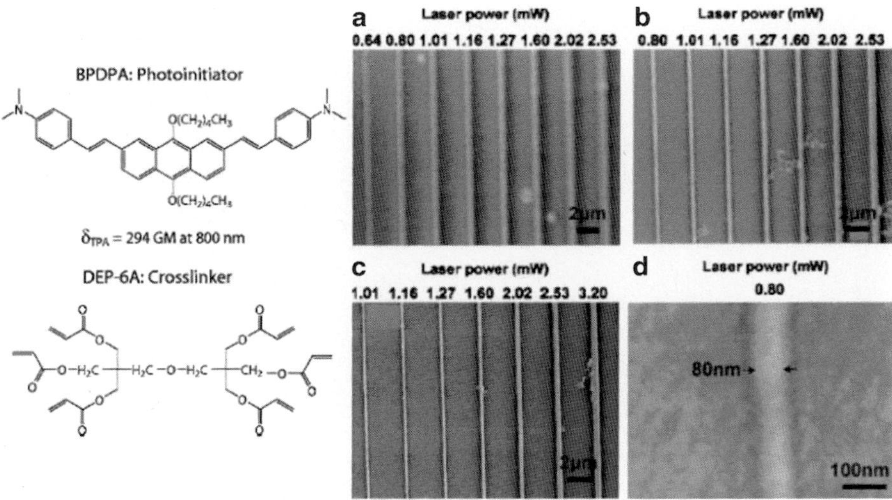

Fig. 4.25 (*Left*) Molecular structures of compound BPDPA used as photoinitiator and the cross-linker in the photocurable resins. (*Right*) SEM images of solidified polymer lines fabricated by using different laser power and linear scanning speed of (**a**) 10 μm/s, (**b**) 30 μm/s, and (**c**) 50 μm/s. (**d**) Magnified SEM image of the polymer line (80 nm) fabricated by using a laser power of 0.80 mW and a scanning speed of 50 μm/s (Reprinted with permission from Ref. [46]. Copyright 2007, American Institute of Physics)

radical quenching moiety and the appropriate photoresist components would lead to high TPP resolution and fine structure fabrication with confining radical diffusion.

Additionally, Farsari et al. presented a method based on quencher diffusion for increasing the resolution of direct femtosecond laser writing by multiphoton polymerization [45]. This method relies on the combination of a mobile quenching molecule with a slow laser scanning speed, allowing the diffusion of the quencher in the scanned area and the depletion of the multiphoton generated radicals. The material used is an organic–inorganic hybrid, while the quencher is a photopolymerizable amine-based monomer which is bound on the polymer backbone upon fabrication of the structures. By using this method, woodpile structures with a 400 nm intralayer period have been fabricated. This is comparable to the results produced by direct laser writing based on stimulated-emission-depletion microscopy; today this method is considered as the state-of-the-art one in 3D structure fabrication. These woodpiles exhibited well-ordered diffraction patterns and stopgaps down to near-infrared wavelengths. The quencher diffusion was further modeled and showed that the radical inhibition was responsible for the increased resolution.

In our previous study, the LSR in multiphoton fabrication was improved to 80 nm by using an anthracene derivative of 9,10-bis-pentyloxy-2,7-bis[2-(4-dimethylamino-phenyl)-vinyl]anthracene (BPDPA) as a highly sensitive and efficient photoinitiator (Fig. 4.25) [46]. Photocurable resin containing 0.18 mol% BPDPA exhibited a low polymerization threshold of 0.64 mW at 800 nm. Furthermore, the theoretical

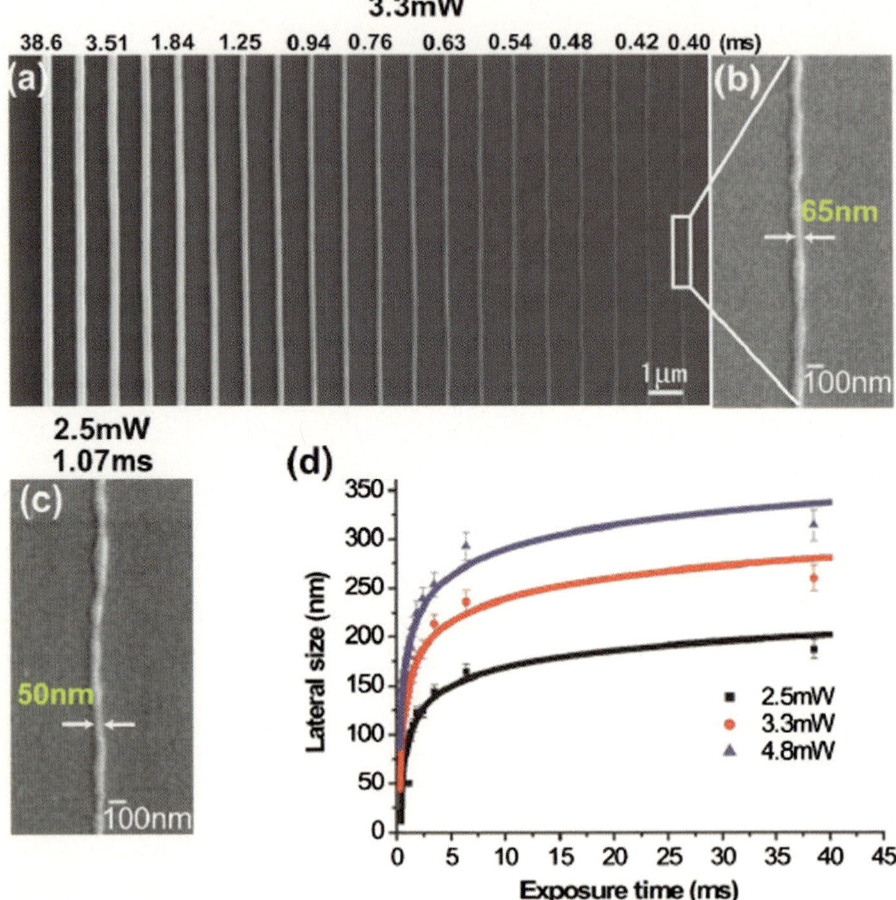

Fig. 4.26 (**a**) SEM images of photocured polymer lines obtained at a laser power of 3.3 mW at various exposure times. (**b**) Magnified SEM image of a line fabricated with the laser power of 3.3 mW and exposure time of 0.4 ms. (**c**) A polymer line with a width of 50 nm. (**d**) LSR versus exposure time under different laser power (lines are the calculated results) (Reprinted with permission from Ref. [29]. Copyright 2008, American Institute of Physics)

calculations showed that the LSR can be increased by reducing the laser power, indicating that the LSR could be improved using more sensitive initiators.

We have achieved an LSR of 50 nm for photocured polymer lines on the surface of a substrate by continuing scanning mode [29]. Figure 4.26 shows the SEM images of photocured polymer lines obtained with a laser power of 3.3 mW at various exposure times. An LSR of 50 nm has been successfully obtained with the laser power of 2.5 mW and the exposure time of 1.07 ms. In addition, the calculation has been carried out, and the corresponding theoretical analysis of the distribution of light intensity indicates that the LSR could be reduced to better than 20 nm.

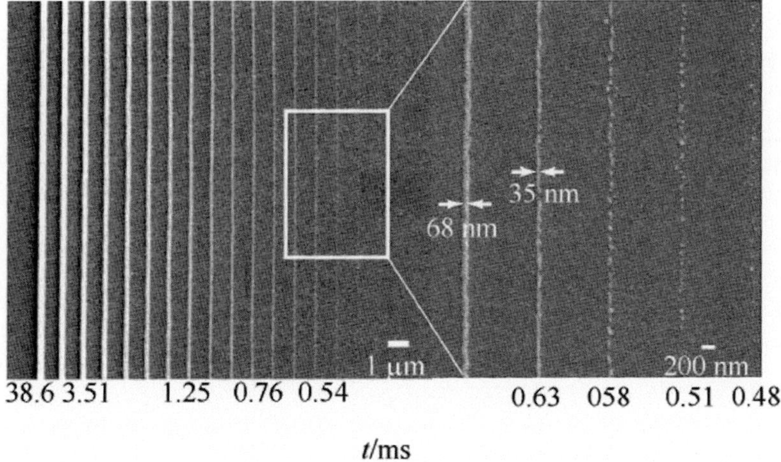

Fig. 4.27   Feature sizes of lines fabricated by SCR500. The average laser power was 3.1 mW

Recently, a polymer line with a resolution of 35 nm has been achieved on the glass substrate by controlling the laser scanning speed, exposure time, and the laser power (Fig. 4.27), which is only 1/22 of the laser wavelength used for the multiphoton fabrication [47]. Among the reported results, it is the highest linewidth resolution achieved by multiphoton fabrication on the glass substrate. It is revealed that the resolution can be improved efficiently with ultra-small exposure time or exposure power which is close to the polymerization threshold.

Furthermore, the feature size of two-photon polymerization using SCR500 was reduced to λ/50 by using femtosecond laser at the wavelength of 780 nm [34]. Lines with sub-25-nm resolution were produced by controlling the incident laser power and the laser focus scanning speed up to 700 μm/s. By fabricating the suspended lines between two closed structures, the feature size has been further reduced to about 15 nm (Fig. 4.28), which demonstrated the potential for fabricating 3D micro-/nanostructures with a high spatial resolution.

4.3.3   Photoreduction for Metal Nanostructure Fabrication

Metal nanostructures are one of the most important components in many micro-/nanodevices, such as nanoelectronic integrated circuits, metamaterials, micro/nanofluidics, and MEMS/NEMS. In the multiphoton photoreduction process, the metal ions absorb two or more photons simultaneously and being reduced into metal nanoparticles. A basic mechanism for photoreduction of metal ions by a dye sensitizer has been proposed. The transfer of an electron from the excited dye to the metal ion leads to the formation of a metal atom which can either (a) react with other metal atoms to nucleate, (b) add onto an existing particle, or (c) undergo

Fig. 4.28 Feature sizes of suspended lines fabricated between two closely transposed supports. The pulse power was 35 mW, the scanning speed was 700 μm/s, and the spacing was 600 nm (Reprinted with permission from Ref. [34]. Copyright 2007, American Institute of Physics)

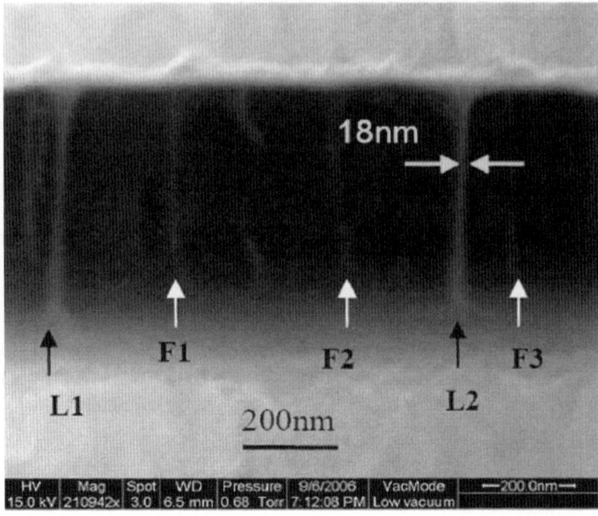

charge recombination. Thus, the formation and growth of metal particles is limited by the competition between the rate of growth or nucleation and charge recombination, as well as local depletion of metal ions. Since growth rates are generally much greater than nucleation rates and depend on the number of nucleation centers, it is reasonable that the introduction of nanoparticle seeds into the composite could significantly enhance the efficiency of formation of a continuous metal phase. Marder and Perry et al. have explored the incorporation of ligand-coated metal nanoparticles into a photoactive material system to provide controllable nucleation centers [48].

In our previous study, the mechanism of multiphoton photoreduction of metal ions with the assistance of surfactant has been proposed. The photoreduction of the metal nanostructures can be divided into three stages: nucleation, growth, and aggregation of metal nanoparticles, which is clearly illustrated in Fig. 4.29a. It was found that the surfactant plays an important role in the growth and aggregation processes. The surfactant molecules made a significant contribution to the control of the diameters of nanoparticles down to tens of nanometers and resulting in the silver stripe with narrow linewidth [49].

More specifically, these three stages in the photoreduction of silver nanopattern include the nucleation of silver in stage 1, growth of nanoparticles in stage 2, and aggregation of silver nanoparticles in stage 3 (Fig. 4.29b). In the absence of surfactant n-decanoylsarcosine sodium (NDSS) which has the alkyl carboxylate, a silver pattern with a lateral resolution of 1 μm consisting of large silver particles has been fabricated [50, 51]. The nucleation process was initiated by laser irradiation, and afterwards the silver nuclei grew up to 1 μm. This ungoverned metal growth leads to large metal particles. The existence of the large particles leads to thick silver lines and prevents the spatial resolution reaching 100-nm scale after the aggregation of the

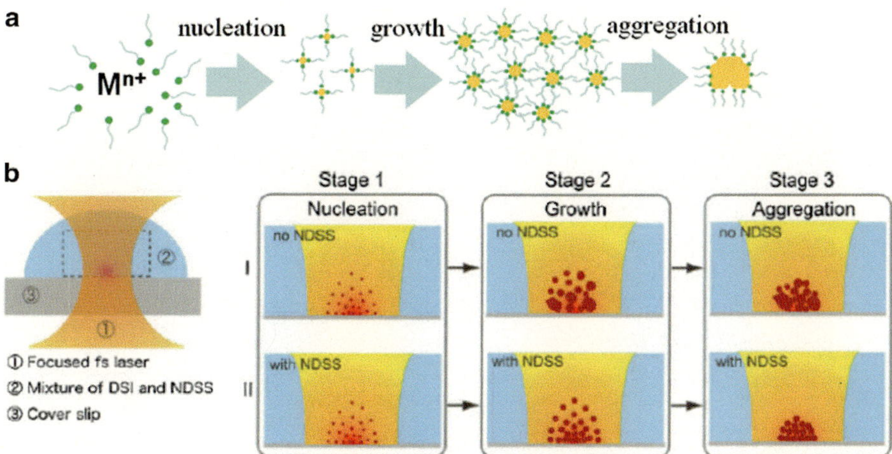

Fig. 4.29 (**a**) Schematic illustration of surfactant-assisted multiphoton photoreduction. (**b**) Schematic illustration of the formation of silver stripe patterns through the multiphoton-induced reduction process. The *left* scheme illustrates the photoreduction system. Routes I and II represent the patterning process of the samples with and without NDSS under the same laser power and exposure time. These two processes start from reduction of silver ions and the creation of silver seeds at stage 1 in the focal laser point. After that, the silver seeds grow up to nanoparticles at stage 2. In the absence of NDSS, varisized particles with different shapes are formed, whereas particles with uniform size and shape are created by adding NDSS. At stage 3, aggregation of the nanoparticles eventually leads to the formation of silver patterns (Reproduced from Ref. [49] by permission of John Wiley & Sons Ltd)

particles (stage 3). On the other hand, the probability of the nucleation is assumed to be the same as that with NDSS. However, the NDSS molecules cover the surface of the silver particles immediately after the nucleation process, and its covered layer eliminates the further metal growth and decreases the particle size down to around 20 nm. In this case, the degree of particle growth suppression increases with the increasing of the surfactant concentration. These growth-suppressed particles aggregate to form silver patterns. In this stage, the concentration of particles is higher at the center of the focused laser spot, since there is a higher nucleation probability associated with the higher laser power. Meanwhile, the higher laser power also helps to break the surfactant layer surrounding the particles [52], which in turn enhances the particle aggregation. These two possibilities lead to the aggregation of silver particles directed to the center of the laser beam.

In order to achieve ideal metallic microstructures, various novel routes have been developed. For example, 3D metallic microstructures were successfully created in the polymer matrix containing homogeneously doped metal ion [53]. Due to the effective support of polymer matrix, continuous microstructures could be retained after developing. Tanaka et al. achieved 3D metallic microstructure through MPR of the metal ions aqueous solution. In this process, ions in a

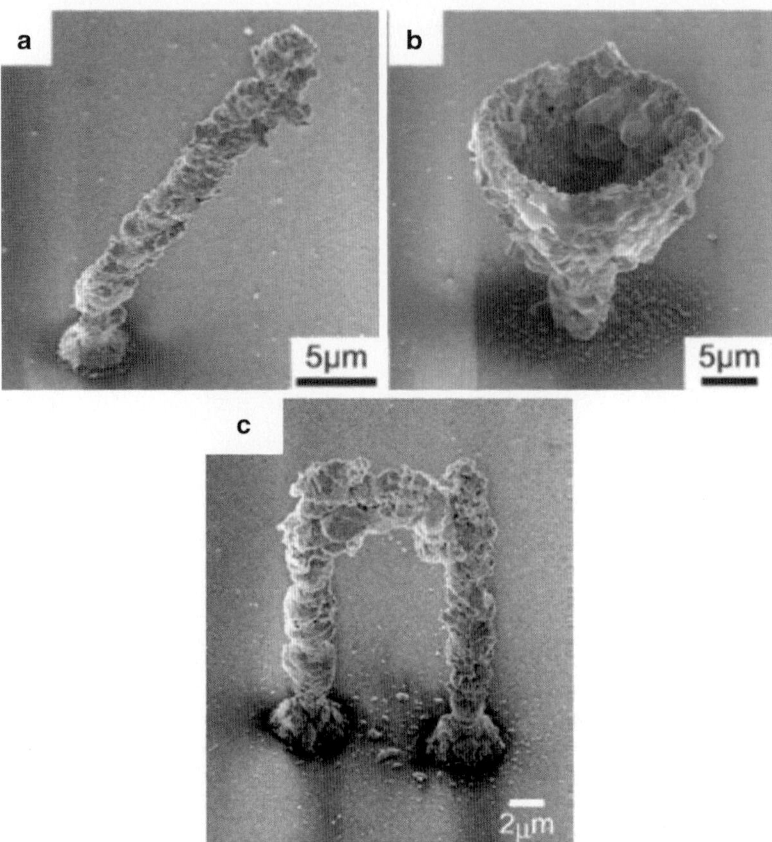

Fig. 4.30 SEM images of a freestanding (**a**) silver tilted rod, (**b**) silver cup (Reprinted with permission from Ref. [54]. Copyright 2006, American Institute of Physics.) and (**c**) microsized 3D silver gate structure (Reprinted with permission from Ref. [50]. Copyright 2006, American Institute of Physics)

metal-ion aqueous solution were directly reduced by a tightly focused femtosecond pulsed laser to fabricate arbitrary 3D structures. The self-standing 3D silver microstructures with arbitrary shapes were fabricated on a glass substrate [50, 54]. SEM images in Fig. 4.30 show a freestanding silver tilted rod, silver cup, and microsized 3D silver gate structure. The length of the rod and the angle relative to the substrate are 34.64 μm and 60°, respectively. The height and the top and bottom diameters of the cup are 26, 20, and 5 μm, respectively. The width, height, and linewidth of the microsized 3D silver gate structure are 12, 16, and 2 μm, respectively. Moreover, the conductivity of the silver gate was only 3.3 times lower than that of bulk silver, while the resolution was at micrometric scale.

4.4 Micro-/Nanostructures Fabricated with Functional Materials

The commercially available photopolymers such as urethane acrylates and epoxies were widely used in the multiphoton microfabrication. In order to achieve certain mechanical and chemical properties of the micro-/nanostructures or devices, many kinds of materials such as the hydrogel and biocompatible materials, photoresists, inorganic–organic polymers, and hybrid polymers containing metal ions have been investigated in the micro-/nanofabrication.

4.4.1 Stimuli-Sensitive Hydrogel and Biocompatible Micro-/Nanostructures

The fabrication of "smart" materials with 3D micro-/nanoscale resolution will extend the utility of these materials to satisfy a broader range of applications. Smart polymeric gel has become a new generation of biomaterials, and a wide range of applications have been developed, such as templates for nanoscale and biomedical devices [55, 56], scaffolds for tissue-engineered prostheses [57, 58], biosensors [59, 60], and actuators [61]. Particularly, stimuli-sensitive hydrogels have been gaining much attention as functional soft and wet biomaterials, since the response to stimuli is a basic phenomenon in living systems. Mimicking the behavior of living systems can offer an opportunity to solve many biomedical problems and is significant to develop biomedical devices. Hydrogels as "chemical muscles" have made significant progress in the field with the conception of artificial muscle-like actuators [62] since Kuhn et al. first identified hydrogels as "chemical muscles" in 1950 [63]. Stimuli-sensitive hydrogels have been proved to response to diverse stimuli, including pH [64], temperature [65], light [66], pressure [67], electric field [68], chemicals [69], ionic strength [70], or any combination thereof [71]. For a smart biomaterial, fast and reversible responsive behavior to the external stimuli signals is needed.

In general, compared to the modes of deformation caused by the whole volume, bending behavior of hydrogels could occur much faster. The hydrogels with specially designed structures are considered to accelerate the bending behavior. Maeda et al. reported a "self-walking" gel actuator through an asymmetric swelling–deswelling change [72]. This asymmetric behavior originates from a gradient structure, resulting in the asymmetric deformation of gel materials. However, conventional stimuli-responsive hydrogels show slow response, which leads to the hysteresis associated with the on and off states [73]. To overcome this drawback, one alternation is to use thinner and smaller hydrogels without significantly deteriorating their mechanical properties. Thus, the fabrication of various microstructures of hydrogels becomes an urgent and desired task for developing biomimetic medical devices.

Multiphoton nanofabrication, a technique that could localize photochemical reactions in 3D based on nonlinear absorption using photoinitiators, is useful to

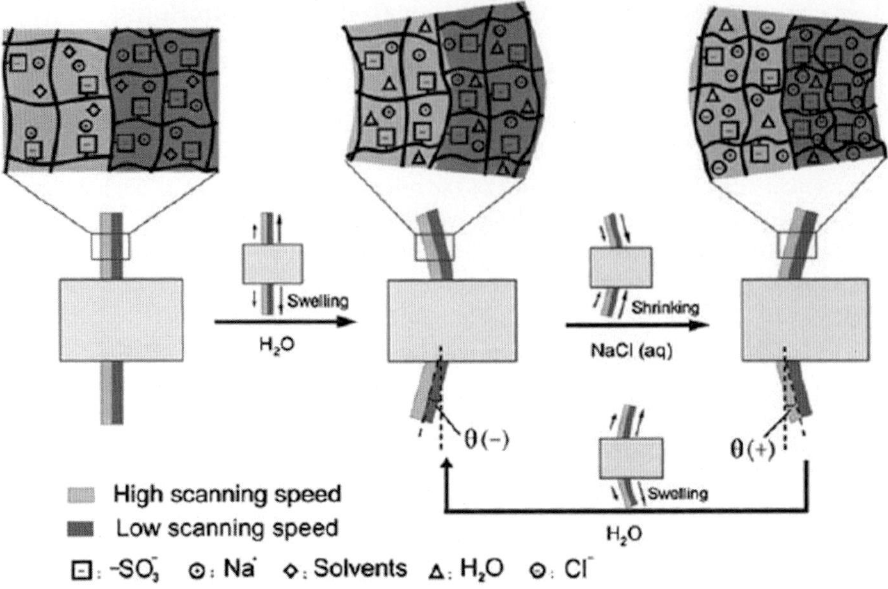

Fig. 4.31 Schematic illustration of designed model for bending controllable ion-responsive hydrogel microcantilever produced by asymmetric-fabricating MPP technique. The microcantilever is fabricated with different laser scanning speed under the same laser power (Reproduced from Ref. [76] by permission of the Royal Society of Chemistry)

create elaborate 3D microarchitectures. It is also a conventional method to fabricate microstructures for gel materials. By using the multiphoton fabrication technique, various micromachines and microdevices with conventional photoresists and polymer nanocomposites have been achieved, such as microneedles, micropumps, microrotors, microgear sets, and 3D photonic crystals. In addition, various microstructures composed of functional materials, including polymeric nanocomposites, photoresponsive and photoluminescent polymers, and stimuli-sensitive hydrogels, such as microrings, microlenses, and microfluidic systems, have been further fabricated and expected to play an important role in the related applications such as biomedical devices. For biomedical applications, especially for those in vivo, microdevices with a size matched with cells are expected. Watanabe et al. realized the fabrication of a 3D hydrogel microstructure through MPP and observed its bending behavior under the irradiation of ultraviolet light caused by the localized swelling [74]. Kaehr et al. reported the fabrication of microstructures composed of protein hydrogel using MPP and demonstrated the hydrogel microelements with bending behavior in prescribed manners [75].

In order to fabricate artificially designable gradient microstructures of hydrogels to achieve controllable fast stimuli-responsive behavior taking the advantage of MPP, we have proposed an asymmetric 3D multiphoton polymerization microfabrication method (Fig. 4.31). The size- and shape-controlled stimuli-responsive asymmetric hydrogel microcantilevers have been fabricated and studied [76]. The

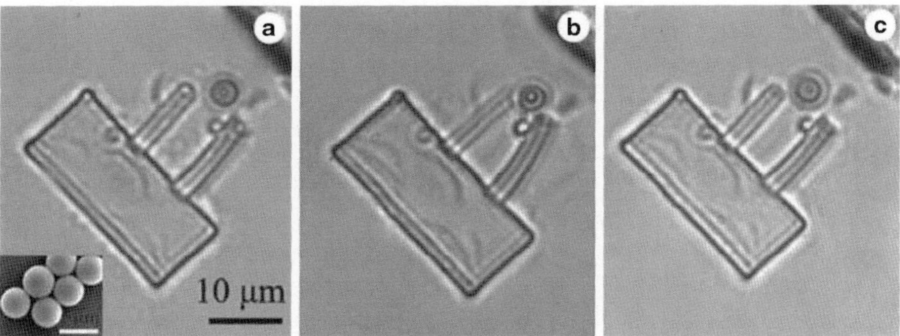

Fig. 4.32 Two-armed hydrogel microcantilever manipulate PS microsphere. (a) The microcantilever was placed in 1 M NaCl. *Inset* was the SEM image of PS microsphere. Scale bar was 5 μm. (b) The covering solution was carefully changed to water. (c) The covering solution was altered to 1 M NaCl again (Reproduced from Ref. [76] by permission of the Royal Society of Chemistry)

reversible ion-responsive hydrogel microcantilevers exhibited controllable bending behavior owing to their asymmetric deformation to external ions. The reversible bending direction of microcantilevers could complete within only 0.133 s when the surrounding environment was alternated from water to 1 M NaCl solution. Furthermore, the microcantilevers exhibited large total bending angles up to 32.9°. The bending behavior can be also strongly influenced by increasing the length of the microcantilevers. The 3D stimuli-responsive hydrogels are demonstrated to be promising for the application of microactuators and micromanipulators as shown in Fig. 4.32. This designable microfabrication technique and the stimuli-responsive behavior of asymmetric microstructure of hydrogel with versatility in the shape would be prospective for developing biomedical microdevices.

Proteins have been integrated as minor components into synthetic hydrogel networks to act as responsive elements as an approach to create smart materials that change volume in response to a chemical signal [75]. For example, hydrogels incorporating proteins such as antibodies and calmodulin have been designed to change hydration degree in response to specific molecular triggers. Natural and engineered proteins offer a diverse physicochemical characteristics and functional properties. In addition, proteins contain a large number of weak acids and bases, similar to polymers used in pH-responsive hydrogels, and, therefore, have the potential to undergo shape and hydration changes in response to chemical triggers. Importantly, matrices composed of functional photo-cross-linked proteins can be localized in 3D microenvironments by using multiphoton fabrication. Thus, multiphoton-fabricated protein structures could provide 3D high-resolution control over the topography of a stimuli-responsive material, facilitating better mechanical functionality and incorporation of responsive hydrogels into more complex 3D devices while potentially providing specificity (e.g., through ligand binding) over volumetric responses. Proteins with differing hydration properties can be combined to achieve tunable volume changes that are rapid and reversible in response to the

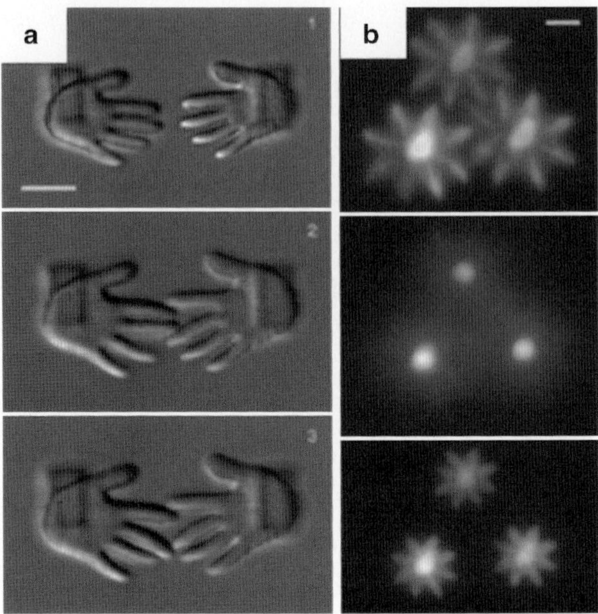

Fig. 4.33 (a) The direct-write process allows BSA hydrogel microstructures to be fabricated with high resolution and arbitrary topography, providing abilities for effecting specific interactions between swelled states. Panels 2 and 3 demonstrate variable interdigitation of middle fingers achieved when the bath solution is cycled between pH 5 and pH 3. (**b**) Fluorescence images (from entrapped photosensitizer) showing stylized BSA microflowers swelled at pH 2.2 (HCl; *Top*) nested ~ 7 μm from the surface on BSA "stems" (*Middle*; focus at coverslip surface), which undergo rapid contraction upon the addition of Na_2SO_4 to a final concentration of 1.0 mM (All of the scale bars are 10 μm) (Reproduced from Ref. [75] by permission of the Royal Society of Chemistry)

changes in chemical environment. Protein matrices having arbitrary 3D topographies and definable density gradients over micrometer dimensions provide the ability to effect rapid (<1 s) and precise mechanical manipulations by means of changes in hydrogel size and shape, and the applicability of these materials to cell biology is shown through the fabrication of responsive bacterial cages.

Multiphoton fabrication of 3D hydrogels composed of photo-cross-linked proteins has been studied and demonstrated their capabilities for achieving chemically responsive micromechanical elements. Appropriate orientation of hand pairs results in interdigitation of individual fingers upon pH-induced expansion (Fig. 4.33a). In addition, microflowers are fabricated atop stems that "bloom" under low ionic strength conditions and rapidly contract on the addition of salts (Fig. 4.33b). Protein microelements that bend in prescribed manners are created by rational incorporation of microscopic thickness and density gradients.

Claeyssens et al. have developed the photopolymerizable resin containing the biodegradable triblock copolymer poly(ε-caprolactone-co-trimethylenecarbonate)-*b*-poly(ethylene glycol)-*b*-poly(ε-caprolactoneco-trimethylenecarbonate) with 4,4′-bis

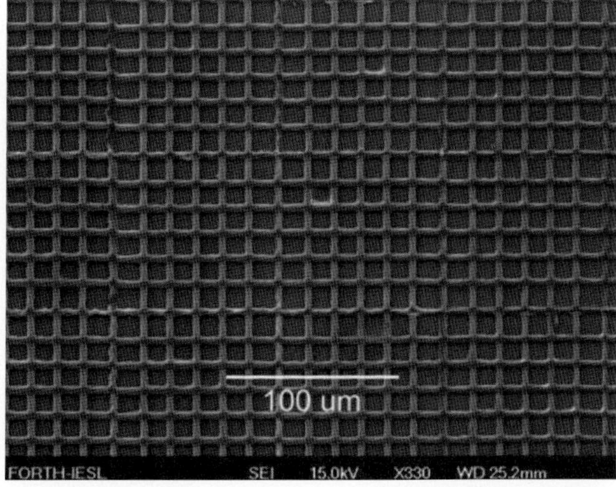

Fig. 4.34 SEM image of a 2D scaffold structure (Reprinted with the permission from Ref. [77]. Copyright 2009 American Chemical Society)

(diethylamino)benzophenone as the photoinitiator [77]. 3D microstructures of good quality have been fabricated by two-photon polymerization technique. As an example, Fig. 4.34 shows a 2D scaffold structure. Furthermore, it was applied to the cell culturing, and the initial cytotoxicity tests showed that the material did not affect cell proliferation. These studies demonstrated the potential of two-photon polymerization as a technology for the fabrication of biodegradable scaffolds for tissue engineering.

Correa et al. reported the two-photon polymerization of an acrylic resin doped with the biocompatible polymer chitosan using a guest–host scheme [78]. The structures fabricated have nanometric surface features (resolution of approximately 700 nm). The fluorescence background in the Raman spectrum indicates the presence of chitosan throughout the structure. Dinca et al. proposed a new method for the precise, 3D patterning of amyloid fibrils (Fig. 4.35) [79]. The technique that combines femtosecond laser technology and biotin–avidin mediated assembly on a polymeric matrix is expected to be applied widely to various fields, from molecular electronics to tissue engineering.

4.4.2   Nanocomposites Used in Micro-/Nanofabrication

4.4.2.1   In Situ Synthesis

Polymer nanocomposites have been widely investigated due to the easy processing originated from polymer and the function from inorganic nanoparticles (NPs). Among them, luminescent semiconductor–polymer nanocomposites are also attractive as luminescent polymeric materials, since various semiconductor NPs show unique tunable light emission properties arising from the quantum size effects. Recently, much attention has been focused on the synthesis and light emission

Fig. 4.35 (**a**) SEM image of a series of 3D micro-/nanostructure, with peptide fibril bridges self-assembled between them. (**b**) One pair of 3D micro-/nanostructure, with peptide fibril bridges self-assembled between them. (**c**) Detail of the self-assembled fiber bridge (Reprinted with the permission from Ref. [79]. Copyright 2008 American Chemical Society)

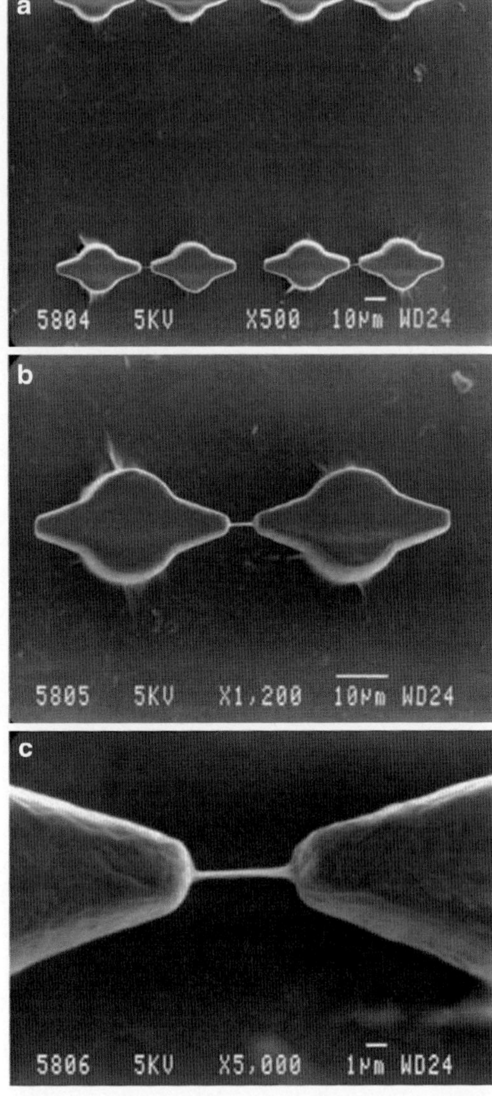

properties of semiconductor–polymer nanocomposites. The resin used in the multi-photon fabrication is required to have good transparency. It is difficult to disperse inorganic NPs into the commercial photoresist resins due to the high viscosity. The aggregation of NPs causes the opaque of photoresist resin, thus decreases the ability for fabricating 3D structures using multiphoton nanofabrication. To realize 3D microstructures of luminescent polymer nanocomposites and to facilitate their use as photonic microdevices, it is essential to develop a suitable method for fabricating 3D microstructures from polymer nanocomposites with size control of the embedded NPs. Such an approach is expected to enable the fabrication of color-tunable microstructures and the development of high-efficiency polymer optoelectronic devices. Thus, we developed the in situ synthesis method for the multiphoton fabrication.

In the in situ synthesis, the metal ions were doped firstly in the photopolymerizable resin that is ready for the multiphoton fabrication. Then, the multiphoton polymerization process will be carried out using the metal-ion-doped resin. After that, the micro-/nanostructure will be developed and the predesigned pattern will be obtained.

In our previous study [80], the titanium (IV) (Ti^{4+}) ions were doped in a commercial photopolymerizable urethane acrylate resin SCR500 that provided by JSR. Titanium (IV) ethoxide was selected as the source of Ti^{4+} ions and was mixed with methacrylic acid to obtain the complexes of titanium (IV) acrylate. 2,2-Diethoxy-acetophenoe was also added into the solution as photoinitiator. Then, the complex solution was introduced into SCR500 resin and mixed to obtain the photopolymerizable urethane acrylate resin doped with Ti^{4+} ions.

Then the developed photopolymerizable resin was dropped on a cover glass. To prevent a reaction between Ti^{4+} ions in resin and water molecules from air, another glass slide was covered to keep the sample. The laser beam was focused by a high NA (1.4) oil-immersion objective lens. Fabrication data files were generated by CAD (computer-aided design) software, and the sample scanning relative to the focal spot was realized by a 3D piezostage translator with an accuracy of less than 1 nm. Two-photon polymerization only occurred at the focal spot and scanning of laser beam allowed for the fabrication of the designed polymer features. After finishing polymerization, the sample was kept in ethanol for 2 h to remove the unpolymerized resin. Then, the sample was dried in air at room temperature.

By using this method, 3D diamond PhC structure can be fabricated by the multiphoton fabrication as shown in Fig. 4.36. Titanium dioxide (TiO_2) nanoparticles were generated in the polymer matrix of micron-sized polymer structures. The diamond photonic crystal structure, which consisted of polymer composite materials of TiO_2 nanoparticles, was successfully fabricated by direct laser writing and its photonic bandgap was confirmed. This is the first report about 3D PhCs of TiO_2-NP-embedded polymer nanocomposites with a confirmed photonic bandgap (PBG).

As reported, we have presented the fabrication of color-tunable 3D microstructures by using multiphoton fabrication technique in conjunction with the in situ synthesis of photoluminescent semiconductor NPs in 3D microstructured polymer matrices [81, 82]. In this study, cadmium methacrylate ($Cd(MA)_2$) was used as precursors for CdS NPs; these species also serve as monomers for photopolymerization. A commercially available oligomer, dipentaerythritol hexaacrylate (DEP-6A), has been selected as the cross-linker. Methacrylic acid (MA) and MMA (Beijing Chemicals) have been used as monomers with 1 wt% of benzyl (Aldrich) as the photoinitiator and 1 wt% 2-benyl-2-(dimethylamino)-4′-morpholinobutyrophenone (Aldrich) as the photosensitizer. First, we have fabricated 3D microstructures of polymers containing CdS precursors by the multiphoton fabrication technique. Subsequently, we have performed the in situ synthesis of CdS NPs in the fabricated 3D polymeric microstructures.

Fig. 4.36 SEM images of the 3D diamond PhCs structure of inorganic nanoparticles–polymer composite materials produced by two-photon polymerization. *Inset* shows magnified image to confirm nanoparticles on surface (Reprinted from Ref. [80], Copyright 2004, with permission from Elsevier)

For the fabrication of 3D microstructures, a mode-locked Ti–sapphire laser was used. The center wavelength, pulse width, and repetition rate were 780 nm, 80 fs, and 82 MHz, respectively. The lasing source was tightly focused by a 100x oil-immersion objective lens with a high NA (1.4, Olympus). The focal point was focused onto the liquid photopolymerizable sample, which was placed on an xyz-step piezostage controlled by a computer. After laser fabrication, the unpolymerized resins were washed away using ethanol. The obtained microstructures were used for characterization or further treatment. The emission wavelength of the micro-/nanostructures can be tuned by the size of CdS nanoparticles (Fig. 4.37). A fluorescent 3D microbull and a 3D microlizard have been successfully fabricated from resins that contained different components. This work has been highlighted and reported by *Nature*.

4.4.2.2 Fabrication for Nanocomposite-Doped Resist

Structures containing metal nanoparticles play an important role in modern nanophotonics. Nanoparticles of noble metal exhibit strong resonant scattering of light through plasmonic oscillation of free electrons. The optical feature of metal nanoparticles, called local surface plasmon resonances (LSPRs), shows the potential as a fundamental platform for novel optical/photonic nanotechnologies. Because the free electrons are confined in the nanoscale volume of metal particles,

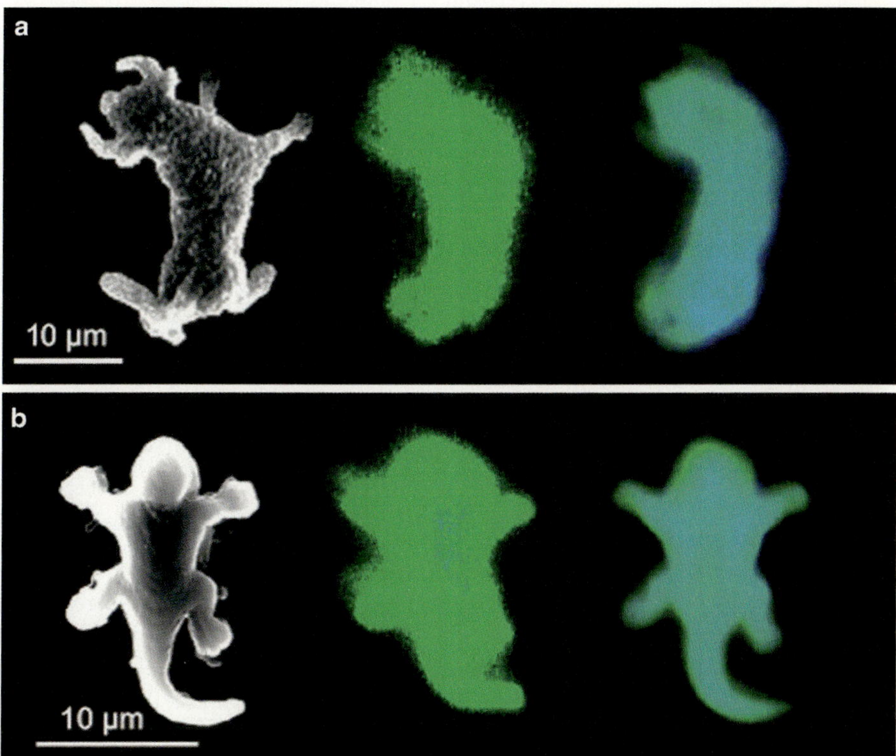

Fig. 4.37 Images of SEM (*left*) and fluorescence microscopy of a 3D microbull (**a**) and a 3D microlizard (**b**) fabricated from resins that contained different components (Reproduced from Ref. [81] by permission of John Wiley & Sons Ltd)

the oscillation of these electrons creates discrete resonant modes, which are strictly defined by the geometry such as size, shape, and the orientation of the particles, as well as various ambient factors that cause refractive index change such as temperature and pH. The excitation of LSPRs could create a strong local electromagnetic field surrounding the nanoparticle, indicating the spatial distribution of the oscillating electrons. The interaction of multiple adjacent nanoparticles causes the conjugation modes of LSPRs. LSPR coupling induces extremely strong electromagnetic field at small gaps between particles and also inter-particle propagation of electromagnetic waves, allowing us to manipulate and utilize photons at nanoscale. Recently, chemical synthesis provides us the method to prepare a variety of noble metal nanoparticles with well-controlled size and shape. Wide applications based on the plasmonic features in assembled metal nanoparticles, including surface-enhanced Raman scattering, chemical and biosensors, left-handed materials, and optical nano-imaging devices, have been proposed.

One of the demands for such plasmonic technologies is to develop novel micro-/ nanofabrication methods to provide arbitrary complex micro-/nanostructures in

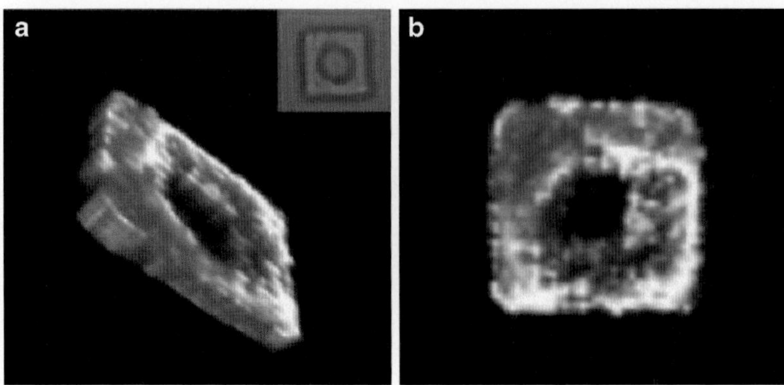

Fig. 4.38 3D TPP microstructure imaged by (**a**) 3D TPL (*Insert*: 2D bright-field image) and (**b**) 2D TPL image of the microstructure in (**a**) at the base (Reproduced from Ref. [84] by permission of Optical Society of America)

which the metal nanoparticles are assembled in desired orientation and arrangement. Up to now, lithographic techniques are widely used to fabricate precisely organized metal nanostructures on substrates. Laser ejection of molten metal nanodroplets is a unique method, in which laser creates and deposits the metal nanoparticles simultaneously to prepare 2D and 3D structures of metal nanoparticle arrangements [83]. There are also reports of the application of TPP lithography for fabricating metal nanoparticles/polymer composite microstructures [84, 85]. The results demonstrate that the developed 3D polymer microstructures containing the AuNRs not only improves efficiency by decreasing the femtosecond laser power but also provides optical properties that can improve the contrast in two-photon luminescence (TPL) imaging. Figure 4.38a shows the TPL image of a hollow 3D microstructure by utilizing the fabrication and imaging conditions as described in reference [84]. This structure has a base area of $20 \times 20\ \mu m^2$ and a height of 5 μm. Finally, the diameter of the hole is 10 μm and the distance between two adjacent axial layers is 0.1 μm. Figure 4.38b is the cross-sectional image of the base of the microfabricated structure of Fig. 4.38a. Figure 4.39a shows the TPL image of a fabricated 3D, AuNRs-doped, woodpile-like bovine serum albumin (BSA) microstructure filled with AuNRs [85]. The two-layer structure has a base area of $24 \times 24\ \mu m^2$ and a height of 4 μm. The width and thickness of BSA sticks are both 2 μm and the distance between the edges of adjacent sticks is 2 μm. In this study, TPL 3D images were excited by the use of 0.13 mW, 100 fs laser at 850 nm, and a scan rate of 50 kHz. Figure 4.39b is the 2D bright-field image of the woodpile-like microstructure (Fig. 4.39a). The SEM zoom in image (Fig. 4.39c) shows clearly that these are AuNRs inside the cross-linked BSA. These results indicated that there was no change to the morphology of the AuNRs after the femtosecond laser fabrication process when 0.75 mW laser power was used in the fabrication. Therefore, cross-linked protein microstructures filled with intact AuNRs can be fabricated by using lower laser power through plasmonic enhancement. This

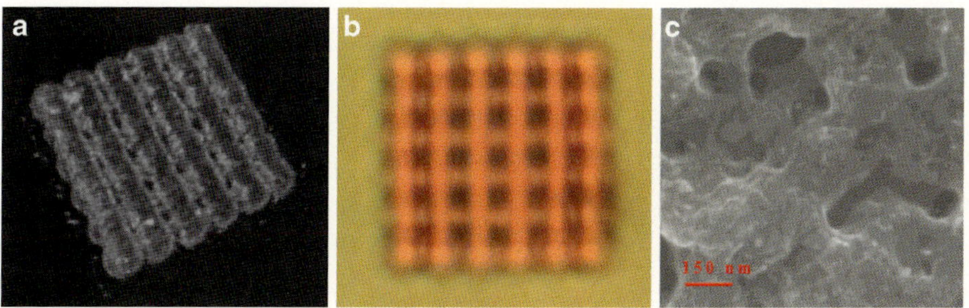

Fig. 4.39 An AuNRs-doped 3D woodpile BSA microstructure imaged with (**a**) 3D TPL, (**b**) 2D bright field, and (**c**) SEM. (Reproduced from Ref. [85] by permission of Optical Society of America)

approach not only can create the freeform 3D cross-linked protein microstructure but also can avoid the use of a high fabrication laser power which will damage the AuNRs. Moreover, the BSA with AuNRs material is biocompatible and, therefore ready, for bioapplications such as 3D scaffolds to cues for cell binding.

We have proposed an approach to dispersing Au nanorods densely into photopolymerizable resin and performed TPP lithography on the colloidal Au nanorods [86–88]. We demonstrated the fabrication of Au nanorod aggregates microstructures by means of a femtosecond near-infrared laser. The laser light was tightly focused into colloidal Au nanorods dispersed in photopolymerizable MMA compound to induce TPP. TPP of MMA gathered the nanorods together to form solid microstructures of aggregates. The laser light excited a local surface plasmon and confined TPP surrounding the nanorods. Consequently, the optical accumulation of nanorods led to the formation of nanorod aggregates with desired microstructures.

We dispersed CTAB wrapped Au nanorods on a glass substrate and poured MMA (mixed with initiator and photosensitizer) over the Au nanorods to prepare for multiphoton fabrication. Figure 4.40a shows the experimental setup. A Ti–sapphire laser (Tsunami, Spectra-Physics, Newport Corp.) at 780 nm, with a pulse width and a repetition rate of 100 fs and 82 MHz, respectively, was focused by an oil-immersion objective lens with NA of 1.4 and irradiated onto the substrate. After laser irradiation, the glass substrate was immersed in ethanol so as to remove the non-polymerized MMA. Figure 4.40b shows the dependence of the thickness of PMMA layer on the average laser intensity. For different laser intensity, the thickness of the formed PMMA layer in longitudinal and perpendicular directions of 10 different particles were measured from high magnification SEM images, and then the mean values were plotted. When the average laser intensity was below 1.3 GW/cm^2, the PMMA layer was not observed on the Au nanorods. The thickness of the PMMA layer was increased nonlinearly until the laser intensity reached the threshold for the TPP of original MMA compound. It was observed that, in the case of conventional TPP with radical polymerization, the size of the polymerized voxel showed nonlinear relation

Fig. 4.40 (**a**) Schematic illustration of the experimental condition. (**b**) The plot of the thickness of PMMA layers formed on Au nanorods as a function of the intensity of Ti–sapphire laser light (Reproduced from Ref. [87] by permission of Optical Society of America)

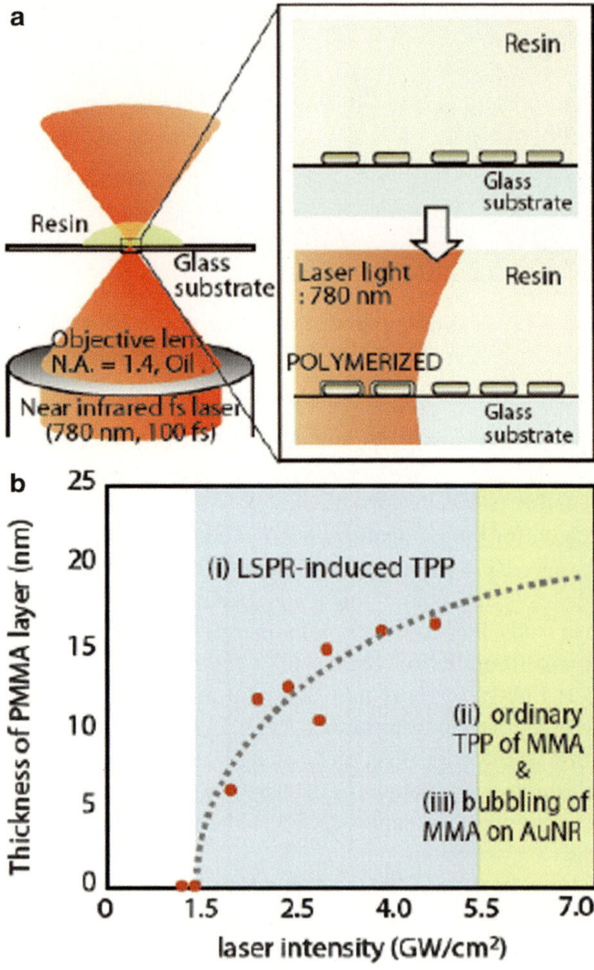

on the laser intensity [31]. The intensity dependence of the thickness of the PMMA layer in Fig. 4.40b is similar to that of conventional TPP.

To define the effect region of plasmonic resonance enhancement on TPP, the dependence of the shapes and thickness of the polymer covered on the isolated single gold nanorods on the laser irradiating intensity was investigated. Figure 4.41a–d shows the SEM images of polymer-coated isolated single gold nanorods obtained by TPP with the laser irradiations under different intensities. With the laser irradiating intensity of 1.7 GW/cm^2, the thickness of the polymerized skin covered the gold nanorod with the length of 62 nm, and the diameter of 22 nm was characterized as 21 and 14 nm at the end and the middle of the gold nanorod, respectively, which formed a polymer/gold nanocomposites "dumbbell" as shown in Fig. 4.41a. The shape of polymer/gold nanocomposites varied from the "dumbbell" to the "ellipsoid" by increasing the laser intensity. When the laser irradiating intensity was increased to

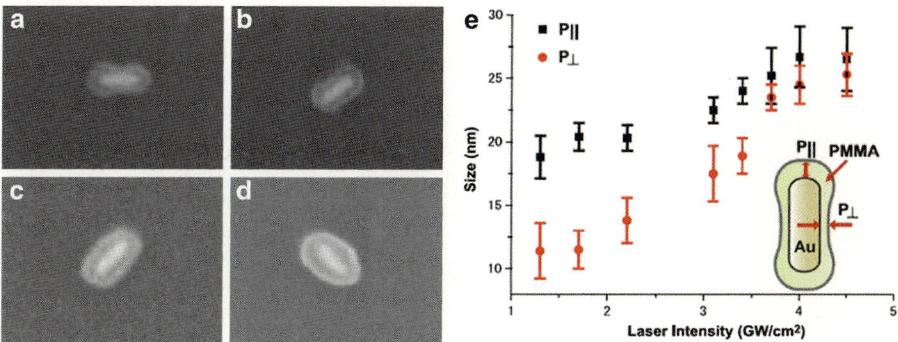

Fig. 4.41 Typical SEM images of the PMMA-coated Au nanorods with different laser intensity, (**a**) 1.7 GW/cm², (**b**) 3.1 GW/cm², (**c**) 3.7 GW/cm², (**d**) 4.5 GW/cm². The scale bar is 100 nm. (**e**) The thickness of polymer layers covered on the Au nanorods in parallel (*black squire*) and perpendicular (*red dot*) directions on the laser intensity. *Inset* is the schematic illustration of P∥ and P⊥ (Reprinted from Ref. [88], with kind permission from Springer Science+Business Media)

4.5 GW/cm², we obtained a polymer/gold nanocomposites "ellipsoid" with the transverse diameter of 111 nm and the conjugate diameter of 74 nm, and a gold nanorod with the diameter of 64 nm and the length of 24 nm was wrapped by the photocured polymer. The parameters of the polymers covered on the gold nanorods under different laser irradiating intensities are summarized in Fig. 4.41b. Thickness changes of the polymer layers covered on the gold nanorods could indicate the effect regions of plasmonic resonance on TPP. It was observed that the thickness of polymer layers covered on the gold nanorods increased along both of the axial (P∥) and radial (P⊥) directions of the gold nanorods when increasing the laser irradiating intensity to 3.7 GW/cm². The thickness of polymer layer at P⊥ direction increased twice of that at P∥ direction, leading to the shape of the polymer/gold nanocomposites from "dumbbell" to "ellipsoid." However, the increase of polymer layer thickness was almost saturated when the laser irradiating intensity was increased to 3.7 GW/cm². Although the laser irradiating intensity of 4.5 GW/cm² was used, only small increase of polymer layer thickness was observed.

The fabrication of arbitrary microstructures of Au nanorods aggregates has been further demonstrated. Figure 4.42a shows SEM images of the letters "GOLD" structures. The letters were fabricated by a single-stroke scan of the laser spot by a step of 50 nm with the exposure time of 3 s at each position. The line width of the structure was 1.2 μm and the total size of each letter was about 10 μm. High magnification images (Fig. 4.42b, c) show that the solid line consisting of closely packed Au nanorods is wrapped by thin PMMA. We evaluated the spatial resolution at different exposure time with a fixed laser intensity. Similar to the typical TPP lithography, we observed an improvement of the spatial resolution when reducing the exposure time. However, further reduction of the exposure time started to cause inhomogeneous linewidth with forming discrete aggregation beyond a certain level. The modulation of the linewidth may be caused by the depletion and consequent

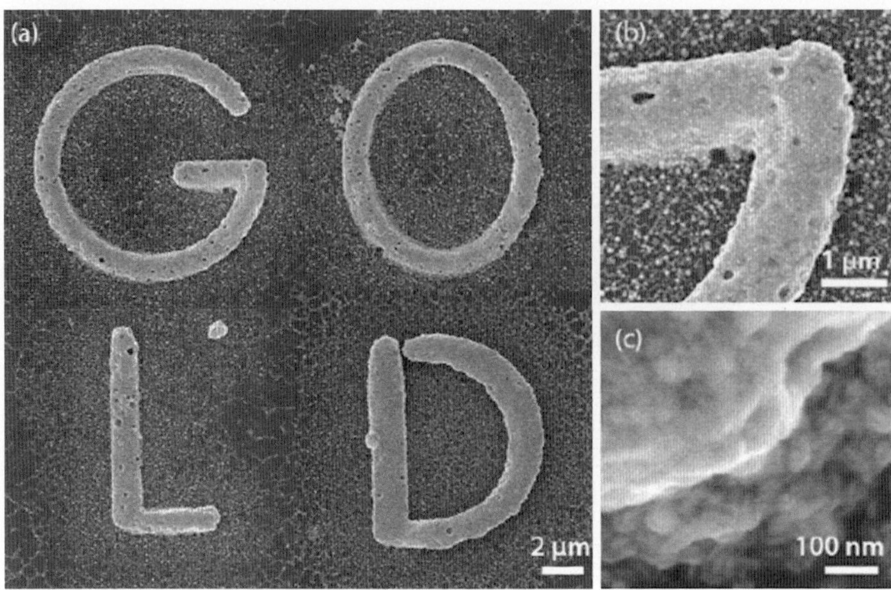

Fig. 4.42 SEM images of letters "GOLD" made of aggregations of Au nanorods. (**a**) Top view, (**b**, **c**) high magnification images. The laser intensity was 1.6 GW/cm^2 and the exposure time was 3 s. The total size of the structure was 10 μm and the line width was of about 1.2 μm (Reproduced from Ref. [87] by permission of Optical Society of America)

temporal change of the local density of nanorods surrounding the laser spot. In this experiment, the smallest linewidth of about 1.0 μm was achieved.

For better understanding the effect of plasmonic resonance of gold nanorod in TPP, finite difference time-domain (FDTD) technique has been performed to simulate the field enhancement of gold nanorod. A gold nanorod with the diameter of 25 nm and length of 62.5 nm and the refractive index of 1.41 in the surrounding of the gold nanorod were used in our calculation model, and an induced irradiating light source with the wavelength of 784 nm was considered. The field distribution is illustrated in Fig. 4.43. The strong field enhancements appeared at two ends of gold nanorod, by which it could be confirmed that the strength of plasmonic resonance enhancement possessed a "dumbbell" shape from the contour lines of field strength shown in Fig. 4.43b, which agreed well with the experimental results of polymer/ gold nanocomposites under low laser irradiating intensity as shown in Fig. 4.43a. It indicated that the polymerized area of TPP was dependent on the strength of plasmonic resonance enhancement. Hence, the ultrafast laser irradiation excited the dipolar longitudinal plasmon mode of the Au nanorods, resulting in the local surface plasmonic resonance-induced TPP.

Furthermore, the mechanism has been induced. Usually, the refractive index of the solidified photoresist would be slightly increased after photopolymerization. Thus, the refractive index around the gold nanorod could be changed after TPP. This change of refractive index from the plasmonic resonance-enhanced

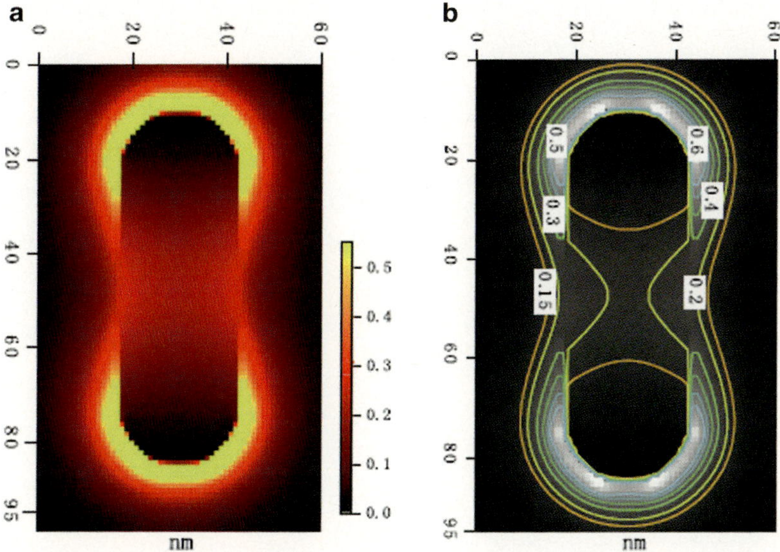

Fig. 4.43 (**a**) Plasmonic field enhancement pattern and (**b**) contour lines of field strength of the Au nanorods calculated at the 784 nm wavelength (Reprinted from Ref. [88], with kind permission from Springer Science+Business Media)

polymerization of photoresist might occur from the surface area of gold nanorod with strong field enhancement and led to the new edge forming between the polymer and the surface of gold nanorod. This new edge could spread the TPP from the ends to the middle of gold nanorod and then change the shape of polymer/ gold nanocomposites from the "dumbbell" to the "ellipsoid" under stronger laser irradiating intensity. Finally, the continued increase of the laser power led to the saturation of the polymerization since the field strength of plasmonic resonance attenuated exponentially from the surface of gold nanorod. Consequently, we could figure out that the effective distance of plasmonic resonance-enhanced TPP was not longer than 30 nm from the obtained experimental results, which agreed with the experimental results described before.

4.4.3   Micro-/Nanofabrication for Hybrid Materials

Hybrid (organic–inorganic) materials are a very popular class of photosensitive materials, since they are easy to prepare, modify, and photopolymerize, and after polymerization, they are optically, mechanically, and chemically stable. In addition, chemical property of the hybrid materials provides the possibility of involving functional groups, such as nonlinear optical molecules and quantum dots [89]. SEM images in Fig. 4.44 shows a micrometer-scale Venus statue fabricated by two-photon microfabrication using a kind of optically high-quality inorganic–organic hybrid

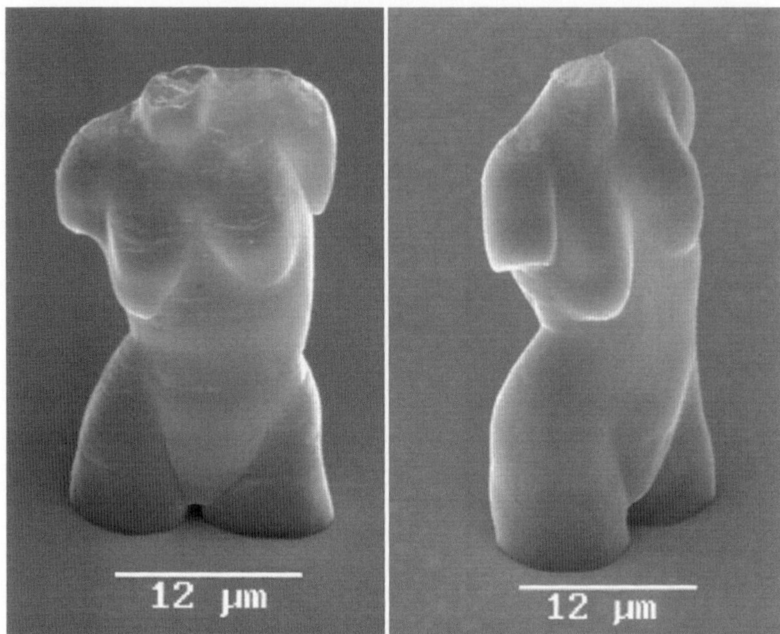

Fig. 4.44 SEM micrometer-scale image of Venus fabricated by two-photon microfabrication. Only the shell was irradiated by femtosecond laser pulses; the inside region was cured with a UV lamp after the liquid resin was developed (Reproduced from Ref. [89] by permission of Optical Society of America)

materials (ORMOCERs). Only the shell was fabricated with femtosecond laser pulses, taking approximately 5 min. The statue was irradiated with UV light for final polymerization of the inner body after that the liquid resin was developed. Aside from the 3D microstructures, photonic crystals fabricated with inorganic–organic hybrid polymers with a structure size down to 200 nm and a periodicity of 450 nm were also achieved.

Chichkov et al. reported the two-photon polymerization of a nonshrinking, photosensitive, zirconium sol–gel hybrid material [90]. This hybrid material can be photostructured even when it contains up to 30 mol% of zirconium propoxide (ZPO). 3D microstructures which were in consistent with the predesigned pattern have been produced by the two-photon polymerization technique. Figure 4.45 shows the 3D photonic crystal structures. No additional effort is required to prevent shrinkage or to improve the structural stability of the fabricated photonic crystals compared to common practice. Its refractive index can be also varied by modifying and tuning the inorganic content in the material. The photosensitivity of the resulting photopolymer was significantly increased by the introduction of ZPO. The fabricated 3D photonic crystal structures demonstrated high resolution and a clear band-stop in the near-IR region. The possibility of tuning this material's optical, mechanical, and chemical properties also makes it suitable for a variety of applications by two-photon polymerization manufacturing.

Fig. 4.45 3D photonic crystal structures produced by the two-photon polymerization technique (Reprinted with the permission from Ref. [90]. Copyright 2008 American Chemical Society)

Furthermore, Farsari et al. reported the nanofabrication of 3D metallic woodpile structures with features below 100 nm, obtained by direct laser writing and selective silver coating [91]. The structures that had fcc geometry and in-layer periodicity 600 nm were fabricated using an organic–inorganic, zirconium–silicon hybrid material doped with a metal-binding monomer, 2-(dimethylamino) ethyl methacrylate. Subsequently, the structures were metalized using electroless plating. The metalized structures exhibited ohmic conductivity, comparable to bulk silver. The author presented experimental and theoretical electromagnetic characterization results of the fabricated structures. It was said that this was the first time that 3D metallic structures with bandgap in the visible range were fabricated, opening thus the way for the design and fabrication of 3D optical components, isotropic optical metamaterials, and optical sensors.

Figure 4.46 shows a metalized woodpile structure with 600 nm periodicity fabricated with a laser power of 1.90 mW. This structure has 3 unit cells (12 layers) thickness, and its features are in the order of 100 nm. For the transmission measurements, a white light source pumped at 1,064 nm (Fianium) was used. The light beam was focused normally onto the sample surface using a 7.5 cm focal length lens, while the transmitted light was collected using an optical spectrometer (Ocean Optics S2000). The half-opening angle of the incident light was 5°. The diameter of the focused beam in the sample was 24 μm, while the measured nanostructure surface normal to the beam was 40 × 40 μm. As the size of the sample's surface was comparable to the beam spot diameter, sample alignment was difficult and critical. The diffraction pattern in the reflected waves produced by the structure was observed, which indicated the quality of the sample. To understand the diffraction pattern, one should note that the woodpile structure presents the {001} family of planes parallel to the crystal surface. In these planes, the structure presents a square symmetry, which is actually revealed by the square symmetry of the diffraction pattern observed. The sharpness of the diffraction spots indicates the long-range order in the crystal.

Fig. 4.46 Perspective view of a 600 nm periodicity and 3-unit-cell (12-layer) thickness woodpile structure. Characteristic cross diffraction pattern of the reflected wave due to the structure geometry (Reproduced from Ref. [91] by permission of John Wiley & Sons Ltd)

4.4.4 Micro-/Nanofabrication for Metallic Structures

4.4.4.1 Micro-/Nanofabrication for Metallic Structures

While the majority of two-photon microfabrication techniques involve the photopolymerizable resin as the fabrication material, studies using metallic materials have also been reported. Up to now, efforts have been devoted to fabrication of metal nanostructures through various nanofabrication techniques. However, it is still a challenging task to obtain 3D structures with nanoscale spatial resolution. For metal nanofabrication via multiphoton process, photoreduction reactions were considered as preferred processes to generate metal building blocks. Perry and Dunn et al. have reported the fabrication of 2D and 3D metal micro-/nanostructures through two-photon fabrication in the gel and polymer systems, respectively (Fig. 4.47) [48, 92]. However, photoreduction usually undergoes the process of isolated metal nanoparticles growing up and aggregation which is uncontrollable and, therefore, usually results in poor metallic microstructures.

The resolution of these 3D metallic structures was still low owing to the presence of unreacted polymers. Typically, the major problem that restricts spatial resolution of metal micro-/nanostructure fabricated via multiphoton process is the uncontrollable growth and aggregation of metal nanoparticles during photoreduction process. If the thermal reduction and rapid growth of metal particles could be efficiently avoided, then metal nanostructures could be directly fabricated in aqueous solutions of metal ions [49, 93, 94].

Marder and Perry et al. have reported the multiphoton fabrication of composite materials containing metal nanoparticles, leading to the growth of nanoparticles into defined continuous metal patterns. Photoactive nanocomposites have been formed using nanoparticles with covalently attached photoreducing dye molecules.

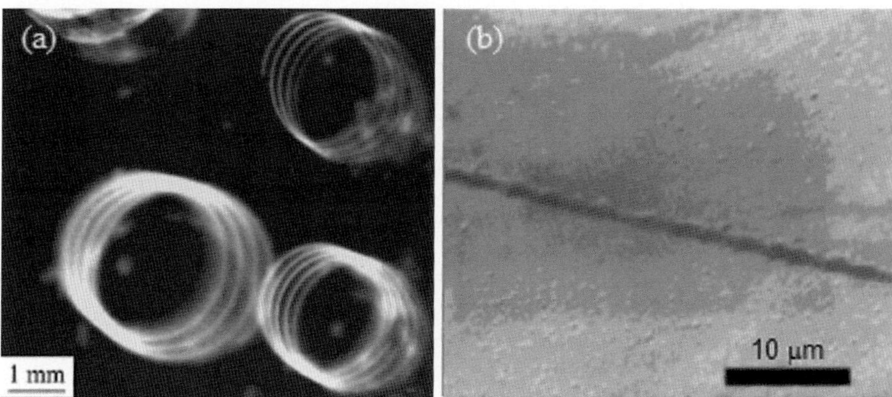

Fig. 4.47 (**a**) SEM image of 3D spiral structure produced within silver-doped sol–gel materials. The latent image was produced using multiple pulse exposures with an energy of 60 µJ per pulse (800 nm, 120 fs pulse) using a 25 mm focusing lens. The development time was 8 h (Reproduced from Ref. [92] by permission of John Wiley & Sons Ltd). (**b**) SEM image of a silver line fabricated on an indium tin oxide-coated substrate by using two-photon fabrication (Reproduced from Ref. [48] by permission of John Wiley & Sons Ltd)

Truly 3D freestanding metallic structures with a height of 100 µm have been microfabricated and characterized (Fig. 4.48). Excitation by one- or two-photon absorption allowed for the one-step laser writing of continuous silver, copper, and gold metal microstructures in 1D, 2D, and 3D patterns. This nanocomposite approach, based on photoreductive-induced growth of nanoparticles, provides a versatile new method for patterning metal features that could have an impact on electronic, optical, and electromechanical technologies [48].

4.4.4.2   Effects of Surfactant on Multiphoton Photoreduction

When we optimize the experimental conditions, such as laser power, scanning velocity of the laser beam spot, and concentration of the surfactant NDSS, the width of the lines can be decreased. Considering the control capability of the growth of the nanoparticles and the particle aggregation contributed to the continuity of the silver stripe in the sample, the sample with the NDSS concentration of 0.099 M was chosen in our previous study [49]. Figure 4.49a and b show the deposition of silver dots on cover slips using MPR microfabrication with assistance of different fatty salts, C_7 and C_9, under the laser power of 1.3 mW and the laser duration of 800 ms. The surface morphology of the silver dots varied with the change of the chain length of the fatty salts. The size and size distribution of silver nanoparticles, which constructed the silver microstructures, also depended on the chain length of the fatty salts, as shown in Fig. 4.49c. When the fatty salt with $n = 5$ was used, the silver nanoparticles showed an average size of 47 nm with a wide distribution from 10 to 130 nm. As n increased to 7, the average size of the nanoparticles was reduced

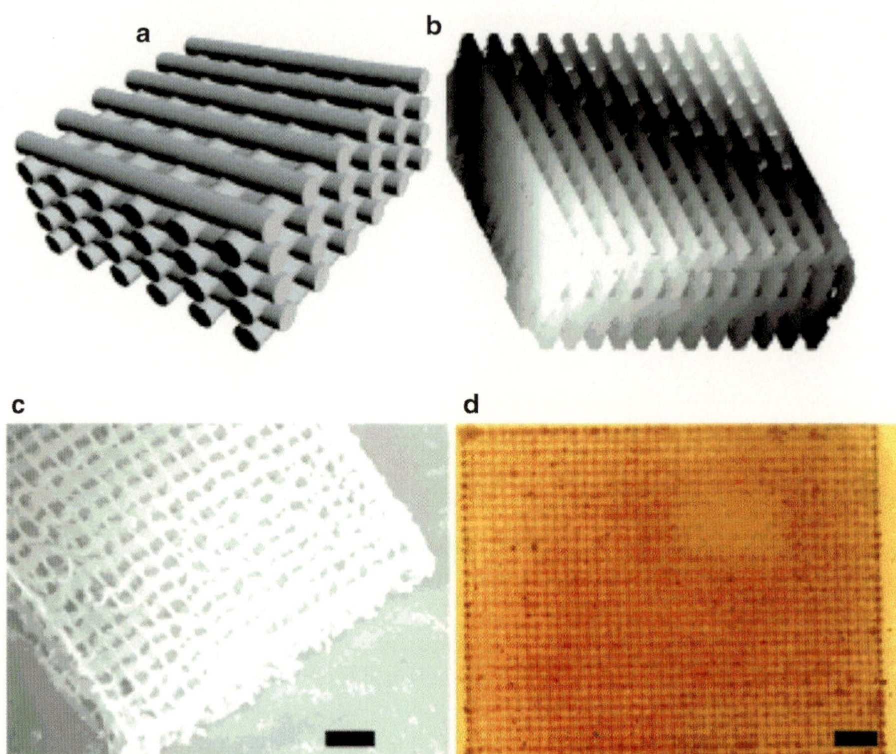

Fig. 4.48 Metallic structures fabricated in nanocomposites by two-photon scanning laser exposure. (**a**) Model of a target 3D structure: a "stack-of-logs" type structure. (**b**) Image of the actual 3D silver structure fabricated in a nanoparticle-seeded polymer nanocomposite by two-photon laser exposure; the image was reconstructed from a series of two-photon fluorescence microscopy images obtained at varying depths in the sample. (**c**) SEM image of the freestanding 3D silver structure in (**b**) after removal of unexposed material using dichloromethane. (**d**) Transmission optical microscopy (TOM) image of the silver microstructure in the polymer nanocomposite after two-photon laser exposure (Reproduced from Ref. [48] by permission of John Wiley & Sons Ltd)

to 21 nm with the narrowing of the size distribution between 10 and 60 nm. When n increased to 9, almost more than 90 % of the silver nanoparticles possessed sizes smaller than 30 nm with a narrow distribution from 10 to 40 nm. These results provide clear evidence that the fatty salts are effective for the size control of silver nanoparticles in the MPR microfabrication process. Longer chain lengths of fatty salts are propitious for obtaining smaller nanoparticles with a narrow size distribution.

The result indicates that this approach not only illustrates an effective process for localized generation and aggregation of metal nanoparticles but also offers a simple way to accurately control the location of nanoparticles at nanometric scale. This finding may open up the way to create arbitrary 3D metal structures and investigate the unique functions of 3D metal structures at nanoscale.

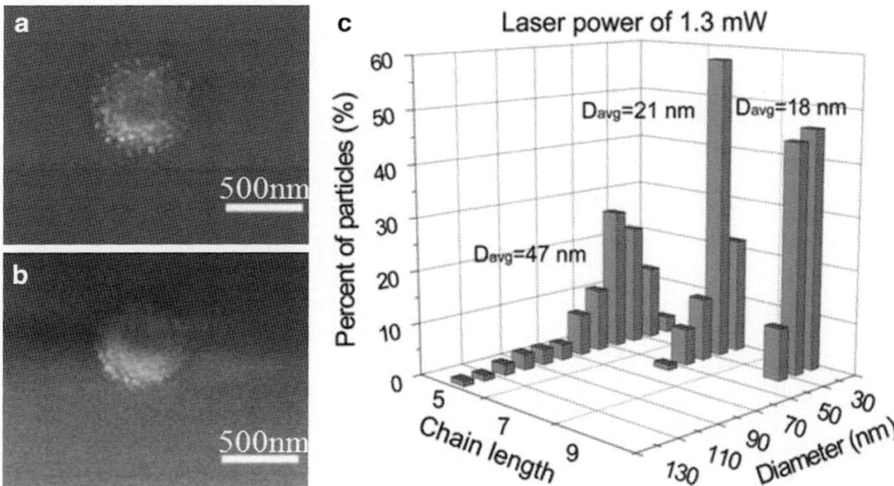

Fig. 4.49 SEM images of the silver dots fabricated in sample solutions with different fatty salts (**a**) C_7 (chain length of 7) and (**b**) C_9 under the laser power of 1.3 mW and the laser duration of 800 ms. (**c**) Size distributions of silver particles on the surface of the silver dots fabricated in the sample solutions with different fatty salts, C_5, C_7, and C_9 (Reprinted from Ref. [49], with kind permission from Springer Science+Business Media)

4.4.4.3   Two-Dimensional Metallic Nanostructures

In our previous studies, we have successfully fabricated 2D and 3D metallic microstructures in both metallic ion-doped polymer matrix [45] and aqueous solutions of metallic salts by using MPR [49, 93]. In particular, the 120 nm linewidth of silver line was successfully achieved by MPR from an aqueous solution of silver ions, which was smaller than the 1/6 of the femtosecond laser wavelength of 800 nm [49]. The silver lines were constructed by the aggregation of silver nanoparticles with surfactants on the surfaces, which probably limited the properties of the silver microstructures for electronic and photonic applications. Thus, the modification of the morphology of metallic microstructures in MPR fabrication process becomes critically important.

We have studied the morphology modification of silver microstructures fabricated by MPR technique [95]. We have demonstrated the modification effect of the laser irradiation time and repeated scanning numbers on the morphology of silver microstructures in MPR technique. *Trans*-4-[4-(dimethylamino)-*N*-methylstil-bazolium] *p*-tosylate (DAST) was used as photosensitizer and effectively reduced the laser power to 0.66 mW. The increase of the irradiation time and repeated scanning induced more reduction in the multiphoton photoreduction microfabrication process, resulting in the optimization of the linewidth.

The effect of repeated laser scanning on the morphology of silver microstructures was confirmed at various exposure times (t). When the repeated scanning N was less than 3, the silver lines were almost constructed by the separated silver particles.

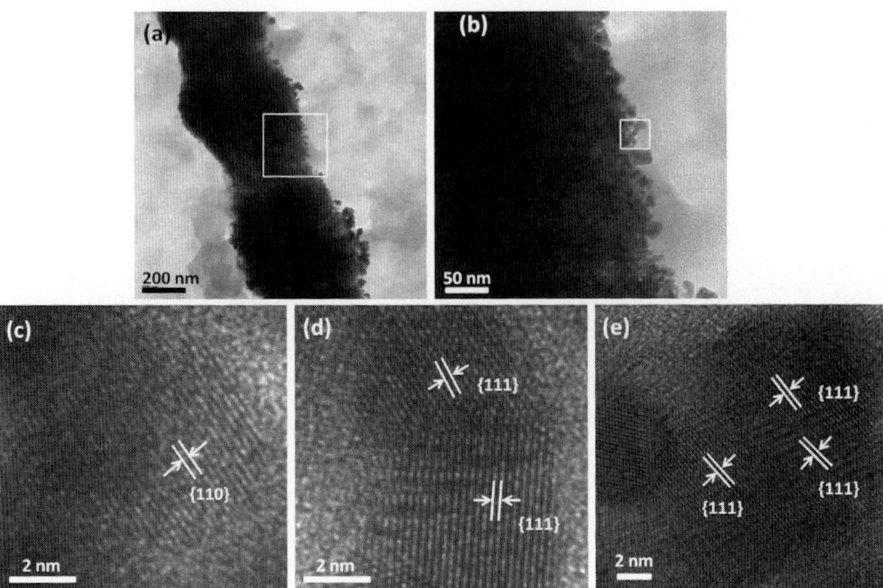

Fig. 4.50 (**a**) TEM image of gold nanoline fabricated using C5-assisted aqueous solution of HAuCl$_4$. (**b**) The magnified TEM images selected from the square region of (**a**). (**c**), (**d**), and (**e**) were the magnified HR-TEM images from the marked region of (**b**)

When N became more than 3, the silver lines were obtained as a continuum. However, the defects of the silver lines were observed even at the case of $N = 4$ while $t = 10$ ms. Then the defects of silver lines were reduced, subsequently, disappeared with the increasing of N and t. Especially in the cases of $t = 15$ ms with $N = 7$ and $t = 20$ ms with $N = 4$, the defects of silver lines were completely disappeared. It was reasonable considering the fusion of silver particles. Therefore, the laser irradiation time and multiple laser scanning play an important role in the fabrication of metallic microstructures without any defects. The fusion of silver nanoparticles led to the morphology change of silver microstructures; therefore, the compact metallic microstructures could be achieved. The results indicated that using the appropriate laser irradiation time and repeated scanning numbers could realize the morphology control of the silver microstructures. It would provide an efficient protocol for fabricating the metallic microstructures and play an important role in the fabrication of metallic microstructures for electronic and photonic applications.

The crystalline morphology of the fabricated metallic nanostructures has been demonstrated by TEM. Figure 4.50 shows the typical TEM images of the gold nanostructure fabricated from the sample solution with ionic liquid under the laser power of 1.57 mW and the scanning speed of 2 μm/s. The metallic line was composed of aggregated gold nanoparticles with the average size less than 10 nm. From high-resolution transmission electron microscopy (HR-TEM) image (Fig. 4.50c–e), we can see that the gold nanostructures obtained were composed of

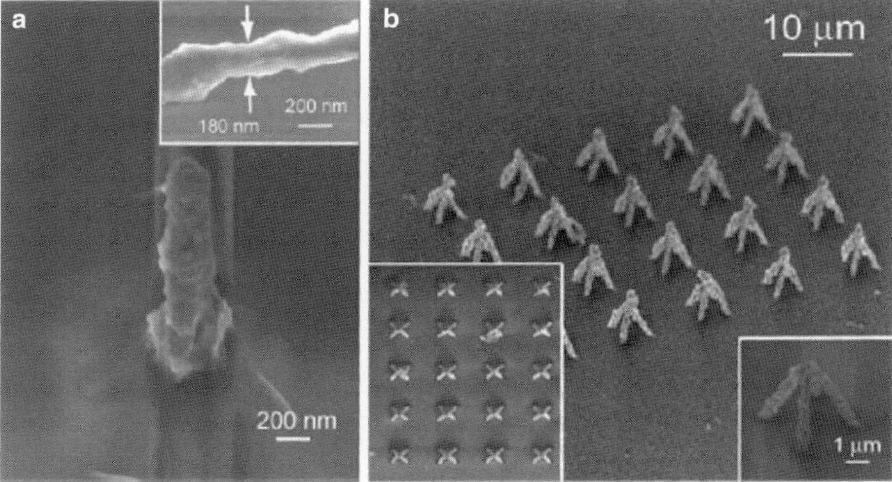

Fig. 4.51 (**a**) SEM image of the freestanding silver pillar on the cover slip, made by using a laser power of 1.14 mW and a linear scanning speed of 3 μm s^{-1}, taken at an observation angle of 45°. The inset is a close-up view of the silver pillar parallel to the substrate, which demonstrates the linewidth of the smallest portion of the silver pillar as 180 nm. (**b**) SEM image of silver pyramids, fabricated with a laser power of 1.3 mW and scanning speed of 2.5 μm s^{-1}, taken at an observation angle of 45°. The *inset* on the *left* is a top view of the silver-pyramid array. The *inset* on the *right* is a close-up view of the silver pyramid (Reproduced from Ref. [49] by permission of John Wiley & Sons Ltd)

crystalline gold nanoparticles with the {111} and {110} facets, possessing either single or joint crystalline structures. These gold nanostructures are expected to provide better electrical and optical properties.

4.4.4.4   Three-Dimensional Metallic Nanostructures

Based on the understanding of experimental parameters of the metal fabrication, the 3D metallic structures were further fabricated [49]. Figure 4.51a shows a freestanding silver pillar on the cover slip, which was obtained by scanning the focused laser spot along the direction normal to the plane of the cover glass. The silver pillar was created with a scanning speed of 3 μm s^{-1} and a laser power of 1.14 mW. SEM images reveal that this pillar has a minimum linewidth of 180 nm. Scanning the focused spot under the computer-controlled preprogram could enable the creation of 3D structures with arbitrary geometry. Figure 4.51b demonstrates the truly freestanding 3D silver pyramids fabricated with a scanning speed of 2.5 μm s^{-1} and laser power of 1.3 mW. The silver particles were closely combined; thus, these silver pyramids structures were strong enough to resist the surface tension in the washing process. The detail of the silver pyramid reveals that the height was 5 μm and the angle for each edge relative to the substrate was 60°. Consequently, the direct photoreduction of the metal ions with surfactants could also lead to 3D metal structures with resolution exceeding the optical diffraction limit.

4.5   Applications

The 3D multiphoton fabrication capability enables the fabrication of 3D arbitrary micro-/nanostructures, which would satisfy the demands for wide applications. Up to now, the multiphoton microfabricated micro-/nanostructures and devices have been widely investigated in photonics, MEMS, as well as the functional MEMS.

4.5.1   Photonic Crystal

Photonic Crystals (PhCs), an interesting class of periodically structured materials, have attracted much attention due to the unique properties for manipulating the propagation of light at a certain frequency. Various PhCs structures have been widely investigated both theoretically and experimentally [96]. Among a variety of structures, PhCs with diamond-lattice geometry are one of the most attractive series of structures, since they possess a full PBG even with low refractive index materials, where differences can be as low as 2.0.

Sun et al. fabricated 3D photonic crystal structures by using laser microfabrication techniques through TPA photopolymerization of resin [97]. Significant bandgap effects in the infrared wavelength region were observed from "layer-by-layer" (LBL) structures. The fabrication quality was demonstrated by transmission analysis of the LBL structures. The proposed technique has merit in either making arbitrary structures or tailoring their optical properties. Figure 4.52a is a top view of the surface of an LBL microstructure from an optical microscope, and Fig. 4.52b is a spatial image of the cross section taken by SEM after the sample being coated with a-few-nanometer-thick Au film. The SEM images showed that a well-defined 3D spatial structure had been achieved, and the rods were arrayed regularly in the same plane and half a period offset in the nearest layers with the same orientation.

In addition, a logpile photonic crystal of CdS nanoparticles–polymer nanocomposites has been successfully fabricated by a novel method combining the TPP technique and the in situ synthesis of CdS nanoparticles in a polymer matrix (Fig. 4.53) [98]. The PBG of the 3D logpile photonic crystal was confirmed and became more effective for CdS nanoparticles–polymer nanocomposites than polymer doped with Cd^{2+} ions, because the nanocomposites possessed a higher-refractive index than the polymer. The proposed concept in the new fabrication method for a 3D microstructure of polymer nanocomposites should be of critical importance in providing a general methodology for functionalizing materials via functional nanocomposites used in the field of laser microstructure fabrication.

By using two-photon photopolymerization technique, Kaneko et al. fabricated diamond-lattice photonic crystal structures with a submicron feature size. Repeating the primitive units in three dimensions produced crystalline structures. Figure 4.54a shows a <100> oriented 8 × 8 × 1 lattice, where the period is 2.5 µm, meaning that the nearest atom distance is 1.1 µm. The diameters of bonded rods and photonic

Fig. 4.52 (**a**) Optical microscope photograph of the top view of a LBL structure. The lateral lattice constant was 1.4 μm, exactly the same value as designed by the CAD. (**b**) SEM image of a cross section of the same structure (Reprinted with permission from Ref. [97]. Copyright 1999, American Institute of Physics)

atoms were 500 and 580 μm, respectively. The shadowed region was magnified under SEM and shown in the inset of Fig. 4.54a, from which a quite smooth rod surface was observed. A vertical two-period structure is shown in Fig. 4.54b, which has the same period as the aforementioned single-layer structure. Viewing from the $\langle 110 \rangle$ direction, the substrate is discernible, showing a complete removal of liquid resin. The single-period power rejection in photonic lattices is approximately 35 %. This study is an important step toward applying laser nanofabrication technology in developing multifunctional, complete-bandgap PhCs, as well as PhC-based polymer optoelectronic devices and integrated systems [99].

We reported a kind of 3D PhC structure consisting of gradient quasidiamond lattices using multiphoton photopolymerization of the commercial negative photoresist SCR500 [100, 101]. The PBG of this 3D PhC was experimentally confirmed by reflection and transmission measurements and simulated with FDTD calculations (Fig. 4.55). The results indicated that the 3D PhC with gradient lattices could effectively expand the width of the PBG and may be beneficial for developing complete-bandgap PhCs with low refractive index materials for applications in

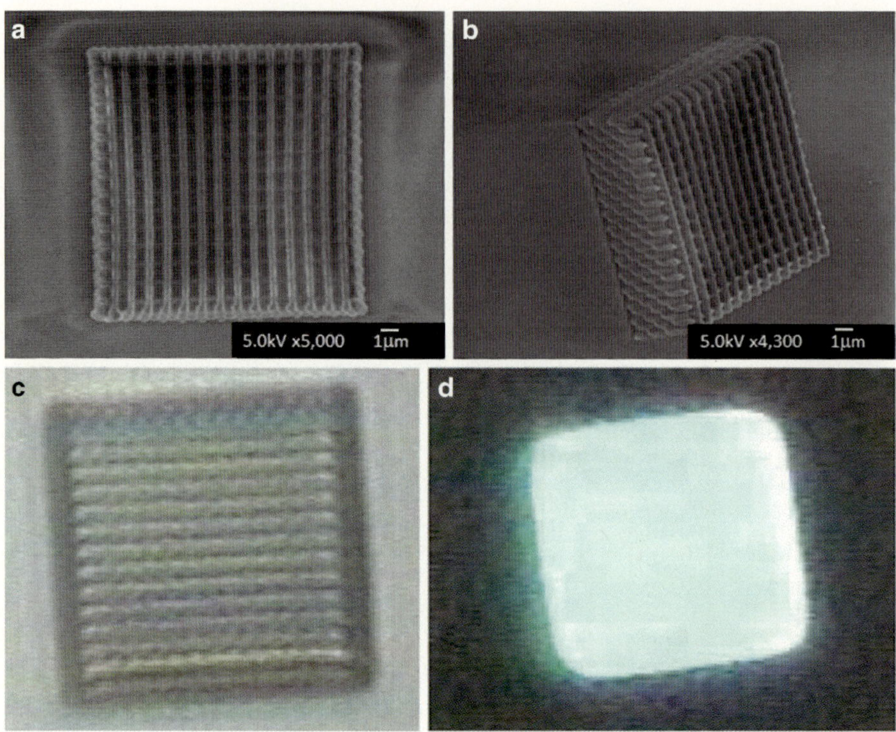

Fig. 4.53 Top (**a**) and side (**b**) view images of SEM for 3D photonic crystal of CdS NP–polymer nanocomposites with size of 20 μm as well as images of optical (**c**) and fluorescence microscopy (**d**) (Reprinted from Ref. [98], with kind permission from Springer Science+Business Media)

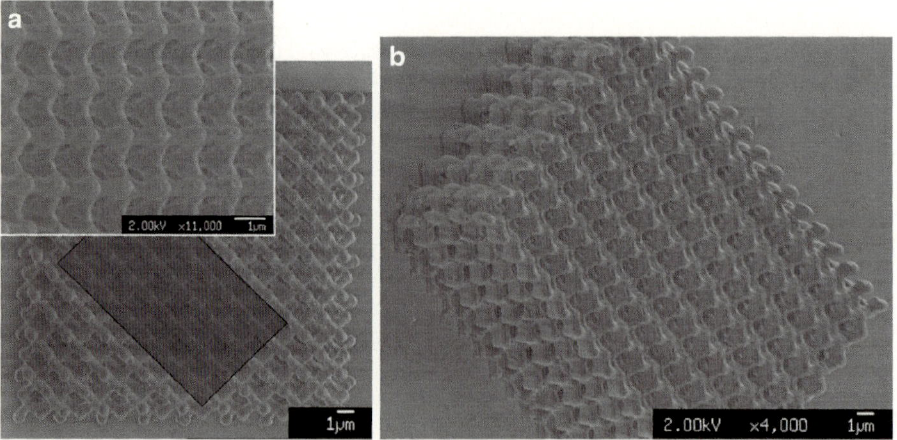

Fig. 4.54 Two-photon photopolymerized diamond PhCs. (**a**) A single-layer 8 × 8 lattice. The shadowed parts are shown by the *top left-hand side inset*. (**b**) A 8 × 8 × 2 period crystal (Reprinted with permission from Ref. [99]. Copyright 2003, American Institute of Physics)

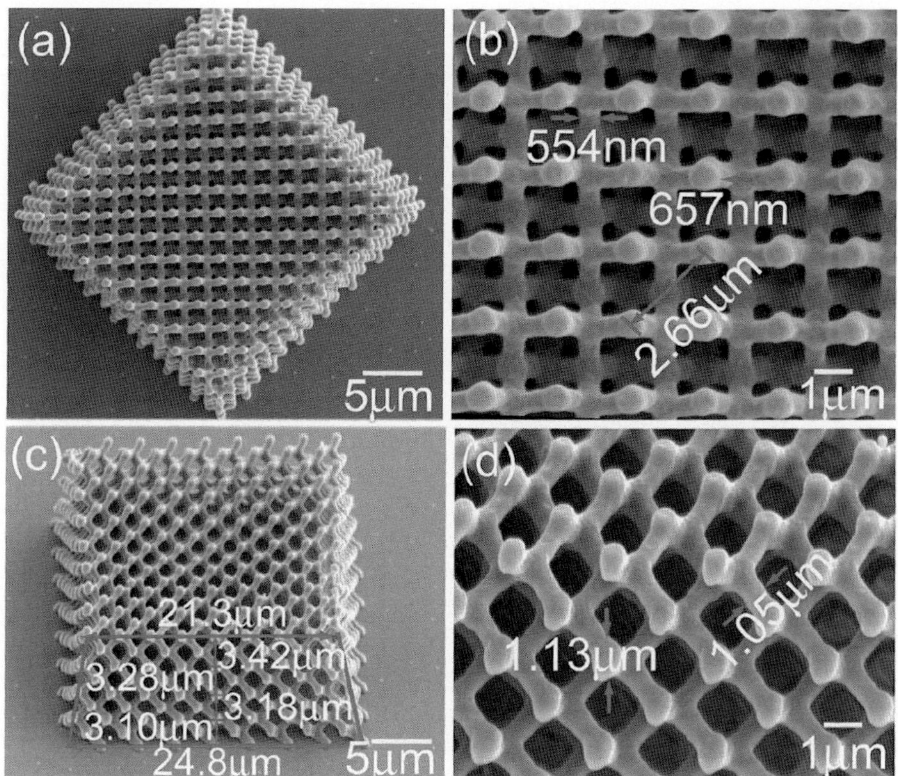

Fig. 4.55 SEM images of the gradient quasidiamond lattice 3D PhC fabricated by MPP. (**a**) Top view of the 3D PhC. (**b**) Magnified image of the top layer. (**c**) Side view (45° angle) of the 3D PhC. (**d**) Locally magnified image of (**c**) (Reprinted with permission from Ref. [100]. Copyright 2008, American Institute of Physics)

polymer-based optoelectronic devices and integrated systems. This work represents an important step toward developing complete-bandgap PhCs in low refractive index materials and for applying PhCs in polymer-based optoelectronic devices and integrated systems.

Misawa et al. reported a kind of 3D spiral-architecture photonic crystals obtained by direct laser writing in SU-8 [102], which exhibited promising photonic bandgap properties. SEM images of the spiral structures are shown in Fig. 4.56. Figure 4.56a shows spiral structures with lattice parameters of a = 1.8 μm, L = 2.7 μm, and c = 3.04 μm, fabricated with pulse energy 0.6 nJ. The vertical edges of the freestanding structure appear to be slightly nonparallel, which typically signifies polymer shrinkage. Individual spirals separated from the structure shown in Fig. 4.56b are mechanically rigid and with smooth turning points. Figure 4.56c shows the sample with two L-shaped waveguides, formed on the walls of the sample by missing parts of the spirals, with the same parameter as that in Fig. 4.56a and extended defects. Figure 4.56d shows the circular-spiral structure

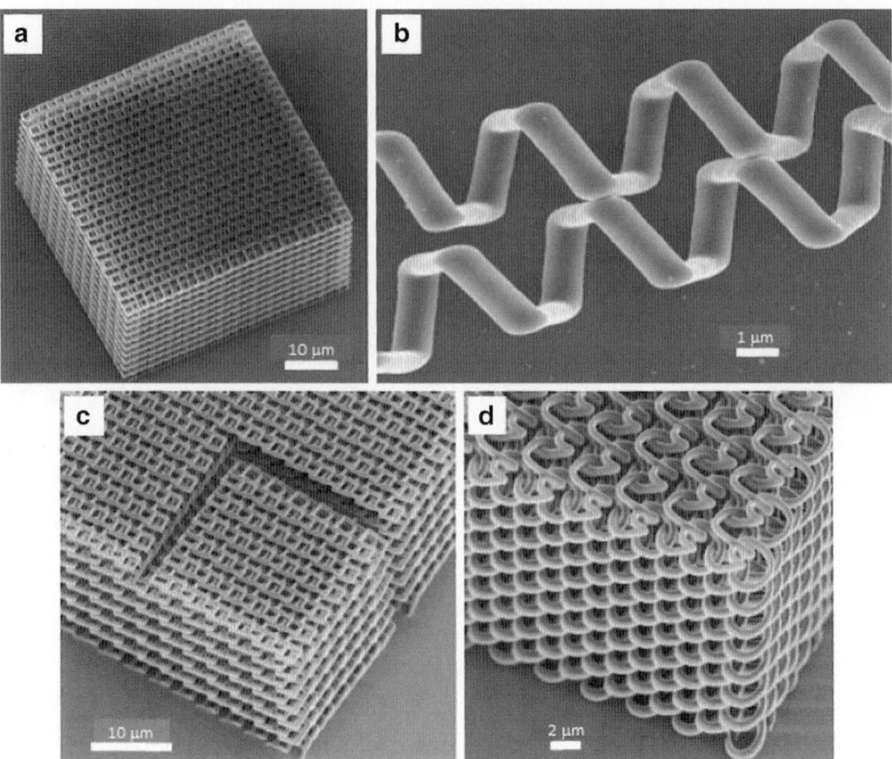

Fig. 4.56 SEM images of the spiral structures. (**a**) Spiral structures with lattice parameters of a = 1.8 μm, L = 2.7 μm, and c = 3.04 μm, fabricated with pulse energy 0.6 nJ. (**b**) Individual spirals separated from the structure. (**c**) Sample with two L-shaped waveguides, formed on the walls of the sample by missing parts of the spirals, with the same parameter as (**a**). (**d**) The circular-spiral structure (a = 1.8 μm, radius is 1.35 μm, c = 3.6 μm) with 180° phase shift between the adjacent spirals (Reproduced from Ref. [102] by permission of John Wiley & Sons Ltd)

(a = 1.8 μm, radius is 1.35 μm, c = 3.6 μm) with 180° phase shift between the adjacent spirals. Self-supporting structures show signatures of photonic stopgaps at infrared wavelengths, which is promising for developing accurate templates for infiltration of higher-refractive-index materials.

4.5.2   *Micro-/Nanoelectromechanical Systems (MEMS/NEMS)*

4.5.2.1   Micro-/Nanomachines

Micromachines have been attracting much interest owing to their extensive applications in integrated circuits, wireless communication, microfluidic pumps, and chemical or biological sensors [39, 103–108]. Femtosecond laser direct writing

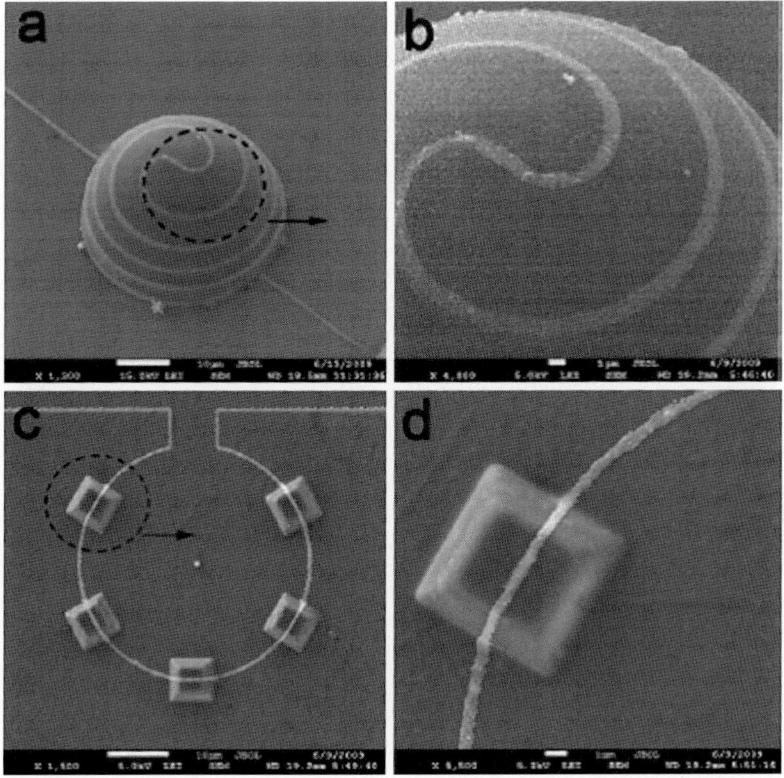

Fig. 4.57 SEM images of silver micropatterns fabricated on substrates with 3D microstructures. (**a, b**) Silver microwinding on a hemisphere; (**c, d**) single silver microwire cycle on the frustum of a pyramid (Reproduced from Ref. [94] by permission of John Wiley & Sons Ltd)

technique enables the fabrication of arbitrary 3D micro-/nanostructures with different purposes for the practical applications. Well-designed true 3D microstructures are preferred to satisfy these requirements. As an emerging micro-/nanofabrication technique, TPP of polymers is promising for fabricating 3D micro-/nanomachines due to its capability for 3D micro-/nanofabrication [40] and high spatial resolution at nanometric scale [39]. Thus, TPP technique has been applied to the fields of micro-/nanodevices and MEMS/NEMS.

Sun's group developed a flexible femtosecond-laser-induced electroless plating technique for metal nanowiring on nonplanar substrates [94]. The fundamental concept consists of immersing the surface where the metal circuitry or electronic interconnection is to be made into a metal-salt solution and then tightly focusing a near-infrared femtosecond-laser beam onto the site to form a wire. The desired metal nanowiring, circuit, or wire micropatterns can be achieved by scanning the laser focus along a preprogrammed trace following the top of either flat or nonplanar surfaces, such as a rough base, spherical surface, or even a sharp corner (Fig. 4.57). The linewidth in the range of 125–500 nm could be easily adjusted through careful

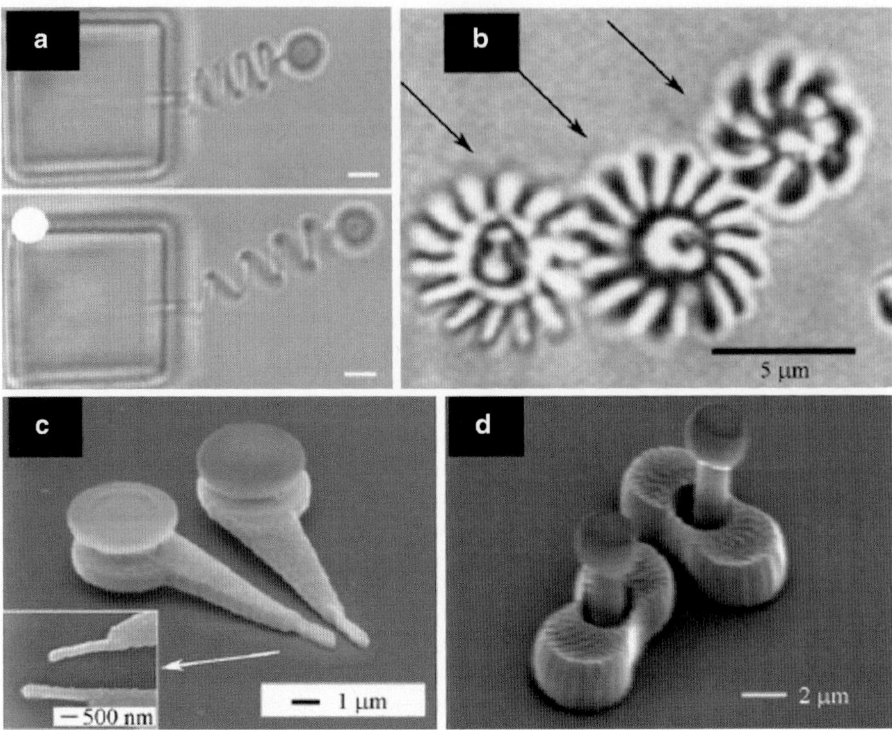

Fig. 4.58 (**a**) Spring (Reprinted by permission from Macmillan Publishers Ltd: Ref. [40], copyright 2001). (**b**) Toothed wheel (Reprinted with permission from Ref. [103]. Copyright 2001, American Institute of Physics). (**c**) 3D microtweezers with submicron probe tips (Reprinted with permission from Ref. [104]. Copyright 2003, American Institute of Physics). (**d**) Micropump (Reprinted with permission from Ref. [105]. Copyright 2006, American Institute of Physics)

control of laser power. Moreover, the patterned silver nanowires maintained a low resistivity, which provided great potential for various circuitry and electronic inter-connection uses.

By using the multiphoton fabrication technique, varieties of arbitrary micros-tructures or devices with attractive properties or functions have been achieved. As shown in Fig. 4.58, movable spring, toothed wheel, 3D microtweezers with submi-cron probe tips, and micropump have been fabricated to realize the optically driven function [40, 103–105].

4.5.2.2   Microfluid Devices

Kumi et al. utilized a novel PAG that with high σ_2 instead of the low concentration of a PAG in an SU-8 photoresist, which is a commercial negative-tone photoresist often employed to create microfluidic masters. Another reason is that there was no refractive

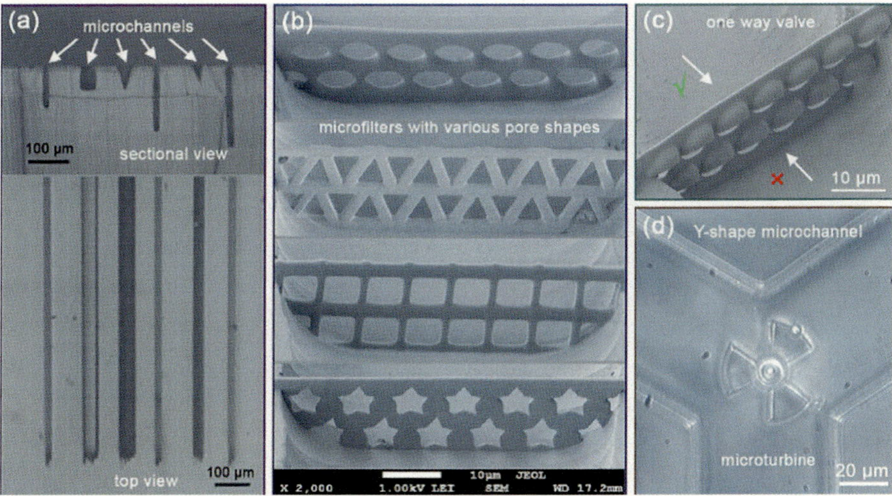

Fig. 4.59 (a) Optical image of microfluid devices with different cross sections and heights (Reproduced from Ref. [106] by permission of the Royal Society of Chemistry). (b) Microfluidic microsieves chip. (c) one-way microvalves were fabricated inside a microchannel for sieving microparticles and control of their flowing direction, respectively. (d) magnetic microturbine fabricated inside a "Y" channel as a mixing microdevice (Reproduced from Ref. [108] by permission of The Royal Society of Chemistry)

index change during exposure [106]. High-aspect-ratio masters with uniform, rectangular cross sections and the low threshold can be created by employing the novel PAG in SU-8 photoresist. Masters with arbitrary, non-rectangular cross sections, and channels with different aspect ratios and cross sections in a single device are able to be fabricated. High-speed fabrication over centimeter-scale distances enables the fabrication of large-scale microfluidic masters on a practical time scale. High-aspect-ratio master structures were therefore attained for the creation of reverse microfluidic channels (Fig. 4.59a). Similarly, microfluidic channels for advanced functions can be fabricated. Figure 4.59b shows the microsieves and one-way microvalve inside a microchannel. Note that the fabrication of these simple structures with vertical arrangement relative to the substrate surface is almost impossible by using simple lithography. It has been demonstrated that particles of size larger than the pore size cannot pass through the sieves experimentally, so the microsieves are expected to perform functions like cell sorting. This fabrication technique significantly increases the diversity of channel architectures available for microfluidic devices and biochip systems [106–108].

4.5.3 Magnetic MEMS/NEMS

Recently, much effort has been made to develop remotely manipulated micro machines or devices to meet the requirement for precisely targeted treatment in

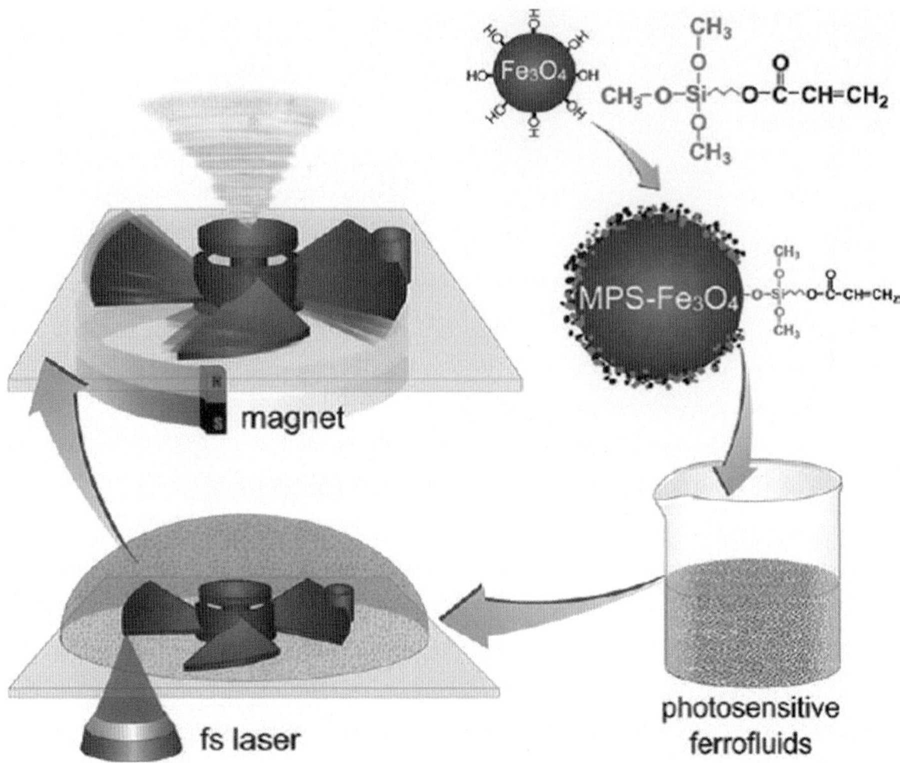

Fig. 4.60 Fabricative procedures of remotely controllable micronanomachines (Reproduced from Ref. [109] by permission of John Wiley & Sons Ltd)

narrow enclosures, harsh environments, and even inside the human body. Magnetically driven micromachines or devices have been proved to be promising in realizing manipulation by remote and noncontact control. An available magnetic force-driven micromachine has been prepared by TPP of ferropolymer, indicating the unique advantages for potential applications in health care (Fig. 4.60) [109].

However, the mechanical performance of the ferropolymer material at micro-/nanoscale, such as hardness and strength, which has been considered important for evaluating micromachines, is still unidentified. Micromachines made of polymers exhibited intrinsically weak mechanical performance originating from the polymer materials, which would limit potential applications in fields that require mechanical strength of a micromachine, for instance, the elimination of thrombus in blood vessels. Moreover, the introduction of the driven force into the polymer material was essential for advanced application of smart micromachines, but it was hard to achieve.

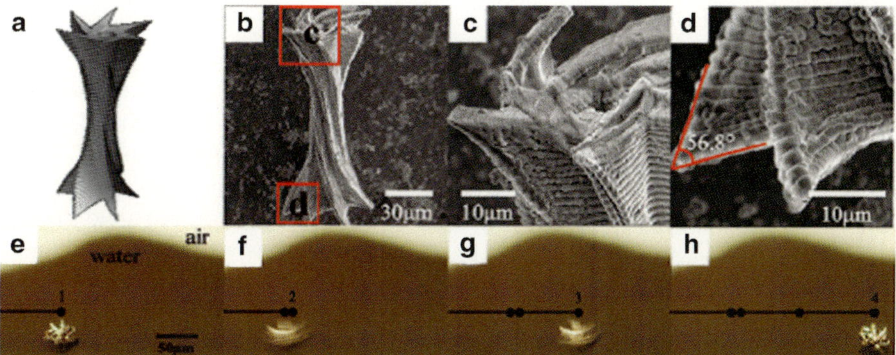

Fig. 4.61 (**a**) Model of the micromachine. (**b**) Polymer micromachine with electroless layer. (**c, d**) Detailed sections of the micromachine. (**e–h**) Continual position change of the micromachine, which rolled to right side (Reprinted with the permission from Ref. [110]. Copyright 2011 American Chemical Society)

Therefore, it is of great importance to exploit a strategy for fabricating precise 3D micromachines with both excellent mechanical strength and manipulated features.

Plating, a facile method to construct composite materials, is used for improving the performance of materials. Electroless plating has been utilized in a wide range of fields since there is no requirement for the conductivity of the substrate. Nickel–phosphorus (Ni–P) alloy deposit has been studied owing to its extremely high hardness, modulus, wear resistance, and corrosion resistance. Furthermore, Ni–P alloy exhibits the characteristics of soft magnetic material, which is sensitive to external magnetic field without any hysteresis and shows no remanence after removal of the external magnetic field. Therefore, preparing polymer microstructures by coating a layer of Ni–P alloy should provide us a great opportunity to develop magnetically driven micromachines.

We have prepared the Ni–P polymer composite and proposed a strategy for fabricating 3D smart micromachines by combining polymer TPP and electroless plating of Ni–P alloy to achieve high mechanical performance and remotely controllable capability, as shown in Fig. 4.61 [110]. The experimental condition of electroless plating was studied to satisfy nanoplating for tiny sections at nanoscale. Ni–P alloy layer exhibited excellent mechanical performance and magnetism to the polymer material, which could be fabricated to 3D micromachine by using TPP. We demonstrated the fabrication of a 3D micromachine combined by TPP and Ni–P alloy nanoplating and the remotely driven motion process of the as-prepared 3D micromachine manipulated by external magnetic field. The strategy proposed here for the development of magnetically driven micromachines with excellent mechanical performance would open up good prospects for the broad applications in MEMS/NEMS.

4.5.4 Metamaterials

Note that another application of multiphoton micro-/nanofabrication is to develop metamaterials. The researches on metamaterials have been largely stimulated by proposals and demonstrations of negative phase velocities, "perfect lenses," invisibility cloaking, etc. Gansel J. K. et al. have investigated the light propagation through a uniaxial photonic metamaterial that composed of 3D gold helices arranged on a 2D square lattice [111]. The nanostructures were fabricated into a positive-tone photoresist via direct laser writing technique, then followed by electrochemical deposition of gold. In order to study the propagation of light along the helix axis, the structure blocks the circular polarization with the same handedness as the helices, whereas it transmits the other, for a frequency range exceeding one octave. The structure is matchable to other frequency ranges and is expected to be used as a compact broadband circular polarizer.

In addition, a 3D invisibility-cloaking structure operating at optical wavelengths based on transformation optics has been designed and fabricated. The metamaterial structure was also fabricated by multiphoton fabrication using direct laser writing technique. A woodpile photonic crystal with a tailored polymer filling fraction was used in the blueprint to hide a bump in a gold reflector. A cloaking operation with a large bandwidth of unpolarized light from 1.4 to 2.7 μm in wavelength was demonstrated when the viewing angles were up to 60° [112].

4.5.5 Other Applications

Other 3D arbitrary micro-/nanostructures have also been studied. An active polymeric distributed feedback (DFB) resonator which consisted of sub-micrometer fiber grating has been fabricated [113]. The DFB resonator was fabricated by using two-photon-induced polymerization technique (Fig. 4.62). Its amplified spontaneous emission (ASE) action with an ultra-low threshold under optical excitation has been observed. The ASE at 556 nm was observed with a threshold of 0.30 μJ/pulse due to the fourth-order diffraction feedback of the optical gain. In this study, a dendrimer was used to modify the laser dye and increased its concentration up to 4.95 wt% in the photocured resin. This protocol for fabricating DFB could be expected to open up a way for the fabrication and application of microscale polymeric mirrorless lasers.

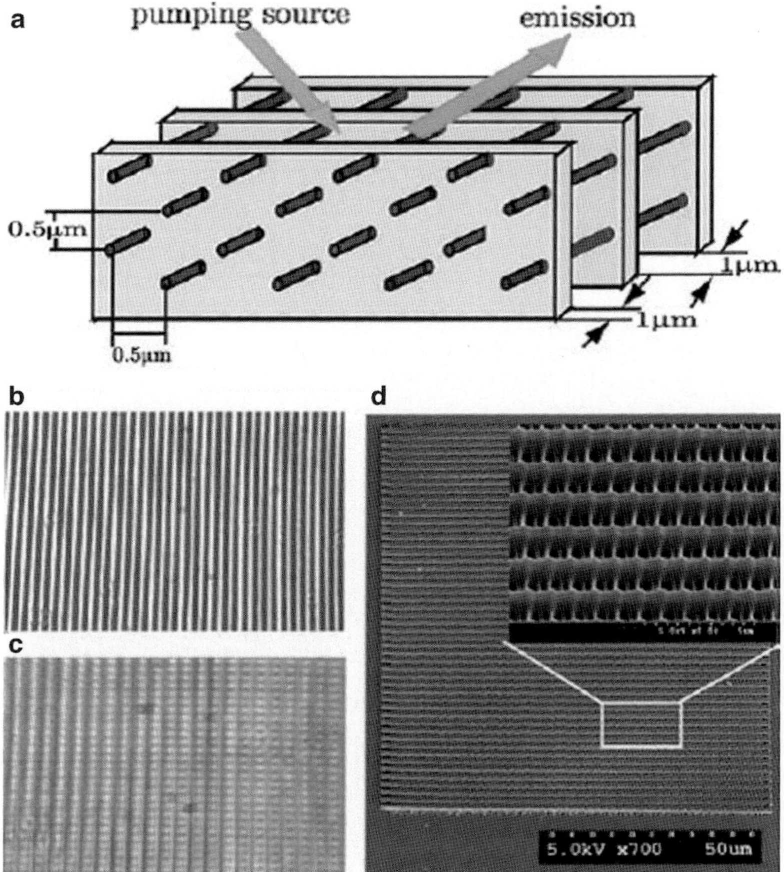

Fig. 4.62 (**a**) Schematic drawing of DFB resonator. Optical microscopy images of the cuboids arrays without (**b**) and with (**c**) the suspended fibers. (**d**) SEM image of cuboids arrays with suspended fibers (Reprinted from Ref. [113], with kind permission from Springer Science+Business Media)

References

1. Lee K-S, Kim RH, Yang D-Y, Park SH (2008) Advances in 3D nano/microfabrication using two-photon initiated polymerization. Prog Polym Sci 33:631–681
2. Totzeck M, Ulrich W, Göhnermeier A, Kaiser W (2007) Pushing deep ultraviolet lithography to its limits. Nat Photon 1:629–631
3. Maruo S, Fourkas JT (2008) Recent progress in multiphoton microfabrication. Laser Photon Rev 2:100–111
4. Dong XZ, Chen WQ, Zhao ZS, Duan XM (2008) Femtosecond laser two photon micronano-fabrication and applications. Science in China Press 53: 2–13

5. Lin T-C, Chung S-J, Kim K-S, Wang X, He GS, Swiatkiewicz J, Pudavar HE, Prasad PN (2003) Organics and polymers with high two-photon activities and their applications. Adv Polym Sci 161:158–190

6. Goeppert-Mayer M (1931) Über Elementarakte mit zwei Quantensprüngen. Ann Phys (Leipzig) 9:273–294

7. Maiman TH (1960) Stimulated optical radiation in ruby. Nature 187:493–494

8. Rumi M, Perry JW (2010) Two-photon absorption: an overview of measurements and principles. Adv Opt Photon 2:451–518

9. Peticolas WL (1967) Multiphoton spectroscopy. Annu Rev Phys Chem 18:233–260

10. McClain WM (1974) Two-photon molecular spectroscopy. Acc Chem Res 7:129–135

11. Xu C, Webb WW (1997) Multiphoton excitation of molecular fluorophores and nonlinear laser microscopy. In: Lakowicz J (ed) Nonlinear and two-photon-induced fluorescence, vol 5, Topics in fluorescence spectroscopy. Plenum, New York, pp 471–540

12. Bhawalkar JD, He GS, Prasad PN (1996) Nonlinear multiphoton processes in organic and polymeric materials. Rep Prog Phys 59:1041–1071

13. Cumpston BH, Ananthavel SP, Barlow S, Dyer DL, Ehrlich JE, Erskine LL, Heikal AA, Kuebler SM, Lee I-YS, McCord-Maughon D, Qin J, Roeckel H, Rumi M, Wu X-L, Marder SR, Perry JW (1999) Two-photon polymerization initiators for three dimensional optical data storage and microfabrication. Nature 398:51–54

14. Chung SJ, Rumi M, Alain V, Barlow S, Perry JW, Marder SR (2005) Strong low-energy two-photon absorption in extended amine-terminated cyano-substituted phenylenevinylene oligomers. J Am Chem Soc 127:10844–10845

15. Zheng S, Beverina L, Barlow S, Zojer E, Fu J, Padilha LA, Fink C, Kwon O, Yi Y, Shuai Z, Van Stryland EW, Hagan DJ, Brédas J-L, Marder SR (2007) High two-photon cross-sections in bis(diarylaminostyryl) chromophores with electron-rich heterocycle and bis(heterocycle) vinylene bridges. Chem Commun 13:1372–1374

16. Xing JF, Chen WQ, Gu J, Dong XZ, Takeyasu N, Tanaka T, Duan XM, Kawata S (2007) Design of high efficiency for two-photon polymerization initiator: combination of radical stabilization and large two-photon cross-section achieved by *N*-benzyl 3,6-bis (phenylethynyl)carbazole derivatives. J Mater Chem 17:1433–1438

17. Xing JF, Chen WQ, Dong XZ, Tanaka T, Fang XY, Duan XM, Kawata S (2007) Synthesis, optical and initiating properties of new two-photon polymerization initiators: 2,7-bis(styryl) anthraquinone derivatives. J Photochem Photobiol A Chem 189:398–404

18. Gu J, Wang YL, Chen WQ, Dong XZ, Duan XM, Kawata S (2007) Carbazole-based 1D and 2D hemicyanines: synthesis, two-photon absorption properties and application for two-photon photopolymerization 3D lithography. New J Chem 31:63–68

19. Xing JF, Zheng ML, Chen WQ, Dong XZ, Takeyasu N, Tanaka T, Zhao ZS, Duan XM, Kawata S (2012) C_{2v} symmetrical two-photon polymerization initiators with anthracene core: synthesis, optical and initiating properties. Phys Chem Chem Phys 14:15785–15792

20. Shi Q, Chen WQ, Xiang J, Duan XM, Zhan X (2011) A low-bandgap conjugated polymer based on squaraine with strong two-photon absorption. Macromolecules 44:3759–3765

21. Huang X, Shi Q, Chen WQ, Zhu C, Zhou W, Zhao Z, Duan XM, Zhan X (2010) Low-bandgap conjugated donor-acceptor copolymers based on porphyrin with strong two-photon absorption. Macromolecules 43:9620–9626

22. Zhou W, Feng J, Huang X, Duan XM, Zhan X (2012) A low-bandgap conjugated copolymer based on porphyrin and dithienocoronene diimide with strong two-photon absorption. Macromolecules 45:7823–7828

23. Wang X, Jin F, Chen Z, Liu S, Wang X, Duan XM, Tao X, Jiang M (2011) A new family of dendrimers with naphthaline core and triphenylamine branching as a two-photon polymerization initiator. J Phys Chem C 115:776–784

24. Drobizhev M, Karotki A, Dzenis Y, Rebane A, Suo Z, Spangler CW (2003) Strong cooperative enhancement of two-photon absorption in dendrimers. J Phys Chem B 107:7540–7543

25. Lee K-S, Yang D-Y, Park SH, Kim RH (2006) Recent developments in the use of two-photon polymerization in precise 2D and 3D microfabrications. Polym Adv Technol 17:72–82
26. LaFratta CN, Fourkas JT, Baldacchini T, Farrer RA (2007) Multiphoton fabrication. Angew Chem Int Ed 46:6238–6258
27. Zhou W, Kuebler SM, Braun KL, Yu T, Cammack JK, Ober CK, Perry JW, Marder SR (2002) An efficient two-photon-generated photoacid applied to positive-tone 3D microfabrication. Science 296:1106–1109
28. Li H, Jin F, Chen WQ, Duan XM (2012) New type positive molecular glass resists for multiphoton lithography. Imaging Sci Photochem 30:347–357
29. Dong XZ, Zhao ZS, Duan XM (2008) Improving spatial resolution and reducing aspect ratio in multiphoton polymerization nanofabrication. Appl Phys Lett 92:091113
30. Sun HB, Tanaka T, Kawata S (2002) Three-dimensional focal spots related to two-photon excitation. Appl Phys Lett 80:3673–3675
31. Sun HB, Tanaka K, Kim M-S, Lee K-S, Kawata S (2003) Scaling laws of voxels in two-photon photopolymerization nanofabrication. Appl Phys Lett 83:1104–1106
32. Tanaka K, Sun HB, Kawata S (2005) Improved spatial resolution and surface roughness in photopolymerization based laser nanowriting. Appl Phys Lett 86:071122
33. Takada K, Sun HB, Kawata S (2006) The study on spatial resolution in two-photon induced polymerization. Proc SPIE 6110:61100A
34. Tan D, Li Y, Qi F, Yang H, Gong Q, Dong XZ, Duan XM (2007) Reduction in feature size of two-photon polymerization using SCR500. Appl Phys Lett 90:071106
35. Lim TW, Park SH, Yang D-Y (2005) Contour offset algorithm for precise patterning in two-photon polymerization. Microelectron Eng 77:382–388
36. Park SH, Lim TW, Yang D-Y, Yi SW, Kong HJ (2005) Direct fabrication of micropatterns and three-dimensional structures using nanoreplication-printing (nRP) process. Sensor Mater 17:065–075
37. Matsuo S, Juodkazis S, Misawa S (2005) Femtosecond laser microfabrication of periodic structures using a microlens array. Appl Phys A 80:683–685
38. Kato J, Takeyasu N, Adachi Y, Sun HB, Kawata S (2005) Multiple-spot parallel processing for laser micronanofabrication. Appl Phys Lett 86:044102
39. Dong XZ, Zhao ZS, Duan XM (2007) Micronanofabrication of assembled three-dimensional microstructures by designable multiple beams multiphoton processing. Appl Phys Lett 91:124103
40. Kawata S, Sun HB, Tanaka T, Takada K (2001) Finer features for functional microdevices. Nature 412:697–698
41. Tanaka T, Sun HB, Kawata S (2002) Rapid sub-diffraction-limit laser microŌnanoprocessing in a threshold material system. Appl Phys Lett 80:312–314
42. Harke B, Dallari W, Grancini G, Fazzi D, Brandi F, Petrozza A, Diaspro A (2013) Polymerization inhibition by triplet state absorption for nanometer-scale lithography. Adv Mater 25:904–909
43. Lu WE, Dong XZ, Chen WQ, Zhao ZS, Duan XM (2011) Novel photoinitiator with a radical quenching moiety for confining radical diffusion in two-photon induced photopolymerization. J Mater Chem 21:5650–5659
44. Lu WE, Chen WQ, Zheng ML, Dong XZ, Zhao ZS, Duan XM (2013) Two-photon induced photopolymerization using photoinitiator with an intramolecular radical quenching moiety for nanolithography. J Nanosci Nanotechnol 13:1343–1346
45. Sakellari I, Kabouraki E, Gray D, Purlys V, Fotakis C, Pikulin A, Bityurin N, Vamvakaki M, Farsari M (2012) Diffusion-assisted high-resolution direct femtosecond laser writing. ACS Nano 6:2302–2311
46. Xing JF, Dong XZ, Chen WQ, Duan XM, Takeyasu N, Tanaka T, Kawata S (2007) Improving spatial resolution of two-photon microfabrication by using photoinitiator with high initiating efficiency. Appl Phys Lett 90:131106

47. Song Y, Dong XZ, Zhao ZS, Duan XM (2011) Investigation into ultimate resolution by femtosecond laser two-photon fabrication technique. High Power Laser Part Beams 23:1780–1784
48. Stellacci F, Bauer CA, Meyer-Friedrichsen T, Wenseleers W, Alain V, Kuebler SM, Pond SJK, Zhang Y, Marder SR, Perry JW (2002) Laser and electron-beam induced growth of nanoparticles for 2D and 3D metal patterning. Adv Mater 14:194–198
49. Cao YY, Takeyasu N, Tanaka T, Duan XM, Kawata S (2009) 3D metallic nanostructure fabrication by surfactant-assisted multiphoton-induced reduction. Small 5:1144–1148
50. Tanaka T, Ishikawa A, Kawata S (2006) Two-photon-induced reduction of metal ions for fabricating three-dimensional electrically conductive metallic microstructure. Appl Phys Lett 88:081107
51. Maruo S, Saeki T (2008) Femtosecond laser direct writing of metallic microstructures by photoreduction of silver nitrate in a polymer matrix. Opt Express 16:1174–1179
52. Abe K, Hanada T, Yoshida Y, Tanigaki N, Takiguchi H, Nagasawa H, Nakamoto M, Yamaguchi T, Yase K (1998) Two-dimensional array of silver nanoparticles. Thin Solid Films 327–329:524–527
53. Kaneko K, Sun HB, Duan XM, Kawata S (2003) Two-photon photoreduction of metallic nanoparticle gratings in a polymer matrix. Appl Phys Lett 83:1426–1428
54. Ishikawa A, Tanaka T, Kawata S (2006) Improvement in the reduction of silver ions in aqueous solution using two-photon sensitive dye. Appl Phys Lett 89:113102
55. Khetan S, Burdick JA (2011) Patterning hydrogels in three dimensions towards controlling cellular interactions. Soft Matter 7:830–838
56. Shin HS, Kim SY, Lee YM (1997) Indomethacin release behaviors from pH and thermore-sponsive poly(vinyl alcohol) and poly(acrylic acid) IPN hydrogels for site-specific drug delivery. J Appl Polym Sci 65:685–693
57. Chang CW, Van Spreeuwel A, Zhang C, Varghese S (2010) PEG/clay nanocomposite hydrogel: a mechanically robust tissue engineering scaffold. Soft Matter 6:5157–5164
58. Lee KY, Mooney DJ (2001) Hydrogels for tissue engineering. Chem Rev 101:1869–1879
59. Kim S, Healy KE (2003) Synthesis and characterization of injectable poly(N-isopropyla-crylamide-co-acrylic acid) hydrogels with proteolytically degradable cross-links. Biomacro-molecules 4:1214–1223
60. Gawel K, Stokke BT (2011) Logic swelling response of DNA-polymer hybrid hydrogel. Soft Matter 7:4615–4618
61. Takada K, Miyazaki T, Tanaka N, Tatsuma T (2006) Three-dimensional motion and trans-formation of a photoelectrochemical actuator. Chem Commun 19:2024–2026
62. Otero TF, Sansinena JM (1998) Soft and wet conducting polymers for artificial muscles. Adv Mater 10:491–494
63. Kuhn PW, Hargitay B, Katchalsky A, Eisenberg H (1950) Reversible dilation and contraction by changing the state of ionization of high-polymer acid networks. Nature 165:514–516
64. Siegel RA, Firestone BA (1988) pH-dependent equilibrium swelling properties of hydropho-bic polyelectrolyte copolymer gels. Macromolecules 21:3254–3259
65. Hu Z, Zhang X, Li Y (1995) Synthesis and application of modulated polymer gels. Science 269:525–527
66. Suzuki A, Tanaka T (1990) Phase transition in polymer gels induced by visible light. Nature 346:345–347
67. Gong JP, Katsuyama Y, Kurokawa T, Osada Y (2003) Double-network hydrogels with extremely high mechanical strength. Adv Mater 15:1155–1158
68. Tanaka T, Nishio I, Sun ST, Ueno-Nishio S (1982) Collapse of gels in an electric field. Science 218:467–469
69. Matsumoto A, Ikeda S, Harada A, Kataoka K (2003) Glucose-responsive polymer bearing a novel phenylborate derivative as a glucose-sensing moiety operating at physiological pH conditions. Biomacromolecules 4:1410–1416

70. Durmaz S, Okay O (2000) Acrylamide/2-acrylamido-2-methylpropane sulfonic acid sodium salt-based hydrogels: synthesis and characterization. Polymer 41:3693–3704
71. Thornton PD, McConnell G, Ulijn RV (2005) Enzyme responsive polymer hydrogel beads. Chem Commun 47:5913–5915
72. Maeda S, Hara Y, Sakai T, Yoshida R, Hashimoto S (2007) Self-walking gel. Adv Mater 19:3480–3484
73. Kataoka K, Miyazaki H, Bunya M, Okano T, Sakurai Y (1998) Totally synthetic polymer gels responding to external glucose concentration: their preparation and application to on-off regulation of insulin release. J Am Chem Soc 120:12694–12695
74. Watanabe T, Akiyama M, Totani K, Kuebler SM, Stellacci F, Wenseleers W, Braun K, Marder SR, Perry JW (2002) Photoresponsive hydrogel microstructure fabricated by two-photon initiated polymerization. Adv Funct Mater 12:611–614
75. Kaehr B, Shear JB (2008) Multiphoton fabrication of chemically responsive protein hydrogels for microactuation. Proc Natl Acad Sci USA 105:8850–8854
76. Xiong Z, Zheng ML, Dong XZ, Chen WQ, Jin F, Zhao ZS, Duan XM (2011) Asymmetric microstructure of hydrogel: two-photon microfabrication and stimuli-responsive behavior. Soft Matter 7:10353–10359
77. Claeyssens F, Hasan EA, Gaidukeviciute A, Achilleos DS, Ranella A, Reinhardt C, Ovsianikov A, Shizhou X, Fotakis C, Vamvakaki M, Chichkov BN, Farsari M (2009) Three-dimensional biodegradable structures fabricated by two-photon polymerization. Langmuir 25:3219–3223
78. Correa DS, Tayalia P, Cosendey G, dos Santos DS, Aroca RF, Mazur E, Mendonca CR (2009) Two-photon polymerization for fabricating structures containing the biopolymer chitosan. J Nanosci Nanotechnol 9:5845–5849
79. Dinca V, Kasotakis E, Catherine J, Mourka A, Ranella A, Ovsianikov A, Chichkov BN, Farsari M, Mitraki A, Fotakis C (2008) Directed three-dimensional patterning of self-assembled peptide fibrils. Nano Lett 8:538–543
80. Duan XM, Sun HB, Kaneko K, Kawata S (2004) Two-photon polymerization of metal ions doped acrylate monomers and oligomers for three-dimensional structure fabrication. Thin Solid Films 453–454:518–521
81. Sun ZB, Dong XZ, Chen WQ, Nakanishi S, Duan XM, Kawata S (2008) Multicolor polymer nanocomposites: in situ synthesis and fabrication of 3D microstructures. Adv Mater 20:914–919
82. Sun ZB, Dong XZ, Chen WQ, Shoji S, Duan XM, Kawata S (2008) Two- and three-dimensional micro/nanostructure patterning of CdS–polymer nanocomposites with a laser interference technique and in situ synthesis. Nanotechnology 19:035611
83. Kuznetsov AI, Kiyan R, Chichkov BN (2010) Laser fabrication of 2D and 3D metal nanoparticle structures and arrays. Opt Express 18:21198–21203
84. Kuo WS, Lien CH, Cho KC, Chang CY, Lin CY, Huang LLH, Campagnola PJ, Dong CY, Chen SJ (2010) Multiphoton fabrication of freeform polymer microstructures with gold nanorods. Opt Express 18:27550–27559
85. Lien CH, Kuo WS, Cho KC, Lin CY, Su YD, Huang LLH, Campagnola PJ, Dong CY, Chen SJ (2011) Fabrication of gold nanorods-doped, bovine serum albumin microstructures via multiphoton excited photochemistry. Opt Express 19:6260–6268
86. Masui K, Shoji S, Ushiba S, Duan XM, Kawata S (2012) Femtosecond laser fabrication of gold nanorod/polymer composite microstructures. Proc SPIE 8457:84571Y
87. Masui K, Shoji S, Asaba K, Rodgers TC, Jin F, Duan XM, Kawata S (2011) Laser fabrication of Au nanorod aggregates microstructures assisted by two-photon polymerization. Opt Express 19:22786–22796
88. Masui K, Shoji S, Jin F, Duan XM, Kawata S (2012) Plasmonic resonance enhancement of single gold nanorod in two-photon photopolymerization for fabrication of polymer/metal nanocomposites. Appl Phys A 106:773–778

89. Serbin J, Egbert A, Ostendorf A, Chichkov BN, Houbertz R, Domann G, Schulz J, Cronauer C, Fröhlich L, Popall M (2003) Femtosecond laser-induced two-photon polymerization of inorganic–organic hybrid materials for applications in photonics. Opt Lett 28:301–303
90. Ovsianikov A, Viertl J, Chichkov B, Oubaha M, MacCraith B, Sakellari I, Giakoumaki A, Gray D, Vamvakaki M, Farsari M, Fotakis C (2008) Ultra-low shrinkage hybrid photosensitive material for two-photon polymerization microfabrication. ACS Nano 2:2257–2262
91. Vasilantonakis N, Terzaki K, Sakellari I, Purlys V, Gray D, Soukoulis CM, Vamvakaki M, Kafesaki M, Farsari M (2012) Three-dimensional metallic photonic crystals with optical bandgaps. Adv Mater 24:1101–1105
92. Wu PW, Cheng W, Martini IB, Dunn B, Schwartz BJ, Yablonovitch E (2000) Two-photon photographic production of three-dimensional metallic structures within a dielectric matrix. Adv Mater 12:1438–1441
93. Cao YY, Dong XZ, Takeyasu N, Tanaka T, Zhao ZS, Duan XM, Kawata S (2009) Morphology and size dependence of silver microstructures in fatty salts-assisted multiphoton photoreduction microfabrication. Appl Phys A 96:453–458
94. Xu BB, Xia H, Niu LG, Zhang YL, Sun K, Chen QD, Xu Y, Lv ZQ, Li ZH, Misawa H, Sun HB (2010) Flexible nanowiring of metal on nonplanar substrates by femtosecond-laser-induced electroless plating. Small 6:1276–1766
95. Jin W, Zheng ML, Cao YY, Dong XZ, Zhao ZS, Duan XM (2011) Morphology modification of silver microstructures fabricated by multiphoton photoreduction. J Nanosci Nanotechnol 11:8556–8560
96. Jia B, Kang H, Li J, Gu M (2009) Use of radially polarized beams in three-dimensional photonic crystal fabrication with the two-photon polymerization method. Opt Lett 34:1918–1920
97. Sun HB, Matsuo S, Misawa H (1999) Three-dimensional photonic crystal structures achieved with two-photon absorption photopolymerization of resin. Appl Phys Lett 74:786–788
98. Sun ZB, Dong XZ, Nakanishi S, Chen WQ, Duan XM, Kawata S (2007) Log-pile photonic crystal of CdS–polymer nanocomposites fabricated by combination of two-photon polymerization and in situ synthesis. Appl Phys A 86:427–431
99. Kaneko K, Sun HB, Duan XM, Kawata S (2003) Submicron diamond-lattice photonic crystals produced by two-photon laser nanofabrication. Appl Phys Lett 83:2091–2093
100. Dong XZ, Ya Q, Sheng XZ, Li ZY, Zhao ZS, Duan XM (2008) Photonic bandgap of gradient quasidiamond lattice photonic crystal. Appl Phys Lett 92:231103
101. Dong XZ, Zhao ZS, Duan XM (2008) Gradient quasidiamond lattice photonic crystal fabricated by two-photon polymerization nanofabrication. Mater Eng 10:118–125
102. Seet KK, Mizeikis V, Matsuo S, Joudkazis S, Misawa H (2005) Three-dimensional spiral-architecture photonic crystals obtained by direct laser writing. Adv Mater 17:541–545
103. Galajda P, Ormos P (2001) Complex micromachines produced and driven by light. Appl Phys Lett 78:249–251
104. Maruo S, Ikuta K, Korogi H (2003) Submicron manipulation tools driven by light in a liquid. Appl Phys Lett 82:133–135
105. Maruo S, Inoue H (2006) Optically driven micropump produced by three-dimensional two-photon microfabrication. Appl Phys Lett 89:144101
106. Zhang YL, Chen QD, Xia H, Sun HB (2010) Designable 3D nanofabrication by femtosecond laser direct writing. Nano Today 5:435–448
107. Wang J, He Y, Xia H, Niu LG, Zhang R, Chen QD, Zhang YL, Li YF, Zeng SJ, Qin JH, Lin BC, Sun HB (2010) Embellishment of microfluidic devices via femtosecond laser micronanofabrication for chip functionalization. Lab Chip 10:1993–1996
108. Kumi G, Yanez CO, Belfieldbc KD, Fourkas JT (2010) High-speed multiphoton absorption polymerization: fabrication of microfluidic channels with arbitrary cross-sections and high aspect ratios. Lab Chip 10:1057–1060

109. Xia H, Wang J, Tian Y, Chen QD, Du XB, Zhang YL, He Y, Sun HB (2010) Ferrofluids for fabrication of remotely controllable micro-nanomachines by two-photon polymerization. Adv Mater 22:3204–3207
110. Wang WK, Sun ZB, Zheng ML, Dong XZ, Zhao ZS, Duan XM (2011) Magnetic nickel-phosphorus/polymer composite and remotely driven three-dimensional micromachine fabricated by nanoplating and two-photon polymerization. J Phys Chem C 115:11275–11281
111. Gansel JK, Thiel M, Rill MS, Decker M, Bade K, Saile V, Freymann G, Linden S, Wegener M (2009) Gold helix photonic metamaterial as broadband circular polarizer. Science 325:1513–1515
112. Ergin T, Stenger N, Brenner P, Pendry JB, Wegener M (2010) Three-dimensional invisibility cloak at optical wavelengths. Science 328:337–339
113. Li CF, Dong XZ, Jin F, Jin W, Chen WQ, Zhao ZS, Duan XM (2007) Polymeric distributed-feedback resonator with sub-micrometer fibers fabricated by two-photon induced photopolymerization. Appl Phys A 89:145–148

Chapter 5
Laser Interference Nanofabrication

5.1 Introduction

Nanoscience and technology have attracted the attention of more and more researchers. Novel material properties and functionalities, such as structural color [1, 2], rely more on the structure than the material itself when the size of the material is at the nanoscale. A number of techniques have been developed to fabricate various nanostructures. And there are mainly three lithography technologies available for direct structure of nano features: ion beam lithography, electron beam lithography, and scanning probe lithography. All of them are using a sequential writing strategy, which is time-consuming and requires high mechanical and electrical stability of the system. Both high costs and low throughput discourage industrial end users from expanding their nanotechnology-related activities. Laser interference lithography (LIL) will play a key role in realizing the full potential of interference nanolithography by combining optical technology, information and communication technology, and micro-/nanotechnology, as current nanofabrication tools are limited to archaic, slow processing rates or do not achieve a competitive cost-effective strategy. It is the aim of LIL to empower laser interference nanolithography technology with a clear focus on industrial use and to drive the rapid development of nanoscience leading to new processes and immediate industrial exploitation. The main advantageous features of the LIL technology in fabrication of nanostructures and devices are high resolution compared with other optical technologies and low cost compared with other beam technologies. As a parallel writing technique, LIL is famous for its high time efficiency. This high efficiency can be further promoted by recording the interference pattern on the target material in one step, so-called direct writing. Photoresist and related processes are not necessary anymore. Local ablation, oxidation, crystallization, or other effects enable LIL to pattern materials directly [3]. Direct writing techniques simplify the processes, thus greatly accelerate the patterning

Q. Liu et al., *Novel Optical Technologies for Nanofabrication*, Nanostructure
Science and Technology, DOI 10.1007/978-3-642-40387-3_5,

process. LIL is a mature technique mainly for fabrication of photonic crystals [4], high-density templates for different applications [5], and periodic structures for MEMS [6] and bioscience [7]. The LIL technology has attracted wide interest on micro- and nanomanufacturing of periodic and quasiperiodic features in recent years [8, 9].

The maximum resolution of LIL is $\lambda/2$ (λ is the incident wavelength). The only way to improve the resolution is to decrease the wavelength λ. There are two ways to achieve small λ: (1) to increase the refractive index of the environment (immersion interference lithography) and (2) to use short wavelength [7, 10]. There are also other techniques which can break the limitation of $\lambda/2$ [8].

This chapter introduces the LIL nanopatterning and its in situ application for epitaxial semiconductor quantum dots (QDs) by LIL-modified molecular beam epitaxy (LIL-MBE). The following parts are included:

Theoretical study shows the relationship between the interference patterns and the key parameters of the laser beams, including intensity, polarization, incident angle, azimuth angle, and difference of the initial phase. The experimental work on LIL was performed with direct writing on semiconductor materials.

Slight differences between angles of incidence may introduce an extraordinary, controllable modulation on a uniform interference field, without changing the periodicity. This complicated interference field leads to variable sizes of features in the final pattern. A new graded-index photonic crystal (GI PhC) structure which can be made by the incidence modification is also discussed in this chapter. It can be applied as miniaturized focusing lens.

There are mainly two ways for epitaxial growth of semiconductor quantum dots (QDs): self-organization on unpatterned and patterned substrates. QDs grown by self-organization on unpatterned substrate can be defect-free but nonuniform in both size and spatial distribution and different from run to run. While, QDs grown on patterned substrate are full of defects including point defects and dislocation, which are induced by the patterning of the substrate. The case will be quite different if we combine LIL and epitaxial growth chamber together. LIL patterning can induce defect-free QD "seeds," and further QD growth on those "seeds" is uniform in both size and spatial distribution. With precise control of both LIL and epitaxial growth, the QDs growth can be identical from run to run.

5.2   Multi-beam Interference

To simplify the simulation of the LIL patterns, we use Gaussian beams for the interference patterns. In the case of small angle of divergence and the pattern area is small, the beams can be considered as plane waves. We set the amplitude of the m-th plane wave as $\boldsymbol{E}^0{}_m$, the wave vector as $\boldsymbol{k}_m$, the initial phase as δ_m, and the position vector of the pattern site P(x, y, z) as $\boldsymbol{r}$. Figure 5.1 shows the beam

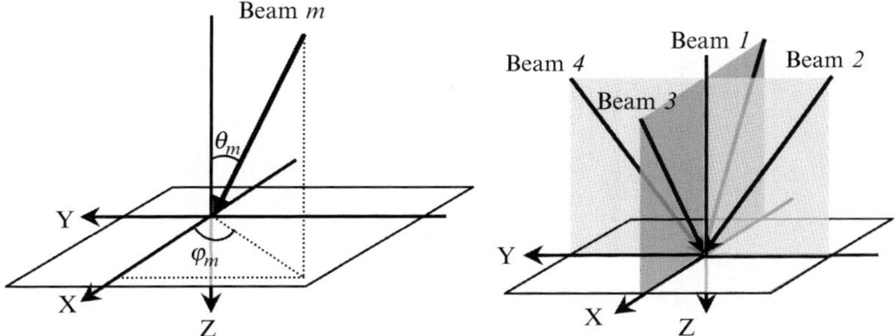

Fig. 5.1 Configuration for beam incident direction and perpendicular symmetric incidence with four beams

configurations. The amplitude of the m-th plane wave can be expressed in the following equation:

$$\begin{aligned}
\boldsymbol{E}_m(\boldsymbol{r}) &= \boldsymbol{E}^0{}_m \exp\left[i(\boldsymbol{k}_m \cdot \boldsymbol{r} + \delta_m)\right] \\
&= \left[\boldsymbol{E}^0{}_{mx}\boldsymbol{E}^0{}_{my}\boldsymbol{E}^0{}_{mz}\right]^* \exp\left\{i[k_m(x\sin\theta_m\cos\varphi_m + y\sin\theta_m\sin\varphi_m + z\cos\theta_m) + \delta_m]\right\}
\end{aligned}$$

$$(5.1)$$

Then the total intensity of the interference field of N coherent light beams is:

$$\begin{aligned}
I &= \left|\sum_{m=1}^{N}\boldsymbol{E}_m\right|^2 = \sum_{m=1}^{N}\sum_{n=1}^{N}(\boldsymbol{E}_m^* \cdot \boldsymbol{E}_n) \\
&= \sum_{m=1}^{N}(\boldsymbol{E}_m^* \cdot \boldsymbol{E}_m) + \sum_{m=1}^{N}\sum_{n=1, n\neq m}^{N}\left(E^0{}_{mx}E^0{}_{nx} + E^0{}_{my}E^0{}_{ny} + E^0{}_{my}E^0{}_{ny}\right)
\end{aligned}$$

$$\exp\left\{i[k(\sin\theta_m\cos\varphi_m - \sin\theta_n\cos\varphi_n)x + k(\sin\theta_m\cos\varphi_m - \sin\theta_n\cos\varphi_n)y\right.$$

$$\left. + k(\cos\theta_m - \cos\theta_n)z + \delta_m - \delta_n]\right\}$$

$$(5.2)$$

Here, the LIL patterns including the direct writing and exposure on photoresist are two dimensional. Reasonably, we can set $z = 0$ for the 2D patterns. While $\boldsymbol{E}_m^* \cdot \boldsymbol{E}_n = \boldsymbol{E}_n^* \cdot \boldsymbol{E}_m$, then,

$$I = \sum_{m=1}^{N}(\boldsymbol{E}_m^* \cdot \boldsymbol{E}_m) + 2\sum_{m=1}^{N}\sum_{n=m+1}^{N}E^0{}_{mn}\cos\left[k(\sin\theta_m\cos\varphi_m - \sin\theta_n\cos\varphi_n)x\right.$$

$$\left. + k(\sin\theta_m\cos\varphi_m - \sin\theta_n\cos\varphi_n)y + \delta_{mn}\right]$$

$$(5.3)$$

where, $E^0{}_{mn} = E^0{}_{mx}E^0{}_{nx} + E^0{}_{my}E^0{}_{ny} + E^0{}_{my}E^0{}_{ny}$, $\delta_{mn} = \delta_m - \delta_n$. The first part of Eq. 5.3 is the uniform background distribution. The second part is the interference field. In the second part, each item "$E^0{}_{mn}\cos[k(\sin\theta_m\cos\varphi_m - \sin\theta_n\cos\varphi_n)x + k$

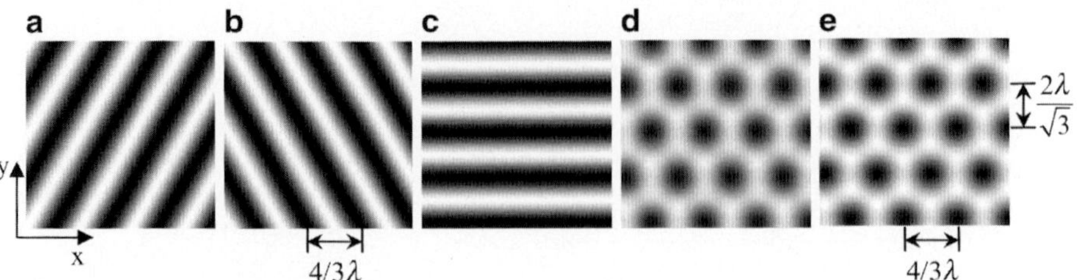

Fig. 5.2 (**a**, **b**) **c**) Patterns of 2-beam interference fringes of beams 1 + 2, 1 + 3, and 2 + 3, respectively; (**d**) superposition of patterns (**a**), (**b**), and (**c**); (**e**) pattern of three-beam interference with beams 1, 2, and 3. All the incident angles are $\pi/6$. The azimuth angles of beams 1, 2, and 3 are 0, $2\pi/3$, and $4\pi/3$, respectively

$(\sin\theta_m\cos\varphi_m - \sin\theta_n\cos\varphi_n)y + \delta_{mn}]$" shows the similar distribution of a 2-beam interference pattern. Therefore, excluding the uniform background field, the multibeam interference pattern is the superposition of the patterns of 2-beam interference.

As an example, the intensity distribution of a three-beam interference pattern field is:

$$I_{1,2,3} = (E_1^* \cdot E_1 + E_2^* \cdot E_2 + E_3^* \cdot E_3) \\ + 2(E_1^* \cdot E_2 + E_1^* \cdot E_3 + E_2^* \cdot E_3) \tag{5.4}$$

While the superposition of the three patterns of 2-beam interference is:

$$I_{1,2} + I_{1,3} + I_{2,3} = 2(E_1^* \cdot E_1 + E_2^* \cdot E_2 + E_3^* \cdot E_3) \\ + 2(E_1^* \cdot E_2 + E_1^* \cdot E_3 + E_2^* \cdot E_3) \tag{5.5}$$

Compare Eq. 5.4 with Eq. 5.5, the difference is the item of the uniform background distribution "$E_1^* \cdot E_1 + E_2^* \cdot E_2 + E_3^* \cdot E_3$." Figure 5.2 shows the three 2-beam interference patterns, the superposition of the 2-beam patterns, and a three-beam interference pattern. Compare (d) with (e), the patterns of the superposition of the 2-beam patterns and three-beam interference are exactly the same except the intensity of the background.

5.2.1 Parameters of the Interference

In the intensity Eq. 5.3, the parameters including E^0_{mn}, θ_m, θ_n, φ_m, φ_n, and δ_{mn} are related to the beam intensity, polarization, incident angle, azimuth angle, and difference of the initial phase. In this section, we will change the parameters one by one to examine their effects on the interference pattern. If not mentioned specially, the laser wavelength is 266 nm, the incident angle of each beam is the same as 60°, the figure area is 3 × 3 μm^2, the polarization of each beam is set to TM (the electric vector lies in the incidence plane), the difference of initial phases $\delta_{mn} = 0$, and the beam configuration is as Fig. 5.1.

5.2.1.1   Beam Intensity

In the case of linear polarization of the laser beams, the intensity ratio of beams by beam splitter varies with the azimuth angle, and the reflection of the mirrors varies with the incident angle. Then, the relative intensity of the interference beams may change from case to case.

Here, we are discussing both 4-beam interference with azimuth setup as $\varphi_1 = 180°$, $\varphi_2 = 90°$, $\varphi_3 = 0°$, $\varphi_4 = 270°$, and 3-beam interference with azimuth setup as $\varphi_1 = 120°$, $\varphi_4 = 240°$, $\varphi_3 = 0°$.

In 4-beam interference, (1) when only one beam changes the intensity ($I_1 = I_0$, $0.1I_0$ and $0.01I_0$) and the others are keeping the same ($I_2 = I_3 = I_4 = I_0$), we found that the interference patterns keep almost the same and only the shape of the interference spots change a little bit to an oval shape even when one beam becomes 100 times weaker than the other beams (Fig. 5.3a–c); (2) when a pair of beams changes the intensity ($I_1 = I_3 = I_0$, $0.1I_0$ and $0.01I_0$) and the other two are keeping the same ($I_2 = I_4 = I_0$), we found that the interference patterns change about 10 times faster than the above case. When one pair of beams becomes 10 times weaker, the spots in the pattern change to the similar oval shape as the case that one beam becomes 100 times weaker than others (Fig. 5.3d–f).

In 3-beam interference, when one beam changes intensity, the pattern changes about 10 times faster than the case when a pair of beams changes intensity in the 4-beam interference. When one beam becomes 10 times weaker than the other two, the intensity change of the pattern looks similar as the case when a pair of beams becomes 100 times weaker than the other two in the 4-beam interference. And when one beam becomes 100 times weaker than the other two, the pattern looks like the case in a 2-beam interference with a small modification.

Normally, for the linear polarized laser beam, when the azimuth angle and the incident angle change, the change of the intensity ratio of the split beams is less than 20 %. Based on the above discussion, this change does not affect the interference patterns much and can be ignored.

Using an attenuator is a good and common choice for altering the intensity of a beam.

5.2.1.2   Beam Polarization

Except the background difference, the pattern of a multi-beam interference is the superposition of the fringes of all 2-beam interferences. The equation of the interference between beam m and n is as following:

$$E_m{}^* \cdot E_n = E^0{}_{mn}\cos\left[k(\sin\theta_m\cos\varphi_m - \sin\theta_n\cos\varphi_n)x + k(\sin\theta_m\sin\varphi_m - \sin\theta_n\sin\varphi_n)y + \delta_{mn}\right]$$

$$(5.6)$$

where $E^0{}_{mn} = E^0{}_{mx}E^0{}_{nx} + E^0{}_{my}E^0{}_{ny} + E^0{}_{mz}E^0{}_{nz}$ is the amplitude of the fringes.

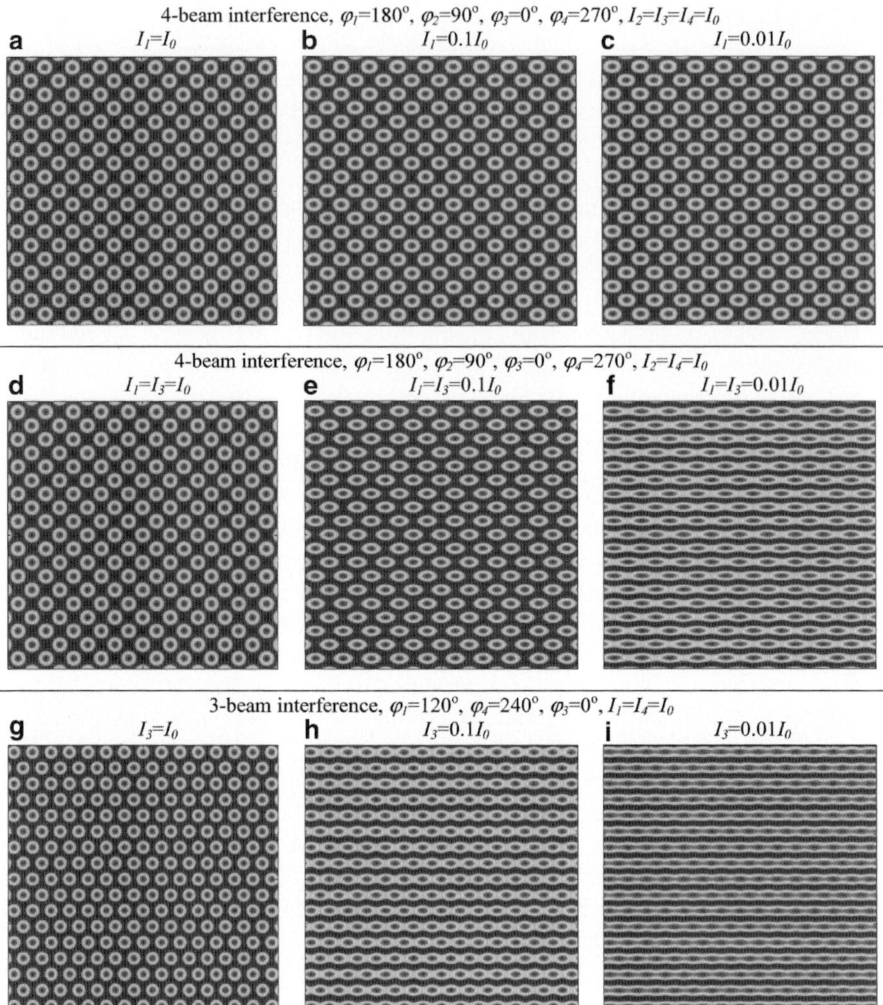

Fig. 5.3 Patterns with different beam intensities in both 4-beam and 3-beam interference

E^0_{mn} changes with the change of the polarization $p = [E_x\ E_y\ E_z]^*$. Only linear polarization is discussed here.

In an isotropic medium, the electric vector is always perpendicular to the wave vector. When a beam irradiates the sample, it is TM polarized if the electric vector lies in the incidence plane and TE polarized if the electric vector is perpendicular to the incidence plane. The interference only happens when the electric vectors of two beams are not perpendicular to each other. It means that beams m and n do not interfere with each other when $p_m \perp p_n$.

Figure 5.4 shows the patterns with different polarization configurations of the 4-beam interference. In Fig. 5.4a, $p_1 = p_{1x} + p_{1z}$, $p_3 = p_{3x} + p_{3z}$, $p_2 = p_{2y} + p_{2z}$,

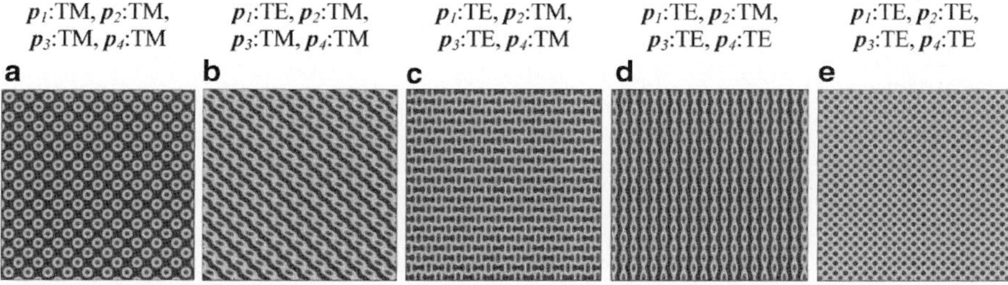

Fig. 5.4 Patterns of 4-beam interference with perpendicular azimuth configuration and different polarization configurations. Especially, the pattern of (**d**) doesn't have any change whether the beam 4 exists or not.

$p_4 = p_{4y} + p_{4z}$, $p_{1z} = p_{2z} = p_{3z} = p_{4z}$, and $p_{1x}//p_{3x} \perp p_{2y}//p_{4y}$, all the four beams interfere with each other and the pattern is symmetric. In Fig. 5.4b, $p_1 = p_{1y}$, $p_3 = p_{3x} + p_{3z}$, $p_2 = p_{2y} + p_{2z}$, $p_4 = p_{4y} + p_{4z}$, $p_{3z} = p_{2z} = p_{4z}$, and $p_{3x} \perp p_1//p_{2y}//p_{4y}$, beams 1 and 3 do not interfere with each other but they interfere with both beams 2 and 4. In Fig. 5.4c, $p_1 = p_{1y}$, $p_3 = p_{3y}$, $p_2 = p_{2y} + p_{2z}$, $p_4 = p_{4y} + p_{4z}$, $p_{2z} = p_{4z}$, and $p_1//p_3//p_{2y}//p_{4y}$, all the four beams interfere with each other and the pattern is symmetric but different from Fig. 5.4a. In Fig. 5.4d, $p_1 = p_{1y}$, $p_3 = p_{3y}$, $p_2 = p_{2y} + p_{2z}$, $p_4 = p_{4x}$, $p_1//p_3//p_{2y} \perp p_4$, and $p_{2z} \perp p_4$, beams 1, 2, and 3 interfere with each other. Beam 4 does not interfere with the other three and only contributes a uniform background.

In Fig. 5.4e, $p_1 = p_{1y}$, $p_3 = p_{3y}$, $p_2 = p_{2x}$, $p_4 = p_{4x}$, and $p_1//p_3 \perp p_2//p_4$, then $E^0_1 = E^0 y$, $E^0_3 = - E^0 y$, $E^0_2 = - E^0 x$, and $E^0_4 = E^0 x$. Therefore, $E^0_{12} = E^0_{14} = E^0_{23} = E^0_{34} = 0$:

$$
\begin{aligned}
I &= 4I_0 + 2(E_1{}^* \cdot E_3 + E_2{}^* \cdot E_4) \\
&= 4I_0 + 2I_0[\cos(2k\sin\theta_0 x) + \cos(2k\sin\theta_0 y)]
\end{aligned} \tag{5.7}
$$

In Eq. 5.7, $E_1{}^* \cdot E_3$ and $E_2{}^* \cdot E_4$ are 2-beam interference. The pattern is more like a double-exposure 2-beam system ($E_1{}^* \cdot E_3$ and $E_2{}^* \cdot E_4$) than a 4-beam system, because there is no interference between beams 1 and 2, 1 and 3, 2 and 3, and 3 and 4.

A half waveplate can easily change the direction of the electric vector on the premise that the laser beam is linear polarized.

5.2.1.3   Incident Angle

Wave vector k is the most important factor influencing the interference pattern. The magnitude of the wave vector is the angular wave number k, which is determined by the laser source used in LIL. The direction of the wave vector is indicated by the angle of incidence and azimuth angle. Here, the azimuth angles of the four beams are set as $\varphi_m = (m-1)90°$ ($m = 1, 2, 3,$ and 4).

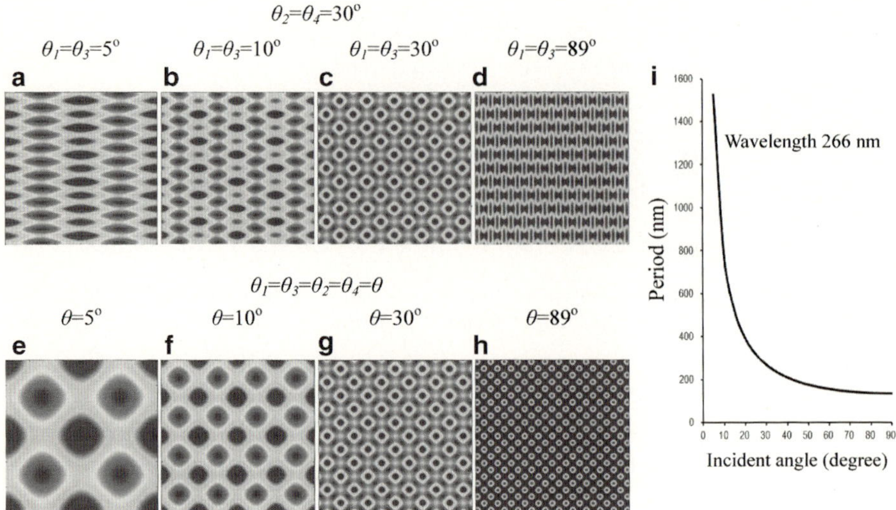

Fig. 5.5 Patterns of 4-beam interference with symmetric azimuth and incidence configuration and different incident angles

If the four beams are in a symmetric arrangement ($\theta_1 = \theta_3$, $\theta_2 = \theta_4$), variation of the angle of incidence only results in a change of period in the x (D_x) or y (D_y) direction.

$$D_x = \frac{\lambda}{(\sin\theta_1 + \sin\theta_3)}, D_y = \frac{\lambda}{(\sin\theta_2 + \sin\theta_4)} \quad (5.8)$$

where, D_x and D_y are the periods of the interference pattern in the x and y directions, respectively. In the symmetric azimuth configuration, there are two kinds of symmetric incidence arrangement: (1) $\theta_1 = \theta_3 \neq \theta_2 = \theta_4$, the periods of two directions are different $D_x \neq D_y$. In Fig. 5.5a–d, if $\theta_2 = \theta_4 = 30°$ keeps unchanged and $D_y = \lambda$, when $\theta_1 = \theta_3 = 5°$, $10°$, $30°$, and $89°$, D_x varies with the incident angles; (2) $\theta_1 = \theta_3 = \theta_2 = \theta_4 = \theta$, when $\theta = 5°$, $10°$, $30°$, and $89°$, both D_x and D_y vary with the incident angles. Figure 5.5i shows the relationship between the period of the pattern and the incident angles. The minimum period is $\lambda/2$ (here 133 nm) with glancing incidence, while it can be very huge with a very small incident angle.

As for an asymmetric case ($\theta_1 \neq \theta_3$ or $\theta_2 \neq \theta_4$), an extra periodical modulation will be introduced in the interference pattern [11]. The period of this modulation is defined as d_x and d_y in the x and y directions.

$$d_x = \frac{\lambda}{|\sin\theta_1 - \sin\theta_3|}, d_y = \frac{\lambda}{|\sin\theta_2 - \sin\theta_4|} \quad (5.9)$$

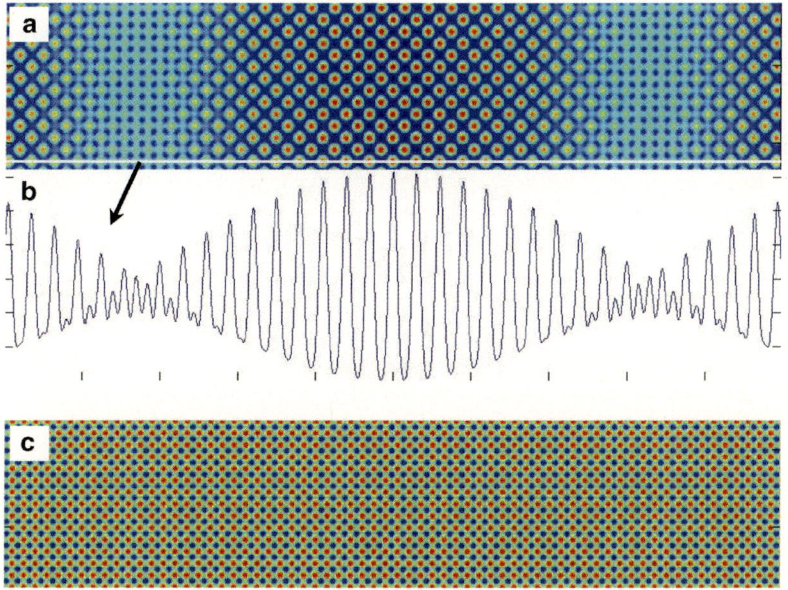

Fig. 5.6 Patterns of 4-beam interference with incidence configuration: $\theta_1 = 65°$, $\theta_2 = \theta_3 = \theta_4 = 60°$. (**a**) TM polarization of each beam; (**b**) intensity profile of the *white line* in (**a**); (**c**) TE polarization of each beam

Figure 5.6 gives an example. Increasing θ_1 from 60° to 65° and keeping other parameters as default, the interference pattern changes from that in Fig. 5.3a to the one in Fig. 5.6a. The intensity profile of the white line in Fig. 5.6a is given in Fig. 5.6b. This modulation can be in any direction, depending on the difference between the four angles of incidence. However, in 4-beam interference with TE polarization of each beam, no modulation will be introduced as claimed above no matter what the incident angles are (Fig. 5.6c). As discussed in Sect. 5.2.1.2, if all the beams are in the TE mode, the pattern is more like a double-exposure 2-beam system than a 4-beam system, because there is no interference between the beams with perpendicular polarization. Figure 5.4e shows a pattern generated in all TE mode at equal angles of incidence at 60°.

The shape of the features corresponding to intensity maxima or minima may change with the angles of incidence, as can be clearly seen from Fig. 5.5. Moreover, the maximum intensity of the pattern also increases with the angles of incidence if all TM-polarized symmetric beam configurations are used.

The asymmetric incidence configuration modulates the pattern. However, it cannot hold any more if all the beams are in the TE mode, because the pattern is more like double-exposure 2-beam interference than 4-beam interference and there is no modulation in a 2-beam interference. Figure 5.6c shows a pattern generated in all TE mode at the same incidence configuration as Fig. 5.6a. No further modulation will be introduced as claimed before no matter what the angles of incidence are.

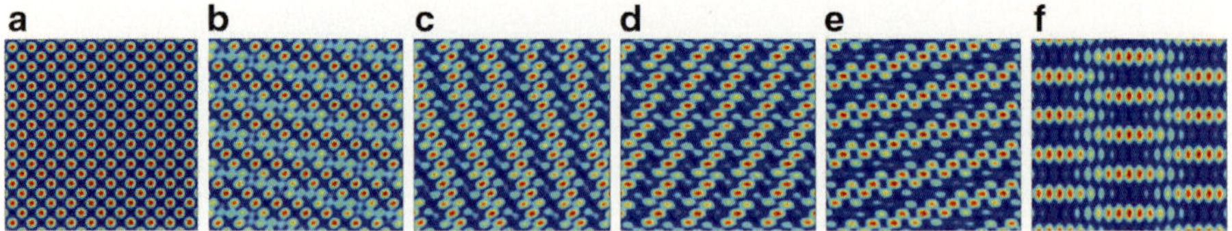

Fig. 5.7 Patterns with different azimuth setup ((**a–e**): $\varphi_1 = 180°$, $\varphi_2 = 90°$, $\varphi_4 = 270°$): (**a**) $\varphi_3 = 0°$; (**b**) $\varphi_3 = 30°$; (**c**) $\varphi_3 = 60°$; (**d**) $\varphi_3 = 120°$; (**e**) $\varphi_3 = 150°$; (**f**) $\varphi_1 = 180°$, $\varphi_2 = 30°$, $\varphi_3 = 0°$, $\varphi_4 = 330°$

Angles of incidence are usually changed by rotating mirrors and the sample. In some experiments, a grating or prism is used for splitting the laser beam, and the angles of incidence are determined by the structure of the grating or prism.

5.2.1.4   Azimuth Angle

k is one of the key parameters influencing the interference pattern. The magnitude of k is the angular wave number k and cannot be modified by changing the beam configuration parameters. The direction of k is indicated by the angle of incidence and azimuth angle, which vary with the beam configuration. The influence of the incident angles on the patterns is discussed in the above section. Here, the incident angles of the four beams are set as $\theta_m = 60°$ (m = 1, 2, 3, and 4), and the azimuth angles will be changed to modify the interference patterns.

Equation 5.3 shows that the interference pattern is very sensitive to the azimuth configuration. Figure 5.7a–e shows different patterns with changing only one azimuth angle φ_3. The multi-beam interference patterns become complicated with the changes of more than one azimuth angles (Fig. 5.7f). All the azimuth angles can be set from $0°$ to $360°$. There are plenty of different multi-beam interference patterns with different azimuth setups.

The change of azimuth angles can be easily achieved with our newly-designed LIL system (see Sect. 5.2 and Fig. 5.10).

5.2.1.5   Phase

The four beams are split from one beam normally by beam splitter [12], Lloyd mirror [13], Dove prism [14], or grating [15]. Finally, they intersect on the sample and end at the same point. They travel along different optical paths. The optical path lengths of the beams are most likely to differ from each other. The initial phase δ_n ($n = 1, 2, 3, 4$) is proportional to the optical path length r: $\delta_n(r) = k_n \cdot r + \delta_n^{\,0}$. The δ_n value of a beam does not have any impact on the pattern, while the phase difference δ_{mn} is important. Then, we can set the initial phase of one beam to be zero, for example, $\delta_1 = 0$, and the others are given by relative values: $\delta_2 = \delta_{21}$, $\delta_3 = \delta_{31}$, and $\delta_4 = \delta_{41}$.

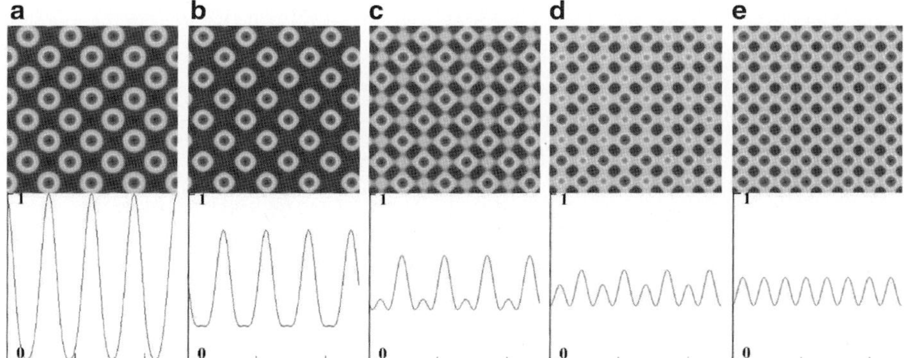

Fig. 5.8 Patterns of 4-beam interference with phase change of beam 1: (**a**) $\delta_1 = 0°$, (**b**) $\delta_1 = 110°$, (**c**) $\delta_1 = 150°$, (**d**) $\delta_1 = 170°$, (**e**) $\delta_1 = 180°$. The curve under each figure is the intensity profile of the *bottom row* of the pattern. Here the background and the maximum intensity of (**a**) are normalized as 0 and 1, respectively, and the intensities of all the other curves are in the same scale of (**a**)

Excluding the uniform background field, the multi-beam interference pattern is the superposition of the patterns of 2-beam interference (Eq. 5.3). In 2-beam interference systems, the variation of phase difference δ_{mn} only causes a shift of the fringe. However, the shifting of the 2-beam interference fringes will cause the change of the superposition of all those 2-beam interferences, and thus the pattern of the multi-beam interference changes. Figure 5.8 shows the pattern changes with phase changing of one beam, and all the other parameters remain the default values. When the phase of beam 1 increases from $0°$ to $180°$, pattern shifts along the x-axis, the background of the patterns increases gradually, another pattern appears, and the intensities of the peaks increase to the same level as the original one.

Figure 5.8e shows the same pattern as Fig. 5.4e, which is the superposition of two sets of perpendicular 2-beam interferences. When $\delta_1 = 180°$,

$$\boldsymbol{E}_1{}^* \cdot \boldsymbol{E}_2 = \cos\left[k\sin\theta_0(x-y) + 180°\right] = -\cos\left[k\sin\theta_0(x-y)\right] = -\boldsymbol{E}^*{}_3 \cdot \boldsymbol{E}_4$$
$$\boldsymbol{E}_1{}^* \cdot \boldsymbol{E}_4 = \cos\left[k\sin\theta_0(x+y) + 180°\right] = -\cos\left[k\sin\theta_0(x-y)\right] = -\boldsymbol{E}^*{}_2 \cdot \boldsymbol{E}_3$$
$$\boldsymbol{E}_1{}^* \cdot \boldsymbol{E}_2 + \boldsymbol{E}^*{}_3 \cdot \boldsymbol{E}_4 + \boldsymbol{E}_1{}^* \cdot \boldsymbol{E}_4 + \boldsymbol{E}^*{}_2 \cdot \boldsymbol{E}_3 = 0$$
$$I_{1,2,3,4} = (\boldsymbol{E}_1{}^* \cdot \boldsymbol{E}_1 + \boldsymbol{E}_2{}^* \cdot \boldsymbol{E}_2 + \boldsymbol{E}_3{}^* \cdot \boldsymbol{E}_3 + \boldsymbol{E}_4{}^* \cdot \boldsymbol{E}_4)$$
$$+ 2(\boldsymbol{E}_1{}^* \cdot \boldsymbol{E}_2 + \boldsymbol{E}_1{}^* \cdot \boldsymbol{E}_3 + \boldsymbol{E}_1{}^* \cdot \boldsymbol{E}_4 + \boldsymbol{E}_2{}^* \cdot \boldsymbol{E}_3 + \boldsymbol{E}_2{}^* \cdot \boldsymbol{E}_4 + \boldsymbol{E}_3{}^* \cdot \boldsymbol{E}_4)$$
$$= (\boldsymbol{E}_1{}^* \cdot \boldsymbol{E}_1 + \boldsymbol{E}_2{}^* \cdot \boldsymbol{E}_2 + \boldsymbol{E}_3{}^* \cdot \boldsymbol{E}_3 + \boldsymbol{E}_4{}^* \cdot \boldsymbol{E}_4) + 2(\boldsymbol{E}_1{}^* \cdot \boldsymbol{E}_3 + \boldsymbol{E}_2{}^* \cdot \boldsymbol{E}_4)$$

$$(5.10)$$

Waveplates can be used for a rigid phase change, while continuous phase change can be realized by many other devices such as Pockels cells, piezoelectric stages, and liquid crystal variable phase retarders. Phase control can also be achieved by clever arrangement of two gratings [16].

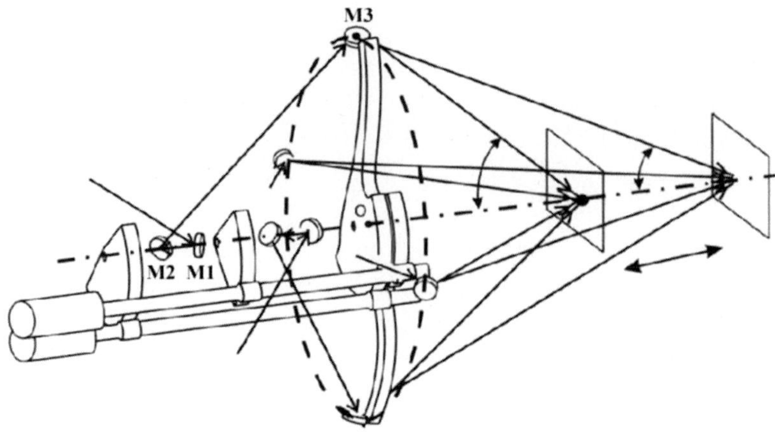

Fig. 5.9 Interference control module: mechanical design and optical path (Reprinted from Ref. [17], Copyright 2009, with permission from Elsevier). *M1*, *M2*, and *M3* are mirrors

5.2.2   Laser Interference Lithography System

There are many azimuth and incident angles to be set in an interference optical setup to obtain a certain pattern. It is a time-consuming process to set those angles. In order to shorten the time of this process, A. Rodriguez et al. designed an LIL system in which the change of the azimuth and incidence angles can be controlled by computer in a very short time [17]. As shown in Fig. 5.9, the laser is split into four beams with identical intensities by beam splitters. Before the interference point on a virtual circumference, last four mirrors reflect the beams onto the sample, which is located on the axis of the circumference and perpendicular to it. A longitudinal displacement of the sample and a rotation movement of each mirror on the circumference define the azimuth and incident angles. The azimuth and incident configuration can be adjusted accurately with this system.

The system enables the automatic selection of two azimuth angles independently, and the other two are fixed. Each of the movable beams is able to rotate around the axis of the system (main axis) without changing the optical path. The laser beam reaches mirror M1 which is fixed in the main axis. Afterwards, the laser beam propagates along the main axis from M1 to M2, then to M3 and finally to the point of interference with the other three laser beams on the sample. The mirrors M2 and M3 rotate synchronously around the main axis to achieve the azimuth angle.

The advantage of this system is that two of the four azimuth angles and all the incidence ones can be modified automatically and quickly. However, the rotation setup is complicated, and only two azimuth angles can be modified.

We designed and manufactured LIL system with beam splitting on the main axis: all the beam splitters are located along the main axis [18]. In Fig. 5.10, S_n ($n = 1$, 2...k−1) (k is number of coherent beams) are beam splitters with reflection/transmission ratio of 1:(k−1)...1:2 and 1:1. The main laser beam 0 propagates along the

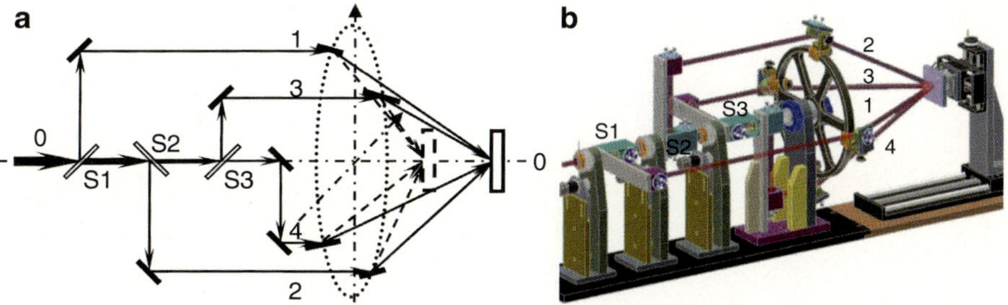

Fig. 5.10 Interference control module designed by us [18]: (**a**) optical path and (**b**) mechanical configuration. *S1*, *S2*, and *S3* are beam splitters with reflection/transmission ratio of 1:3, 1:2, and 1:1. *0*, *1*, *2*, *3*, and *4* are the initial laser beam and the four split beams

main axis and is split into k beams with the same intensities. All the beams reflect from the last mirrors located on a wheel and are focused on the sample. The main axis is perpendicular to the wheel and through the center of the wheel. The sample is located along the main axis. The incident and azimuth angles can be easily tuned by adjusting the sample position and rotating the mirrors respectively.

Compared to the system developed by A. Rodriguez et al. [17], our system is simplified and reduces the number of mirrors and beam splitters, which are the main factors of the reduction of the beam quality. In addition, no laser beam will be blocked by mechanical parts; the entire azimuth configuration can be realized. The optical paths of all the beams are the same and are much shorter than those in Fig. 5.9.

5.3   Patterns Created on Semiconductor Wafer

Si and GaAs are widely used semiconductor substrates and were chosen as the recording material in our experiments. High-power pulsed 308 nm XeCl excimer laser was used in the 4-beam LIL system. Both Si and GaAs epi-ready wafers were used for direct writing by high-power LIL patterns.

We discovered that the temperature difference of the peak and valley of the temperature distribution by the LIL patterns can be over 1,000°C (Fig. 5.11a). It is indicated that we can achieve temperature difference at the range of 0–1,000°C for the LIL patterns by adjusting the laser power. We even achieved temperature higher than the melting point of Si (1,410°C) (Fig. 5.11b).

As we know, there is no reaction when the laser interacts with materials if the laser power is smaller than a special value, which is so-called threshold (T). Sometimes, there are more than one threshold for semiconductors, especially compound semiconductors. We used GaAs epi-ready wafers for LIL direct patterning. The area of the patterning was about 1 cm^2. The laser we used was pulsed (single 10 ns width pulse) 308 nm XeCl excimer laser. The laser power was tuned to about 60, 100, 140, and 180 mJ. Figure 5.12 shows the results. Clearly, four

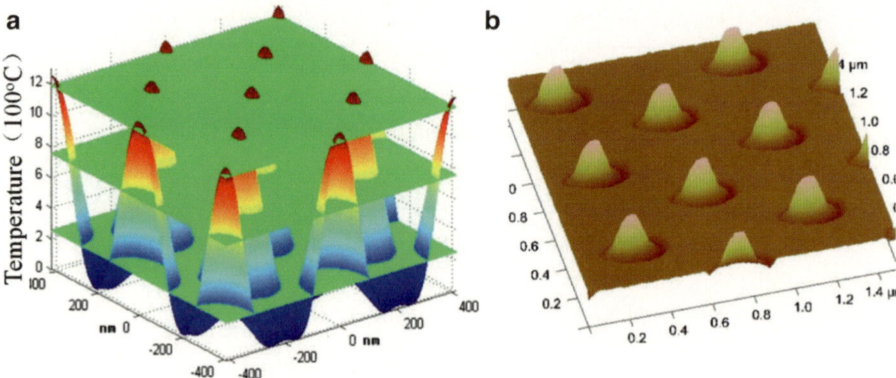

Fig. 5.11 (**a**) Temperature distribution by LIL patterns, (**b**) LIL burned islands on Si epi-ready wafer by high-power 308 nm XeCl excimer laser

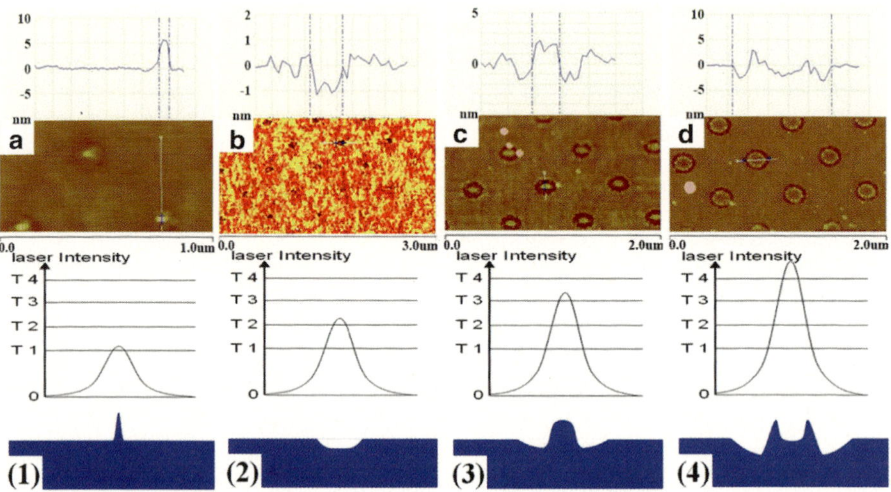

Fig. 5.12 LIL direct writing on GaAs epi-ready wafer by pulsed (single 10 ns width pulse) 308 nm XeCl excimer laser at different laser intensities of (**a**) ~60 mJ, (**b**) ~100 mJ, (**c**) ~140 mJ, (**d**) ~180 mJ with beam sizes of 1 cm^2. (*1–4*) are the diagrams of the shapes of the features and laser burning thresholds of the GaAs wafer for (**a–d**), respectively

different thresholds were observed: (1) at ~60 mJ, GaAs started to melt and islands formed; (2) at ~100 mJ, GaAs started to evaporate and holes formed. Besides the applications for photonic crystals, the nanoholes could also be used as templates for the controlled growth of periodic quantum dots [19]; (3) at ~140 mJ or more, GaAs further melted and island formed in the holes during the solidification after the pulse; (4) big holes were burned, and small island formed around the big holes when the laser power reached 180 mJ.

Fig. 5.13 AFM micrograph
of interference patterns
generated by the laser
beams directly writing.
White square is the selective
area for Fig. 5.14
(Reprinted from Ref. [20],
© Inderscience
Enterprises Ltd)

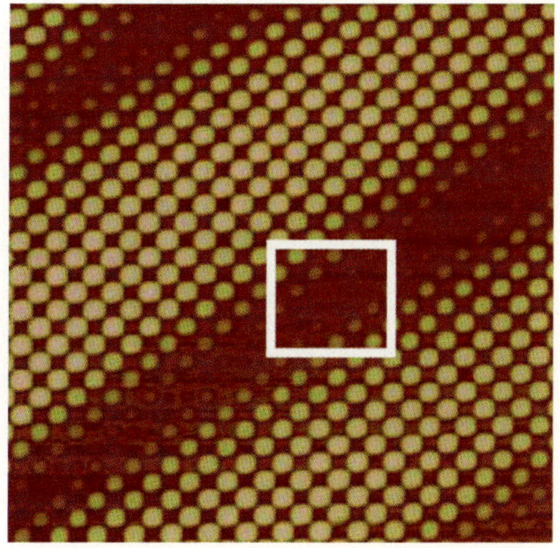

The above patterns were formed/burned by a single laser pulse. As shown in
Fig. 5.6, by tilting the sample, extra periodical modulation with larger period than the
patterns is superpositioned on the periodic patterns. In the modulated area (e.g., inside
the white frame in Fig. 5.13), interference dots of the pattern with different laser
power burned the GaAs material, and relevant different thresholds were observed. A
SiO_2-coated GaAs wafer was directly patterned by LIL. The LIL patterns were
periodic arrays of nanoholes on the GaAs wafer covered with SiO_2 bubbles.

A 5 nm thick SiO_2 layer was grown on a GaAs wafer by plasma-enhanced
chemical vapor deposition. A pulsed XeCl excimer laser ($\lambda = 308$ nm) was used to
create the patterns. The beam was split into four equal-intensity beams and then
converged on the wafer. The incident angles of the four beams were $\theta_1 \approx 25.9°$,
$\theta_2 \approx 27.3°$, $\theta_3 \approx 27.9°$, and $\theta_4 \approx 30.3°$, which was derived from the simulation
with Eq. 5.3 based on the pattern (Fig. 5.13). The azimuth angles were $\varphi_1 = 0°$,
$\varphi_2 = 90°$, $\varphi_3 = 180°$, and $\varphi_4 = 270°$ (Fig. 5.1). After exposure, the patterns were
modulated by periodic dark lines due to the nonidentical value of θ_n ($n = 1, 2, 3, 4$).

The surface patterns generated by laser beam exposure were bubble-like dots
[12]. Wet or dry chemical etching was used to remove the SiO_2 and GaAs materials,
respectively. The area (2×2 μm^2) in the white square of Fig. 5.13 was selectively
for a detailed analysis. Figure 5.14 shows atomic force microscopy (AFM) images
of the white square area before and after etching. There were bubbles in Fig. 5.14a
which indicate that the temperature at the bubble sites was high enough to evaporate
the GaAs and melt the SiO_2 blow into bubbles. While there is no bubble at the
sites 1, 2, 3, and 4, Fig. 5.14 (1) shows simulated threshold of bubble formation.
SiO_2 was selectively etched by dry etching in Fig. 5.14b. No GaAs evaporation
was observed at sites 2, 3, and 4. Figure 5.14 (2) indicates the threshold of
evaporation of GaAs. A 15 nm of GaAs was wet etched by buffer HF in
Fig. 5.14c. It was observed that there were defects induced at sites 2 and 3 while
nothing was observed at site 4. Figure 5.14 (3) indicates the threshold of defect
formation on GaAs.

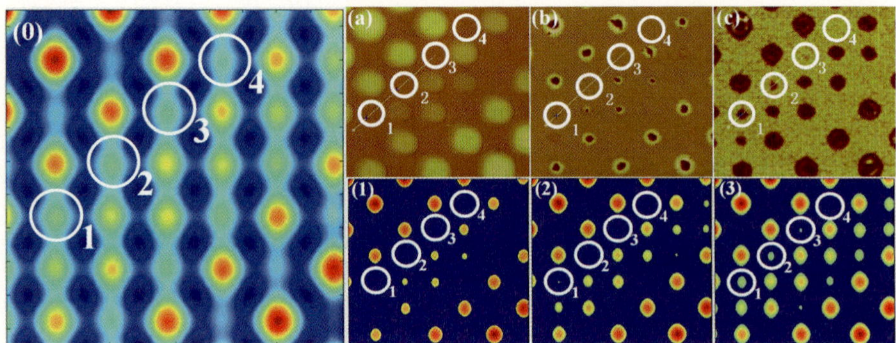

Fig. 5.14 Patterns of 4-beam LIL on SiO_2(5 nm)/GaAs and theoretical simulation. (*0*) Theoretical simulation for the threshold = 0; (**a**) modified LIL patterns before SiO_2 was selectively etched; (*1*) the simulation for the threshold of bubble formation; (**b**) SiO_2 was selectively etched by dry etching; (*2*) the simulation for the threshold of GaAs evaporation; (**c**) 15 nm of GaAs was wet etched by buffer HF; (*3*) the simulation for the threshold of defect formation

The smallest feature could not be observed by AFM with resolution lower than 5 nm. This indicates that the smallest feature of the pattern was less than 5 nm. The time taken to fabricate the LIL nanopatterns was less than 5 min including processing.

LIL offers a simple and cost-effective method to prepare large-area periodic surface patterns well below 5 nm in size without resist and elaborate processing.

5.4   Gradient Index Lens

Photonic crystals (PhCs) are able to realize subwavelength imaging in the near-field optics. However, this structure with uniform rods or holes does not work for plane wave focusing. Graded-index (GI) PhC structure is a practical approach for this purpose. The effective refractive index varies slowly in the direction perpendicular to incident light in the GI PhCs. The effective refractive index can be achieved by changing the lattice spacing, holes/rods sizes, or refractive indices of the materials of the rods or filling in the hole in the same direction [21, 22]. Most of the reported GI PhCs are fabricated by the time-consuming processing, such as electron beam and ion beam lithography, which hinders high-yield fabrication of GI PhCs. LIL is an impressive technology with a high efficiency and low cost for GI PhCs fabrication.

Figure 5.6a shows the modification of the LIL patterns by varying the incidence configuration. PhCs fabricated with those patterns show amazing ability to manipulate light, especially for subwavelength imaging. In this section, four-beam LIL is shown to be a proper technology for GI PhCs fabrication with a square lattice. Numerical simulation by the finite-difference time-domain method (FDTD) was carried out to study the lens effects.

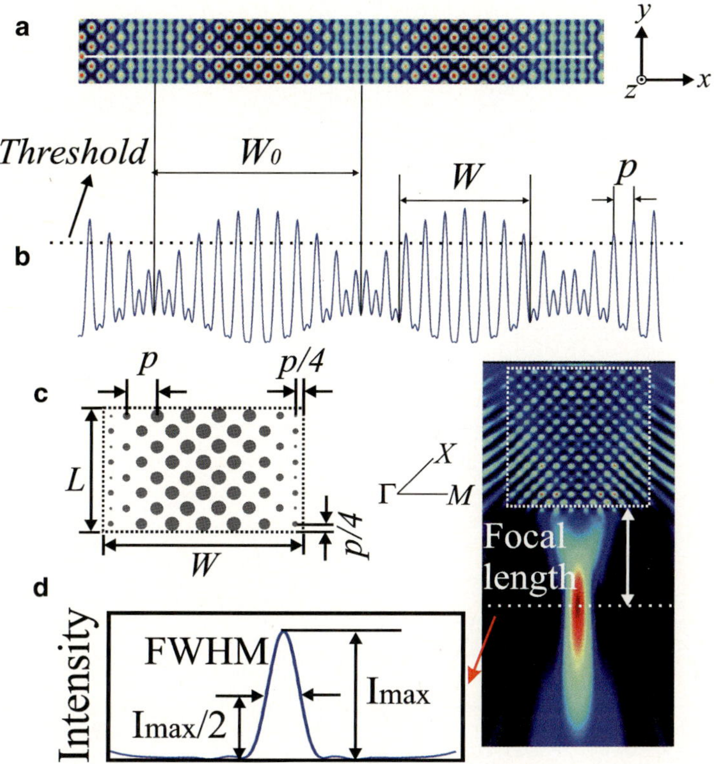

Fig. 5.15 Simulations on a modulated interference pattern: (**a**) interference pattern generated by four beams with different incident angles. (**b**) Intensity profile along the *white line* in (a). (**c**) A sample of GI PhC structure fabricated by LIL. (**d**) Focusing phenomenon of a GI PhC simulated by FDTD. Intensity profile on focal plane is given in the *left*. The *dash frame* indicates the position of GI PhC used in the simulation (Reprinted from Ref. [25], Copyright 2011, with permission from Elsevier)

Figure 5.15a shows the simulation of 4-beam laser interference with modulated interference patterns by slightly different incident angles. The modulation period, W_0, is modulated by the incident angles. No pattern appears in the area where laser interference intensity is lower than the "threshold" (T) of a recording material. The GI PhC structure is formed by those features above the T value. The GI PhC structure consists of rods with varying sizes in square array, and all the rods are made of the same materials, thus have the same refractive index. W is the width of the PhC clab, which consists of those features whose intensity is higher than T in the LIL pattern. High T yields a small W (Fig. 5.15b and c). In our simulation, W reaches its maximum W_0 when the structures are generated at a low T value of 6.1 [23]. A sample of GI PhC is shown in Fig. 5.15c, where parameters p, W, and L are defined as period on x-axis, width, and length of the PhC, respectively. The relations between the rod size (the diameter is Φ) and the position (x) can be expressed approximately as $(\Phi - \Phi_0)^2/A^2 + (x - x_0)^2/B^2 = 1$, where Φ_0, A, and B are variables determined by the incident angles of laser beams and the relative threshold T [23].

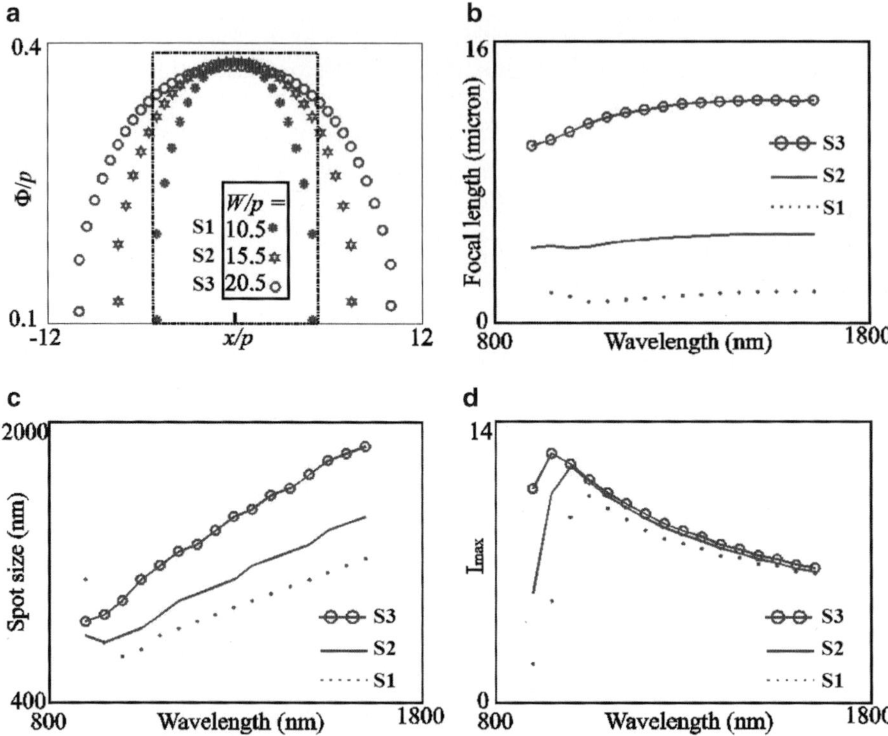

Fig. 5.16 (**a**) Diameter–position relations of the rods in three structures with different width; dependence of focal length (**b**), focusing spot size (**c**), and focusing spot intensity (**d**) of the three structures with wavelength (Reprinted from Ref. [25], Copyright 2011, with permission from Elsevier)

FDTD was used to study the focusing properties. The effective dielectric constant approximation was adopted to treat the curved surface of the rod in the squared unit cell of the FDTD grid [24]. Fully light absorbing materials were used as absorption boundaries. TM-polarized (E field along the rod) plane wave at unit intensity was taken into account in the simulation. The frequency ranges from *900 nm* to *1,650 nm*, corresponding to the mostly used semiconductor lasers. Take the refractive index of all the rods as 1.5 in the simulation. The focus properties here refer to the focal length, spot size (FWHM of the intensity curve on the focal plane), and maximum intensity (I_{max}) on focal plane. The definitions of them can be found in Fig. 5.15d. Focal plane is the plane that includes the point of highest intensity. Focal length here is defined as the distance between focal plane and the back face of GI structure.

Three GI PhC structures by modulated LIL patterns with the same length of $L = 16\,p$ but different widths were studied. The width W ($= W_0$) can be increased from *10.5 p* to *20.5 p* by decreasing the difference of the incident angles in LIL. The rod diameter $\Phi(x)$ curve is given in Fig. 5.16a. The focal lengths keep almost constant at long-wavelength range but behave differently at short wavelengths. The slopes of the three focal-length (wavelength) curves in Fig. 5.16b change from negative to positive in the short-wavelength range as the PhC-structure width increases. The size of focus spot changes in the similar way in the short-wavelength range, while the sizes of all three spots increase with wavelength in the long-wavelength range (Fig. 5.16c). The intensity of the focus spot reaches the maximum at the wavelength range of 950–1,050 nm (Fig. 5.16d).

Four-beam LIL opens an attractive way to fabricate GI PhCs-type planar lens.

5.5 Long-Range Ordered Quantum Dots

5.5.1 Semiconductor Quantum Dots

The theoretical study shows that the quantum devices built from quantum dots (QDs) with 3D confinement [26, 27] have great future applications in nanoelectronics, photonics, life sciences, and quantum computation, which are due to the special optoelectronic properties of QDs. The achievement of quantum Hall effect and fractional quantum Hall effect won the Nobel Prizes in 1985 and 1998 respectively. Semiconductor QDs and QD devices are very hot topics in the semiconductor research. Especially in the application of optoelectronics, QD has intrinsic advantages compared to other gain materials [28]:

1. *Solar cells:* Based on multi-exciton generation (MEG), besides the photons fit the band gap, QD solar cells can utilize the higher energy photons to produce multiple excitons. Therefore, the conversion efficiency can be increased to 66 % [29]. This kind of solar cells are significant in both theory and application [30]. Conventional solar cells can utilize efficiently only the photons with energies close to the band gap E_g. A large amount of the photons with energies higher than E_g are absorbed before they reach the inner field of the pn junction and turned into heat energy, while those reach the pn junction can only generate single photons and waste the extra energy of the photons. The MEG effect attracts great attention and the research develops very fast. The long-range-ordered (LRO) QDs can increase the MEG efficiency greatly due to the resonance effect between the QDs.
2. *Semiconductor lasers:* There is a large amount of reports on QD laser diodes (LDs) [31, 32]. They have a lot of merits such as very low threshold, very high thermo stability, very high efficiency, high wavelength stability, and low absorption loss. Especially for communication laser operation at 1.3 and 1.5 µm, QD LDs play a more important role than other LDs. Liu et al. reported 100 mW InAs QD LDs with threshold of 1.5 mA [33]. The threshold of QD LDs is only 1/5 ~ 1/10 of the other similar LDs. However, the nonuniformity is very harmful for the QD LDs. Miyamoto et al. [34] reported theoretically that the threshold density of the QD LDs can be as low as 14 A/cm^2. But the threshold density of the real QD LDs is at the scale of 10^2 A/cm^2 due to nonuniformity in both size and spatial distribution.

3. *Infrared detectors:* Infrared detectors have very wide application in the fields of night vision, tracking, medical diagnostics, environmental monitoring, and space exploration [35]. The GaAs/AlGaAs quantum well infrared photodetector (QWIP) develops very fast due to the development of molecular beam epitaxy (MBE). QWIP is used in infrared camera and large-area focus plane array. The most disadvantage of QWIP devices cannot detect the vertical incident photons due to the polarization selection rules. Compared to QWIP, quantum dot infrared photodetector (QDIP) has advantages: (1) QDIP can detect the vertical incident photons. It does not need the complicated grating; (2) The difference of the quantum sub-states of the QDs is around 50–70 meV. The electron lifetime of the sub-states is longer than that of the QWs due to the phonon bottleneck effect. QDIP can work at higher temperature; (3) 3D carrier confinement low down the thermo emission and dark current; (4) QDIP does not need cooling. This decreases significantly the size and cost of the infrared camera. QDIP becomes the frontier of the photon detecting and achieved a significant progress. However, the nonuniformity seriously impedes the advantage of QDIP.
4. *Single quantum emitter (SQE):* The emission of the classic light source follows a Poisson distribution or super Poisson distribution. SQE can emit single photon stably. It will have important application prospect in quantum cryptography communication [36]. SQE normally can only emit single photon from a single defect-free QD. The separation of the single QD becomes very important, and this is almost impossible for self-organized random-distributed QDs.
5. *Quantum computation (QC):* Quantum effect dominates more and more in the properties of materials and devices, while the lithography size of the component goes down to nanoscale. It is obvious that QC will replace the classic computation. In December of 2001, IBM realized the computation of $3 \times 5 = 15$ by one molecular with five fluorine atoms and two carbon atoms. However, they could not control more atoms. Large-scale QC can be achieved only in the solid-state nanostructured materials such as semiconductors or superconductors. Two-state exciton in QD can achieve QC [37]. However, QC devices cannot work in disorder and noncontrollable QDs or with high density of defect.

The above devices require defect-free (or low defect density), order, and uniform in both size and spatial distribution. Otherwise, it is useless that several usable ones with different performance are found out in thousands of devices.

Both ordered and disordered QDs have been reported widely [38]. At present, the QDs used for working devices are normally those by S–K epitaxial self-organized growth "disordered" QDs. The main advantage of these QD is that they are defect-free and can fabricate working devices which have been demonstrated that they are much better than those fabricated by other kinds of materials. However, many important parameters, such as the size and spatial distribution, are random and uncontrollable; the reproducibility is very bad; and it is impossible for commercialization. Such QDs cannot be used for quantum information devices. The fabrication process of the reported "long-range-ordered" QDs is as follows: at first, the GaAs substrate is nanopatterned by conventional nanolithography; then, QDs are grown on this nanopatterned substrate. Conventional nanolithography always induces large amount of defect which seriously reduce the performance of devices. Such devices even do not work.

In conclusion, theoretically, QD materials have unparalleled advantages compared with other materials, and they are the ideal materials for the next generation of high-performance devices. However, both the size and spatial distribution are random and uncontrollable for the QDs grown by self-organization on smooth substrates, while the nanopatterning in conventionally patterned QDs induces large amount of defect inevitably. Both of those separate characters are intrinsic and cannot be overcome by the development of technology. The huge advantages predicted by theory cannot be realized due to those characteristics. It is the bottleneck of the commercialization of QDs.

In this chapter, a new idea is discussed to solve the above problems. Uniform in both size and spatial distribution, defect-free, controllable, reproducible "perfect QDs" (P-QDs) may be realized.

5.5.2 Long-Range Ordered QDs

MBE is one of the key technologies for epitaxial growth of QDs. As discussed above, there are two conventional techniques for epitaxial growth of QDs: self-organization on smooth surface and patterned surface.

Self-organization on smooth surface: In lattice mismatch epitaxial growth, misfit strain is induced during the layer-by-layer (2D) growth. The strain will relax when the accumulation of the strain is beyond its threshold (th_1) by clustering (3D) the atom on the surface. The initial states of these 3D clusters are the "seeds" of QDs. Further growth after QDs are formed can accumulate strain again till it reaches another threshold (th_2), and then, misfit dislocation will be induced to relax the strain. Therefore, defect-free QDs can be grown if the strain accumulation is between th_1 and th_2. This is the so-called S–K growth mode. However, the formation of "seeds" on the smooth surface is the result of statistical fluctuations of thermodynamics. The time, sites, speed, and size of each seed formation are totally random. Therefore, the size and the spatial distribution of self-organized QDs on the smooth surface are totally random, uncontrollable, and nonreproducible.

Epitaxial growth on patterned surface: The strain distribution is according to the surface energy, and the distribution of surface free energy is according to the patterns after the patterning of the smooth substrate. Therefore, the distribution of the strain is according to the patterns. At last, the "seeds" of the QDs are induced according to the patterns, and thus the QD distribution is according to the patterns. Such QDs are both controllable with sites and size. However, dry and/or wet etching has to be involved in the patterning, and hence huge amount of defects is induced in the QDs.

Advantages and disadvantages: QDs grown on smooth surface can be defect-free, but both size and spatial distribution are totally random, uncontrollable, and

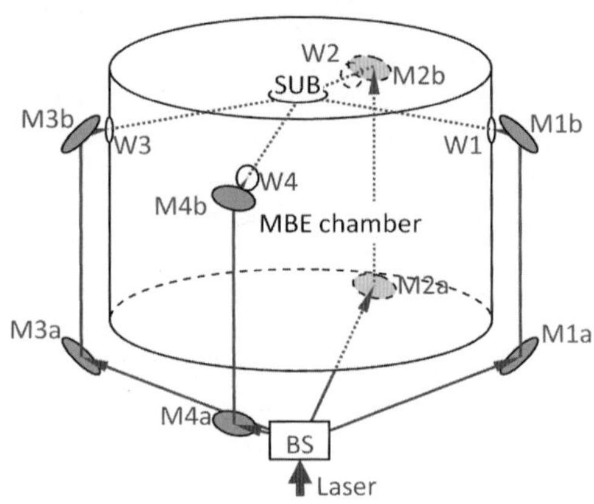

Fig. 5.17 LIL-MBE system. Incident laser is split into four beams by a beam splitter (BS). Eight mirrors (Mkx (k = 1, 2, 3, 4, x = a, b)) reflect the beams into the MBE chamber through windows (W1, W2, W3, W4) and focus on the substrate (SUB) to form LIL patterns

nonreproducible. Vice versa, the QDs grown on patterned surface can be very uniform in both size and spatial distribution but are full of defects. Those disadvantages are intrinsic and cannot be overcome by just improving the conventional techniques. Thus the huge advantage of the QDs cannot be realized. High-quality epitaxial growth is the bottleneck of the ODs mass production.

In lattice mismatch epitaxial growth, misfit strain is induced during the 2D growth. Then the strain will relax by 3D growth mode. The critical thickness from 2D to 3D growth mode is so-called the threshold thickness (TT). Kim et al. reported that the growth temperature (GT) is a key parameter to decide the TT value. TT value at lower GT is bigger than that at higher GT [39]. We invented a new system in which LIL is combined in situ with MBE [40] (Fig. 5.17). By controlling the power of the laser, the GT, and the growth thickness ($\leq th_1$), the growth in the area at low temperature in the LIL pattern will be in 2D growth mode, while in the peak temperature sites of the pattern the growth will be in 3D growth mode, and the threshold of those sites at high temperature is lower than th_2. Then, uniform, periodic-distributed, controllable, and reproducible QDs can be achieved.

These QDs can be grown defect-free because the threshold of 3D growth is controlled to lower than th_2. Therefore, P-QDs can be achieved with LIL-MBE technology due to the following scientific issues:

1. Numerous studies demonstrated very clearly that the growth temperature dominates the threshold thickness of the InAs QDs grown on GaAs. The threshold thickness at low growth temperature (e.g., <430°C) is thicker than that at high growth temperature (e.g., >480°C) [39].
2. We discovered that the temperature difference between the peak and valley of the temperature distribution during the LIL patterning can be over 1,000°C (Fig. 5.11). It indicates that we can achieve temperature difference in the range of 0–1,000 °C for LIL patterning by adjusting the laser power [12].

3. Our research results [12] show that LIL patterning with high-power laser can evaporate the GaAs at the high-temperature sites. If we adjust the laser power to a proper level, the growth mode at the temperature peak site is controlled between th1 and th2, while the growth mode at the temperature valley site is controlled below th1 (e.g., the site 4 in Fig. 5.14). Thus, defect will not be induced in the P-QDs growth by LIL.

In conclusion, there is no technology barrier for LIL-MBE system to grow P-QDs. It is expected that we can break through the bottleneck of achieving defect-free, long-range-ordered, uniform, and controllable QDs. Then, the commercialization of the QD materials and devices can be realized.

5.6   Outlook

At present, the main lithography technologies for fabricating ordered nanostructures are electron beam lithography (EBL), X-ray lithography (XRL), LIL, scanned tunneling microscopy (STM), and atomic force microscopy (AFM). High accuracy and resolution are the main advantages of EBL, STM, and AFM. However, low efficiency, high cost, and complicated processing are the critical disadvantages of these three methods and of XRL. The Institute of Information Optics Engineering at Soochow University can make up to 60 in. master with nanostructures for roll-to-roll nanoimprint lithography by using LIL, a technology with the advantage of being mask-free and offering direct writing, high efficiency, and high controllability. Further, it is non-contaminating, non-contacting, environment friendly, suitable for large-area fabrication at low cost.

Industrial end users are currently discouraged from expanding their nanotechnology-related business activities by either unacceptably high costs or the inability to control production processes on a nanometric scale. It will play a key role in realizing the full potential of interference nanolithography by combining optical technology, ICT, and micro-/nanotechnology. LIL will empower the interference nanolithography technology with a clear focus on industrial use, and to drive the rapid development of nanoscience leading to new processes and immediate industrial exploitation. The main advantages of the LIL system in the fabrication of nanostructures and devices are high resolution (better than 10 nm) compared with other optical technologies and low cost and high efficiency compared with other beam technologies.

Since the concept of semiconductor superlattice and quantum well was first introduced in 1970 [41], it has opened a new field of research in low-dimensional quantum structures, which includes two-dimensional quantum well, one-dimensional quantum wire, and zero-dimensional QDs. In QDs, carriers are confined in all the three dimensions. We can change the electron states by changing the shape and size of QDs to modify the electronic and optic properties. QD materials offer unparalleled advantages over other materials, and they are the ideal materials for the next generation of high-performance devices. By combining LIL with MBE technologies, QDs that are uniform in both size and spatial distribution, periodic distributed, defect-free, controllable, and reproducible can be achieved. The huge advantages predicted by theory can be realized, and the chief bottleneck in the commercialization of QDs can be removed with the help of LIL-MBE technology.

References

1. Vukusic P, Sambles JR (2003) Photonic structures in biology. Nature 424:852–855
2. Vukusic P, Sambles JR, Lawrence CR (2003) Structural colour: colour mixing in wing scales of a butterfly. Nature 404:457–457
3. Venkatakrishnan K, Sivakumar NR, Tan B (2003) Fabrication of planar gratings by direct ablation using an ultrashort pulse laser in a common optical path configuration. Appl Phys A 76(2):143–146
4. Campbell M, Sharp DN, Harrison MT, Denning RG, Turberfield AJ (2000) Fabrication of photonic crystals for the visible spectrum by holographic lithography. Nature 404:53–56
5. Konkola PT, Chen CG, Heilmann RK, Joo C, Montoya JC, Chang C-H, Schattenburg ML (2003) Nanometer-level repeatable metrology using the nanoruler. J Vac Sci Technol B 21 (6):3097–3101
6. Shibata S, Che Y, Sugihara O, Okamoto N, Kaino T (2004) Fabrication of high-resolution periodical structure in photoresist polymers using laser interference technique. Jpn J Appl Phys 43:2370–2371
7. Bloomstein TM, Marchant MF, Deneault S, Hardy DE, Rothschild M (2006) 22-nm immersion interference lithography. Opt Express 14(14):6434–6443
8. Chang C-H, Zhao Y, Heilmann RK, Schattenburg ML (2008) Fabrication of 50 nm-period gratings with multilevel interference lithography. Opt Lett 33(14):1572–1574
9. Rodriguez A, Ellman M, Ayerdi I, Perez N, Olaizola SM, Zhang J, Ji Z, Berthou T, Peng CS, Verevkin YK, Wang Z (2009) Interference lithography processes with high-power laser pulses. Proc SPIE 7201:72010R
10. Solak HH, David C, Gobrecht J, Golovkina V, Cerrina F, Kim SO, Nealey PF (2003) Sub-50 nm period patterns with EUV interference lithography. Microelectron Eng 67–68:56–62
11. Tan C, Peng CS, Petryakov VN, Verevkin YK, Zhang J, Wang Z, Olaizola SM, Berthou T, Tisserand S, Pessa M (2008) Line defects in two dimensional four-beam interference patterns. New J Phys 10(2):023023-1–023023-8
12. Tan C, Peng CS, Pakarinen J, Petryakov VN, Verevkin YK, Zhang J, Wang Z, Olaizola SM, Berthou T, Tisserand S, Pessa M (2009) Ordered nanostructures written directly by laser interference. Nanotechnology 20(12):125303-1–125303-5
13. Cho KS, Mandal P, Kim K, Baek IH, Lee S, Lim H, Cho DJ, Kim S, Lee J, Rotermund F (2011) Improved efficiency in GaAs solar cells by 1D and 2D nanopatterns fabricated by laser interference lithography. Opt Commun 284(10–11):2608–2612
14. Lei M, Yao B, Rupp RA (2006) Structuring by multi-beam interference using symmetric pyramids. Opt Soc Am 14(12):5803–5811
15. Xu D, Chen KP, Ohlinger K, Lin Y (2011) Nanoimprinting lithography of a two-layer phase mask for three-dimensional photonic structure holographic fabrications via single exposure. Nanotechnology 22(3):035303-1–035303-8
16. Hendrik J, Wielea K, Simon P (2003) Fabrication of periodic nanostructures by phase-controlled multiple-beam interference. Appl Phys Lett 83(23):4707–4709

17. Rodriguez A, Echeverría M, Ellman M, Perez N, Verevkin YK, Peng CS, Berthou T, Wang Z, Ayerdi I, Savall J, Olaizola SM (2009) Laser interference lithography for nanoscale structuring of materials: from laboratory to industry. Microelectron Eng 86(4–6):937–940
18. Peng CS, Dong X, Zhang W, Gu X.-Y, Zhou Y, Liu W.-P (2011) A laser interference lithography system. Chinese Patent application, No. 201110178877.6 June 2011
19. Alonso-González P, González L, González Y, Fuster D, Fernández-Martínez I, Martín-Sánchez J, Abelmann L (2007) New process for high optical quality InAs quantum dots grown on patterned GaAs(001) substrates. Nanotechnology 18(35):355302-1–5
20. Peng CS, Tan C, Zhang W, Gu X-Y, Liu W-P (2012) Nano fabrication by laser interference. Int J Nanomanufact 8(3):212–215
21. Wu Q, Gibbons JM, Park W (2008) Graded negative index lens by photonic crystals. Opt Express 16(21):16941–16949
22. Kurt H, Colak E, Cakmak O, Caglayan H, Ozbay E (2008) The focusing effect of graded index photonic crystals. Appl Phys Lett 93(17):171108-1–171108-3
23. Tan C, Peng CS, Petryakov VN, Verevkin YK, Zhang J, Wang Z, Olaizola SM, Thierry B, Stéphane T (2009) Fabricate planar photonic crystal gradient index lens by laser interference lithography. In: Proceeding of the IEEE Nanotechnology, IEEE Nano 2009, pp 450–453. Genoa, 26–30 Jul 2009
24. Kaneda N, Houshmand B, Itoh T (1997) FDTD analysis of dielectric resonators with curved surfaces. IEEE Trans Microwave Theory Tech 45(9):1645
25. Tan C, Niemi T, Peng CS, Pessa M (2011) Focusing effect of a graded index photonic crystal lens. Opt Commun 284(12):3140–3143
26. Bimberg D, Grundmann M, Ledentsov NN (1999) Quantum dot heterostructures. Wiley, Chichester
27. Masumoto Y, Takagahara T (2002) Semiconductor quantum dots: physics, spectroscopy and applications. Springer, Berlin/Heidelberg
28. Bimberg D, Kuntz M, Laemmlin M (2005) Quantum dot photonic devices for lightwave communication. Appl Phys A 80(6):1179–1182
29. Rabami E, Baer R (2010) Theory of multiexciton generation in semiconductor nanocrystals. Chem Phys Lett 496(4–6):227–235
30. Schaller RD, Sykora M, Pietryga JM, Klimov VI (2006) Seven excitons at a cost of one: redefining the limits for conversion efficiency of photons into charge carriers. Nano Lett 6 (3):424–429
31. Ghosh S, Pradhan S, Bhattacharya P (2002) Dynamic characteristics of high-speed $In_{0.4}Ga_{0.6}As$/GaAs self-organized quantum dot lasers at room temperature. Appl Phys Lett 81(16):3055–3057
32. Hopfer F, Kaiander I, Lochmann A, Mutig A, Bognar S, Kuntz M, Pohl UW, Haisler VA, Bimberg D (2006) Vertical-cavity surface-emitting quantum-dot laser with low threshold current grown by metal-organic vapor phase epitaxy. Appl Phys Lett 89(6):061105
33. Liu HY, Badcock TJ, Groom KM, Hopkinson M, Gutierrez M, Childs DT, Jin C, Hogg RA, Sellers IR, Mowbray DJ (2006) High performance 1.3-mum InAs/GaAs quantum dot lasers with low threshold current and negative characteristic temperature. SPIE Proc 6184 (43):618417–618418
34. Miyamoto Y, Miyake Y, Asada M, Suematsu Y (1989) Threshold current density of GaInAsP/InP quantum-box lasers. IEEE J Quantum Electron QE-25(9):2001–2006
35. Rogalski A (2002) Infrared detectors: an overview. Infrared Phys Technol 43:187–210
36. Michler P, Imamolu A, Mason MD, Carson PJ, Strouse GF, Buratto SK (2000) Quantum correlation among photons from a single quantum dot at room temperature. Nature 406:968–970
37. Zrenner A, Beham E, Stufler S, Findeis F, Bichler M, Abstreiter G (2002) Coherent properties of a two-level system based on a quantum-dot photodiode. Nature 418:612–614
38. Stangl J, Holy V, Bauer G (2004) Structural properties of self-organized semiconductor nanostructures. Rev Mod Phys 76(3):725–783

39. Kim MD, Kim TW, Woo YD, Kim SG, Hong JS (2005) Formation process of and lattice parameter variation in InAs/GaAs quantum dots dependent on the growth parameters. J Crystal Growth 278(1–4):125–130
40. Peng CS (2011/2012) A fabrication device and method for quantum dots. Chinese patent application & PCT patent application, No. 201110224270.7 Aug 2011 & PCT/CN2012/078013 Jul 2012
41. Esaki L, Tsu R (1970) Superlattice and negative differential conductivity in semiconductors. IBM J Res Dev 14(1):61–65

Chapter 6
Super-Resolution Patterning and Photolithography Based on Surface Plasmon Polaritons

6.1 Introduction

With the developments of the nanoscale science and technology, the demand for fabrication of nanoscale patterns has been significantly increased. Photolithography has still been the most widely used microfabrication technique because of its ease of repetition, effective cost, and suitability for large-area fabrication. The diffraction limit, however, restricts single-exposure resolution to features with periods no smaller than half wavelength of the illuminating sources. To achieve nanometer feature sizes within the regime of diffraction physics, one straightforward method is to reduce the working wavelength by employing light sources of higher photon energy, such as extreme ultraviolet light (EUV), soft X-rays, or atomic wave packet [1–4]. The main drawbacks, however, are the drastic increase of complexity and cost for instrumentation and processing, including the developments of new sources, photoresist, and the optical components. Several alternative techniques, such as electron beam lithography [5–9], focused ion beam lithography [10–14], dip-pen lithography [15–18], and nanoimprint lithography [19–22], can also achieve nanoscale feature sizes; however, these methods require the introduction of a new infrastructure of tools, materials, and processing technologies, which costs huge expenses.

Recently, a new photolithographic scheme, which is based on the unique properties of surface evanescent waves [23–26] or surface plasmon polaritons (SPPs) [27–37] induced at the interface between a metal and a dielectric material, to achieve sub-diffraction-limit nanopatterns was proposed and demonstrated. The wave vector of SPPs can be significantly larger than that of the free-space illuminating light at the same frequency, which results in extraordinary "optical frequency but X-ray wavelength" property. As a result, the resolution of this surface plasmon nanolithography can go far beyond the free-space diffraction limit of the light. Different versions of the implementation of the SPPs-based super-resolution nanolithography have been proposed and conducted both theoretically and experimentally in recent years, which all demonstrate that SPPs-based nanolithography has the potential to satisfy the requirements of high resolution, low cost, and high throughput simultaneously.

Q. Liu et al., *Novel Optical Technologies for Nanofabrication*, Nanostructure Science and Technology, DOI 10.1007/978-3-642-40387-3_6,
© Springer-Verlag Berlin Heidelberg 2014

In this chapter, we will review the development and progress of the SPPs-based nanolithography technique, in which the SPPs nanolithography is categorized into four major types in terms of different mechanisms and implementation techniques, namely, prism-coupled SPPs nanolithography, grating-coupled SPPs nanolithography, superlens imaging nanolithography, and SPPs direct writing nanolithography, respectively. In the following sections, we will review these SPPs-based nanolithographic techniques separately from both theoretical and experimental aspects.

6.2   Introduction to Surface Plasmon Polaritons (SPPs)

Surface plasmon polaritons (SPPs) are the electron plasma oscillations existing on the interface of good metal and dielectric. Under normal circumstances, however, a surface plasmon polariton cannot be excited directly with light on a smooth metal interface as its wave vector is greater than the photon wave vector. The respective dispersion relation is shown in Fig. 6.1 together with the photon dispersion in the free-space dielectric medium. Since the SPPs dispersion curve lies to the right of the light line of the dielectric, SPPs cannot be excited with conventional illumination unless special techniques for phase-matching are employed.

Phase-matching condition of SPPs can be achieved in the total internal reflection geometry or diffraction effects [38]. The most common configuration of the total internal reflection geometry is prism coupling; there are two major implements, including Kretschmann and Otto configurations. In the Kretschmann configuration [39], the metal film is illuminated through the dielectric prism at a resonant angle θ, as shown in Fig. 6.2a, where the in-plane component of the photon wave vector in the prism coincides with the SPPs wave vector on an air–metal surface, resonant light tunneling through the metal film occurs, and light is coupled to the SPPs:

$$k_{SPPs} = \frac{\omega}{c}\sqrt{\varepsilon_{prism}}\sin\theta \qquad (6.1)$$

And in the Otto configuration [40], the prism is separated from the metal film by a thin air gap; total internal reflection takes place at the prism/air interface, exciting SPPs via tunneling to the air/metal interface, as shown in Fig. 6.2b.

Another way to provide wave vector conservation for the SPPs excitation is to use grating coupling [41]; components of the diffracted light whose wave vector coincides with the SPPs wave vector will be coupled to SPPs (Fig. 6.2c), and one can excite the SPPs by coupling the illumination light with the grating momentum:

$$k_{SPPs} = k\sin\theta \pm m_x G_x \pm m_y G_y \qquad (6.2)$$

where k_{SPPs} is the surface plasmon wave vector, $k\sin\theta$ is the in-plane component of the incident wave vector, G_x and G_y are the reciprocal lattice vectors in the

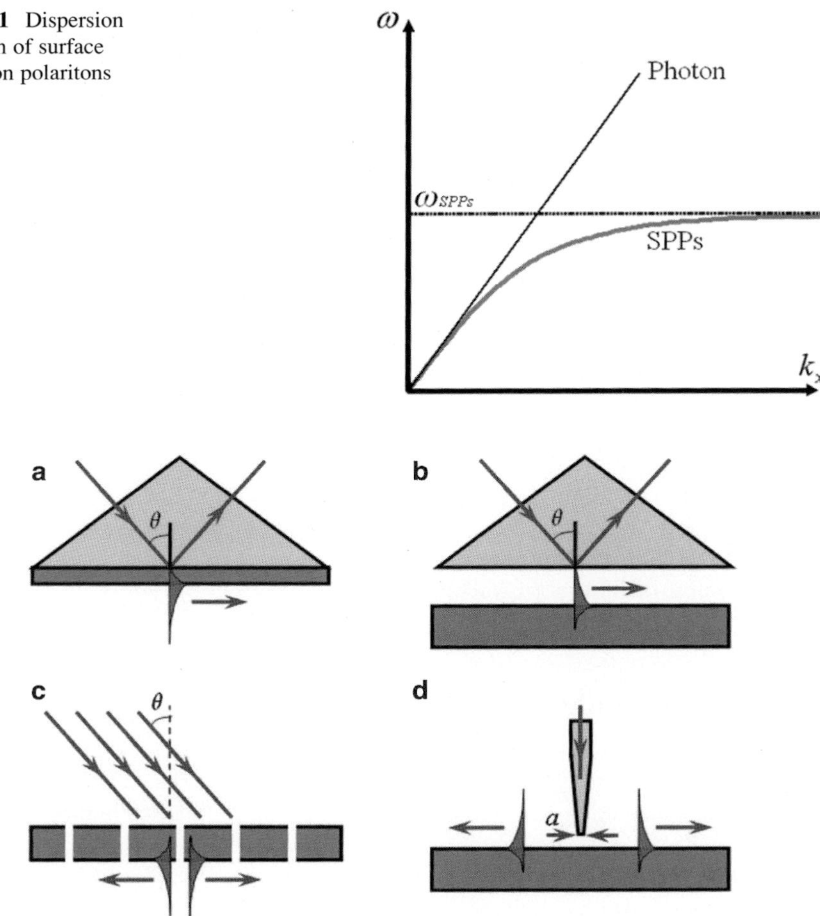

Fig. 6.1 Dispersion relation of surface plasmon polaritons

Fig. 6.2 SPPs excitation configurations: (**a**) Kretschmann geometry; (**b**) Otto geometry; (**c**) diffraction structure; (**d**) excitation with an SNOM probe

x and y directions, and m_x and m_y are integers. With the help of reciprocal lattice vectors, incident light with long wavelength can excite SPPs. For normal incidence, $k_{SPPs} = \pm\, m_x G_x \pm m_y G_y$, which means that $k_{SPPs} = \pm\, m_x G_x$ and $k_{SPPs} = \pm\, m_y G_y$ may be excited simultaneously. That is to say, the SPPs waves on the mask have opposite propagating directions along the metallic films. The interference of the SPPs waves results in standing waves along the metallic film.

Excitation schemes such as prism or grating coupling excite SPPs over a macroscopic area defined by the dimensions of the spot of the coupling beam of wavelength λ_0. In contrast, near-field optical microscopy techniques allow for the *local* excitation of SPPs over an area $a << \lambda_0$ and can thus act as a point source for SPPs [42, 43]. Figure 6.2d sketches the typical geometry of a scanning near-field optical microscopy (SNOM) applications in SPPs studies: a small probe tip of aperture size $a < = \lambda_{SPP} < \lambda_0$ illuminates the surface of a metal film in the near field. Due to the small aperture size, the light ensuing from the tip will consist of wave vector components $k > = k_{SPPs} > k_0$, thus allowing phase-matched excitation of SPPs. Due to the ease of lateral positioning of such probes in scanning near-field optical microscopes, SPPs at different locations of the metal surface can be excited. This configuration can be treated as either a diffraction or tunneling mechanism of the SPPs excitation.

There are many significant advantages of SPPs, such as short wavelength, transmission enhancement, and super-resolution imaging. When the incident light transmits through a single subwavelength aperture or subwavelength hole arrays consisting of metal and dielectric, components of the diffracted light will be greatly enhanced with the excitation of SPPs [44–46]; thus, for SPPs-based nanolithography, resolution and contrast can be improved obviously due to the extraordinary enhancement of transmittance. With the Pendry's concept of superlens [47], it is able to realize super-resolution imaging; both propagating and evanescent waves contribute to the resolution of the image [48, 49]. Actually, a conventional lens is only capable of transmitting the propagating components; the evanescent waves that carry subwavelength information about the object decay exponentially and cannot be collected in far field to reconstruct the image.

6.3   Prism-Coupled SPPs Nanolithography

Based on the essential physics behind the excitation of the SPPs with prism coupling, different implementation arrangements of the SPPs interference were proposed in the past few years. In 2006, Guo et al. proposed a large-area surface plasmon polariton interference lithography (LSPPIL) [50]. Figure 6.3 shows the physical arrangement. The uppermost layer is an isosceles triangle or semispherical prism with a high refractive index, and a thin metal film is coated on the bottom surface of the prism, and then brought into intimate contact with a thin photoresist coated on a substrate. When two mutually coherent plane waves with p-polarizations are incident on the base of the prism in the vicinity of the resonance angle, there arise multiple counterparts of the SPPs everywhere on the interface. As a result, SPPs interference fringes are formed in the photoresist.

Figure 6.4 shows the electric field distributions of the interference pattern with the incidence angle of $59.9°$ at a wavelength of 441 nm. The metal material is silver with the thickness of 30 nm. Figure 6.4a, b represents the E_x and E_z components of the SPPs interference, respectively. The period of the interference pattern is approximately 120 nm, i.e., the critical dimension of 60 nm, smaller than $\lambda/7$, and it has the potential to realize a 32 nm node periodic nanostructure if a 193 nm source is employed [51].

In 2010, Sreekanth et al. experimentally demonstrated a maskless single-step two- and four-beam SPPs interference lithographic configuration [52, 53]. A schematic of the experiment is shown in Fig. 6.5. A 20 mm × 20 mm test sample is

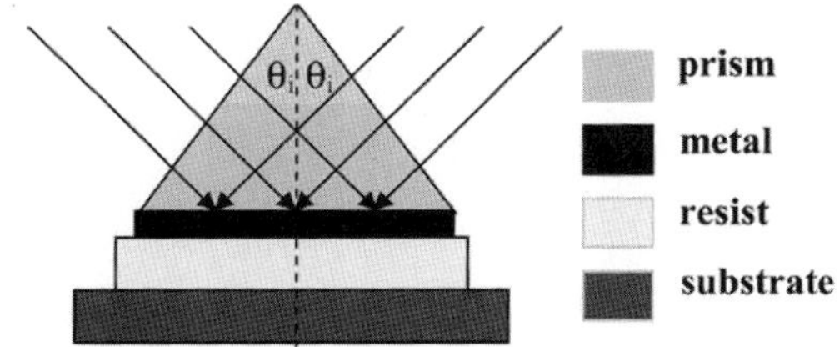

Fig. 6.3 Schematic of the LSPPIL process (Reprinted with permission from Ref. [50], © 2006 Optical Society of America)

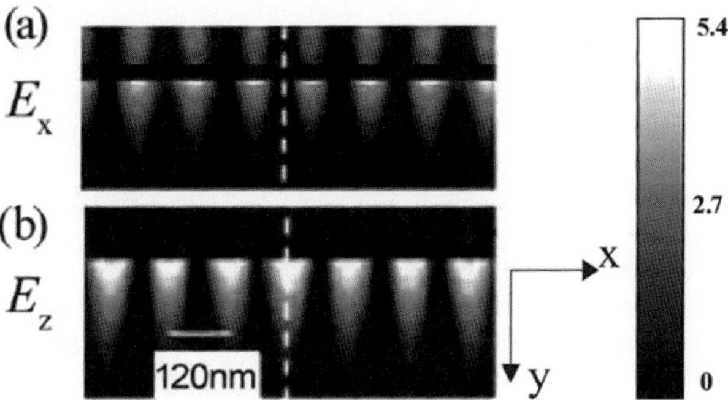

Fig. 6.4 Electric field distribution of the interference patterns with TM incident wave (Reprinted with permission from Ref. [50], © 2006 Optical Society of America)

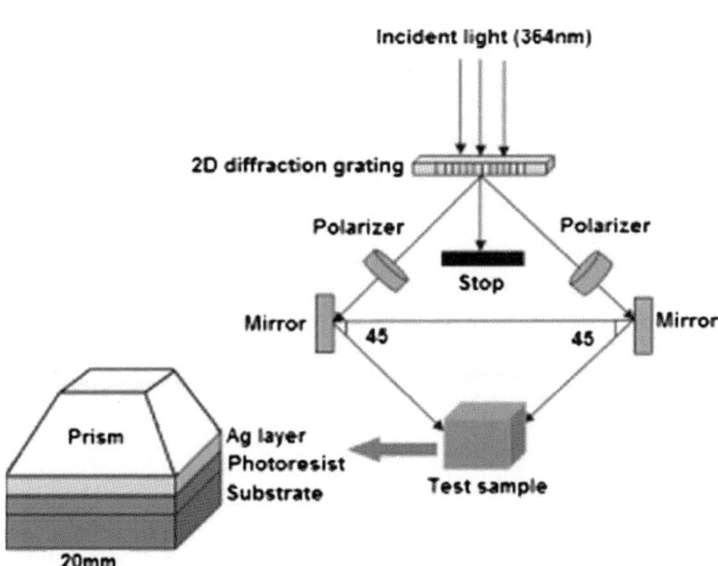

Fig. 6.5 Schematic of the experimental setup (Reprinted with permission from Ref. [52], © 2010 Optical Society of America)

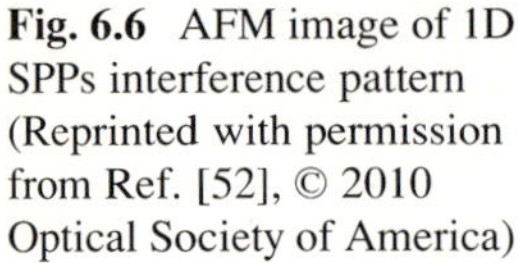

Fig. 6.6 AFM image of 1D SPPs interference pattern (Reprinted with permission from Ref. [52], © 2010 Optical Society of America)

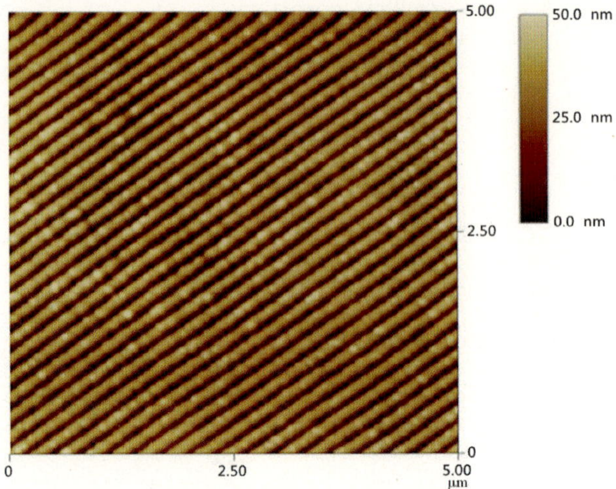

designed for interference lithography at 364 nm wavelength in which the upper layer is a 60° triangular prism of refractive index 1.745 (NLAK 8 glass). A 50 nm thick silver film is coated on the bottom surface of the prism, the silica substrate coated with a 100 nm thick photoresist (AZ7220) layer, and the prism is brought into intimate contact (using index-matching liquid). In order to obtain thin photoresist thickness, AZ7200 is diluted with EBR solvent (AZ Electronic Material) at 1:4 proportions (1 ml AZ7200 and 4 ml EBR solvent) and spin coated at 1500 rpm.

In detail, the illumination source is vertically polarized light from a continuous-wave UV argon ion laser. An electronic shutter is placed in front of the laser head to provide control over the exposure time. The incident light is allowed to be incident normal to the axis on a custom-made 2D diffraction grating to obtain various diffraction orders. The 2D diffraction grating is designed in a way that it can suppress the zeroth-order beam and generate four high-diffraction-efficiency (17 %) first-order beams. When 45° linear polarized light is incident on the grating surface, the diffracted light is unpolarized. The beam is then allowed to pass through the polarizer to get p-polarized light. The central beam is blocked with a stop, and four counter-propagating first-order beams are made to be incident on the test sample.

The p-polarized incident beams at the prism surface has a targeted power density of 334 mW/cm^2, and exposure time of the samples is 1 s. The development time is taken as 2 min. An AFM image of the lithographic pattern of the SPPs interference of the two beams is shown in Fig. 6.6. Large-area uniform periodic grating lines were obtained at this exposure dose and exposure time. A four-beam SPPs interference lithography experiment is also carried out to pattern 2D features, as shown in Fig. 6.7. It is noted that large-area grating lines and dot arrays with feature size 90 nm can be achieved with this approach.

To improve the performance and increase the flexibility of the prism-based SPPs interference, many other implementation methods were demonstrated by different

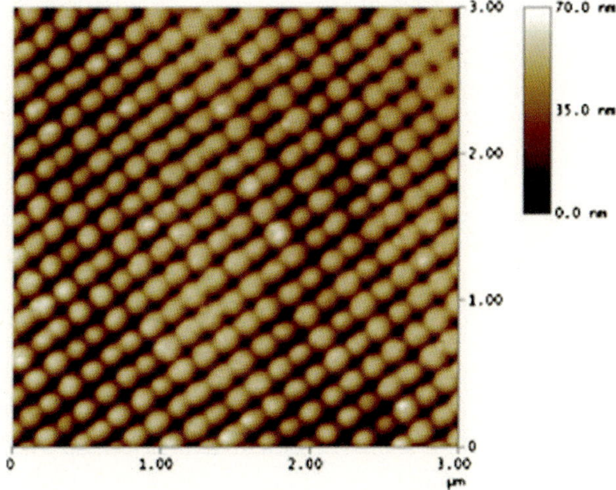

Fig. 6.7 AFM images of 2D SPPs interference pattern (Reprinted with permission from Ref. [52], © 2010 Optical Society of America)

groups. Lim et al. proposed a thin-film patterning method based on Kretschmann configuration where the surface plasmon mode and plasmon waveguide modes coupled into a target photosensitive layer are generated [54]. The key point of this method is that there are three resonant angles, 33.5°, 56.3°, and 90°, respectively. Different SPPs interference patterns can be generated at different resonant angles, and the field intensity and visibility of the interference is considerably high. The periods of the interference patterns at different resonant angles are different, which are, respectively, 294.5, 233.8, and 195.4 nm for the angles of 33.5°, 56.3°, and 90° with a 633 nm light source. It is apparent that when multiple beam paths coupled into the PMMA layer are generated, various interference patterns can be made. In addition, Guo et al. employed a metal-bounded dielectric structure to excite symmetric mode or antisymmetric mode of SPPs at different resonant angles changing with the thickness of dielectric [55], i.e., it has the potential to develop a new tunable SPPs lithography technique.

In all those above introduced implementation schemes, the photoresist layer and the metallic thin film are arranged in intimate contact; this may result in the damage or pollution of surfaces. It is worth to mention that He et al. introduced a backside-exposure technique to fabricate nanostructures [56] that can avoid the drawbacks of the contact exposure scheme.

The structure of the backside-exposure SPPs interference lithography (SPPIL) is shown in Fig. 6.8, which composes a prism, matching fluid layer and glass substrate, silver film, resist layer, and air layer. According to Maxwell equations and boundary conditions, the wave vector k_{SPPs} can be expressed as [57]:

$$k_{SPPs} = k_x^{halfspace} + k_x^{photoresist} + k_x^{metal} \tag{6.3}$$

where $k_x^{halfspace}$ describes the dispersion relationship of SPPs at the boundary of the

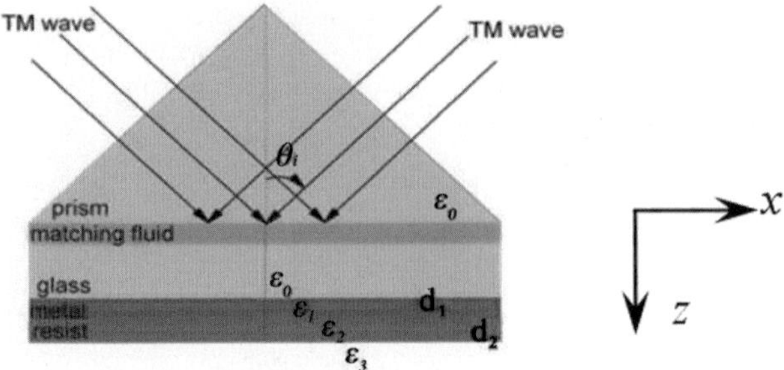

Fig. 6.8 Schematic of the backside-exposure SPPIL structure (Reprinted with permission from Ref. [56], © 2010 Optical Society of America)

silver halfspace and air halfspace, the second term $k_x^{photoresist}$ gives the influence of the resist to SPPs, and it is considered in the third term k_x^{metal} that the thickness of the silver film is finite.

Being different from the former SPPs interference lithography [50] which requires a high refractive prism to be used, a low refractive prism could also be used to excite SPPs in the backside-exposure SPPIL device; however, due to the limited transmission distance of SPPs, it can pass through a resist layer only when it is thinner than 100 nm. The resonant condition of surface plasmon and resolution of interference patterns are varied with different thickness of the resist layer. When the thickness of the resist layer (d_2) varies from 30 to 100 nm, the resolution decreases from 88 to 64 nm and corresponds to the prism with n = 1.89, and when d_2 reaches a certain thickness, the resolution does not change anymore, which is equivalent to the situation when the thickness of the resist layer is infinite. When d_2 varies from 30 to 70 nm, the resolution decreases from 89 to 69 nm and corresponds to the prism with n = 1.527, and when d_2 > 70 nm the enhancement of SPPs disappears.

Relevant experiment was developed. The thicknesses of silver film and resist are 40 nm and 50 nm, respectively; a laser of wavelength 441.6 nm is employed as a light source. Figure 6.9 is the SEM images of the patterns; the period of the patterns is about 150 nm; the only difference between (a) and (b) is that the material of the prime and substrate was dense flint (n = 1.89) and k9 glass (n = 1.527), respectively. It is obviously that the backside-exposure method can obtain fringes with high quality easily. In theory, prism with high refractive index can be used to achieve patterns with high resolution. But in experiment, matching fluid with high refractive index is hard to find. Figure 6.9a shows that the scattering of light in the matching fluid layer leads to a low contrast of the lithography patterns, as the matching fluid with low refractive index does not match the high refractive prism and high refractive substrate well. When a low refractive index prism and substrate is used, prism can match well with the matching fluid, so the better quality lithography patterns can be obtained, as shown in Fig. 6.9b. It indicates that using prism and substrate with low refractive index for backside-exposure SPPIL is more practical and feasible.

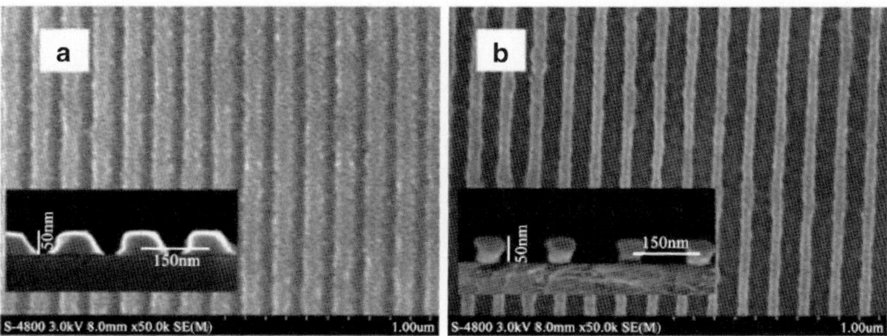

Fig. 6.9 The SEM images of the pattern using back exposure SPPIL (**a**) prism with high index; (**b**) prism with low index (Reprinted with permission from Ref. [56], © 2010 Optical Society of America)

In summary, prism coupling has several advantages, including maskless, high transmission, soft contact, and large-area fabrication capability. It is expected that prism-coupled SPPs nanolithography can provide a convenient route for patterning high-throughput nanometer structures without employing expensive equipment or complex mask. The superiority of the prism-coupled SPPs nanolithography may be in the fabrication of 1D or 2D subwavelength gratings.

6.4   Grating-Coupled SPPs Nanolithography

6.4.1   Single Metallic Grating Lithography Scheme

Grating-coupled SPPs nanolithography is a method to use different metallic grating masks associated with appropriate structures to excite SPPs and forms corresponding nanoscale features. Many efforts have been devoted to the exploitation of this technique in recent years.

In 2004, Luo et al. experimentally demonstrated a method for obtaining subwavelength resolution based on the interference of SPPs in the near field with a metallic grating mask [58, 59]. A schematic is shown in Fig. 6.10, a silver mask, which was fabricated on 2 mm thick quartz by electron beam lithography, and lift-off process is illuminated from the top with a collimated light.

The features on the silver mask have a periodicity of 300 nm, and the opening is 1/5 of the periodicity. A photoresist (TSMR series) was spun onto silicon substrates with a subwavelength thickness. Then, a vacuum chuck is used to pull the mask onto the resist-coated substrate. The incident light tunnels through the mask via SPPs and reradiates into the photoresist. Exposure is carried out using g-line; a 12 s exposure followed by a 60 s development in the developer (NMD-W 2.38 %) was used. The experiment result is presented in Fig. 6.11; 50 nm wide lines were

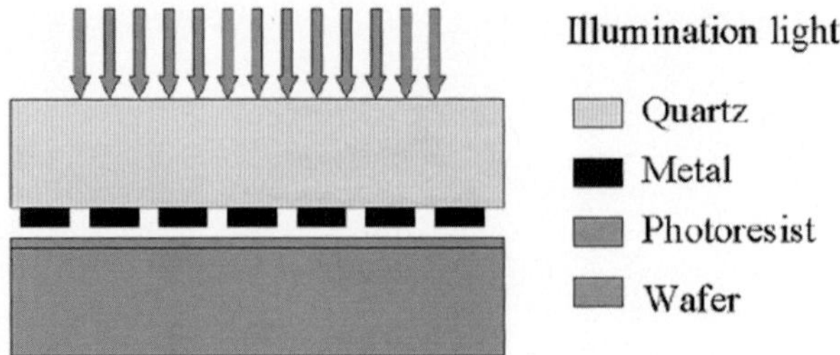

Fig. 6.10 Schematic of a single metallic grating lithography (Reprinted with permission from Ref. [58], ©2004 American Institute of Physics)

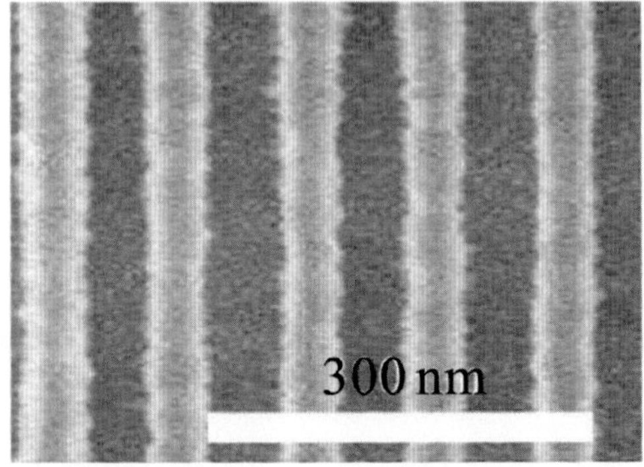

Fig. 6.11 SEM picture of the pattern using a single metallic grating lithography (Reprinted with permission from Ref. [58], ©2004 American Institute of Physics)

obtained. The edge roughness of the pattern shown in Fig. 6.11 is attributed to edge roughness on the mask. With a better electron beam lithography tool, the pattern fidelity would be improved.

In 2007, Doskolovich et al. designed a diffraction grating with a metal film coated onto the substrate to produce a high-contrast interference pattern [60, 61]. Figure 6.12 shows the schematic of the 1D binary grating with a metal film.

This work suggests that the surface plasmon interference pattern can be obtained using a diffraction grating made of the dielectric material with a metal film applied in the substrate region. At normal incidence of light onto the grating, the orders $\pm m$ directly contribute to the interference pattern. The other orders after reflection from the metal film collide with the grating again. Upon secondary reflections from the grating, they contribute to the useful orders $+m$ and $-m$, thus enhancing the interference pattern. So, one can suppose a quasiwaveguide propagation of the other orders in the grating substrate, accompanied by energy transfer to the $+m$th and $-m$th diffraction orders and to the surface plasmon interference pattern, respectively.

Furthermore, the same group discussed the generation of 2D surface plasmon interference patterns using a 3D metal–dielectric diffraction structure [62]. The considered structure consists of a 3D binary diffraction grating (DG) and a metal film applied in the substrate region (Fig. 6.13).

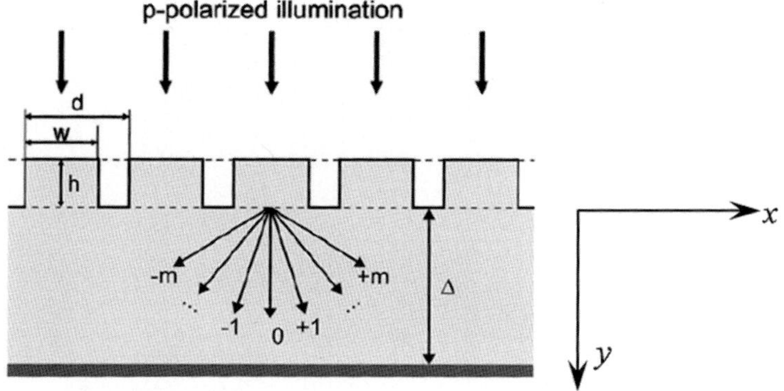

Fig. 6.12 Binary grating with a metal film on the substrate (Reprinted with permission from Ref. [60], © 2007 IOP Publishing Ltd)

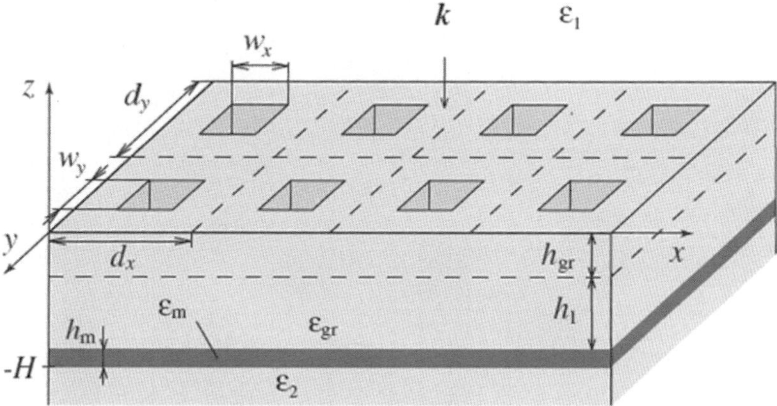

Fig. 6.13 Geometry of a 2D SPPs interference structure (Reprinted with permission from Ref. [62], © 2010 Elsevier B.V.)

Figure 6.14 shows a calculated interference pattern generated underneath the metal film in the case of TM wave incidence with a wavelength of 550 nm; ψ is the polarization angle between the direction of the electric field vector E and the x-axis. In this case, SPPs are excited by the orders $(\pm 3, 0)$. The pattern period is 154 nm. The values of the contrast and the field enhancement factor are 0.73 and 10.7, respectively. In Figs. 6.15 and 6.16, the calculated interference patterns for the cases of the linear and circular polarization of the incident wave are given. The periods of the patterns are 218 and 154 nm, respectively, and the values of the contrast and the field enhancement factor are (0.99, 19.5) and (0.73, 10.8), respectively. In general, the angle ψ can be tuned from 0° to 90°, and the phase

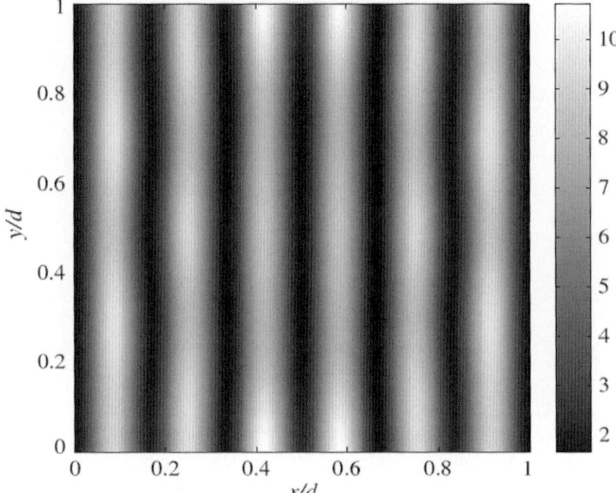

Fig. 6.14 Electric field distribution underneath the metal film with TM wave incidence ($\psi = 0°$) (Reprinted with permission from Ref. [62], © 2010 Elsevier B.V.)

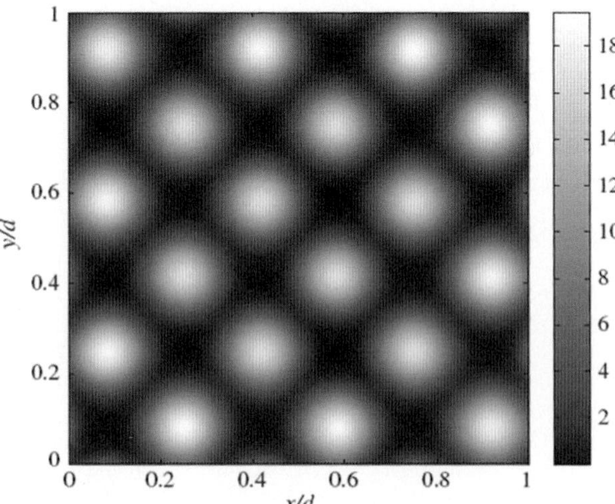

Fig. 6.15 Electric field distribution in the case of linear polarization wave ($\psi = 45°$, $\delta = 0°$) (Reprinted with permission from Ref. [62], © 2010 Elsevier B.V.)

difference δ between TE and TM components can be also different from $0°$ and $90°$. The arbitrary choice of the angles ψ and δ gives further possibilities in changing the interference pattern configuration. It is shown that high-contrast interference patterns with the period 2.5–3.5 times smaller than the incident wave length can be produced. At the interference maxima, electric field intensity exceeds incident wave intensity by an order of magnitude. The interference pattern and period can be controlled by ψ, δ, and the length of the incident wave.

Fig. 6.16 Electric field distribution in the case of circular polarization wave incidence (Reprinted with permission from Ref. [62], © 2010 Elsevier B.V.)

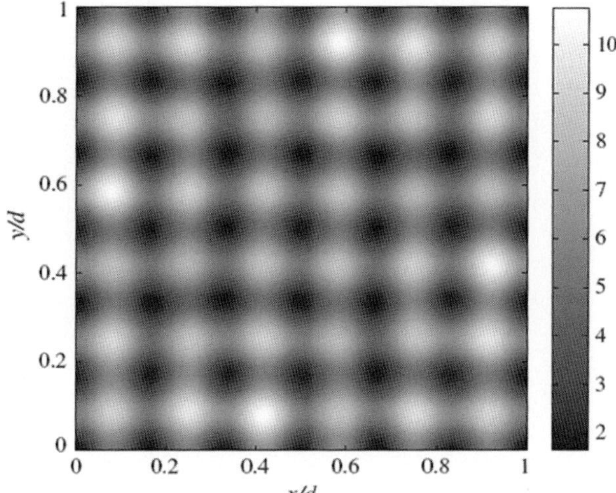

Fig. 6.17 Schematic of the proposed SAMS structure (Reprinted with permission from Ref. [63], © 2008 Optical Society of America)

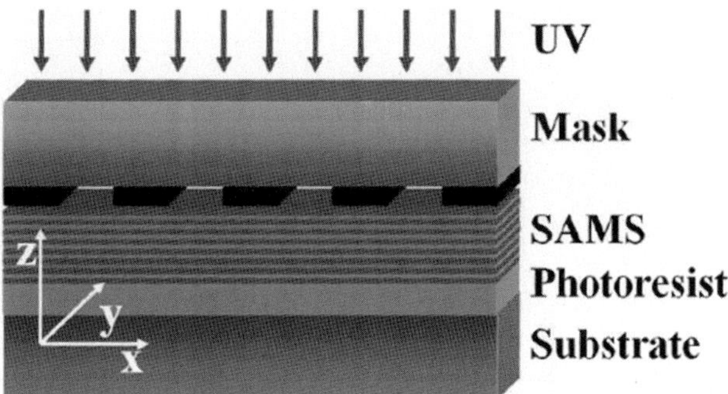

6.4.2   Multilayer Metal/Dielectric Structure

Following the experimental demonstration in Refs. [58] and [59], many works on the SPPs photolithography based on metallic grating coupling were proposed. In 2008, Luo's group theoretically presented an interference lithography technique beyond the diffraction limit achieved by positing an anisotropic metamaterial under the conventional lithographic mask [63]. The proposed structure consists of a Cr grating mask affiliated with a sandwich anisotropic metamaterial structure (SAMS), a layer of photoresist, and a silica substrate, as shown in Fig. 6.17.

The SAMS is constructed with the periodical unit cell which is composed of 30 pairs of silver and fused silica with thicknesses of 20 and 30 nm, respectively. A normal plane wave of 442 nm in p-polarization is incident on the top side of the mask. Figure 6.18 gives the distributions of total electrical intensity; two Cr gratings with periodicities 160 and 320 nm (duty-cycle 0.5) are posited on the

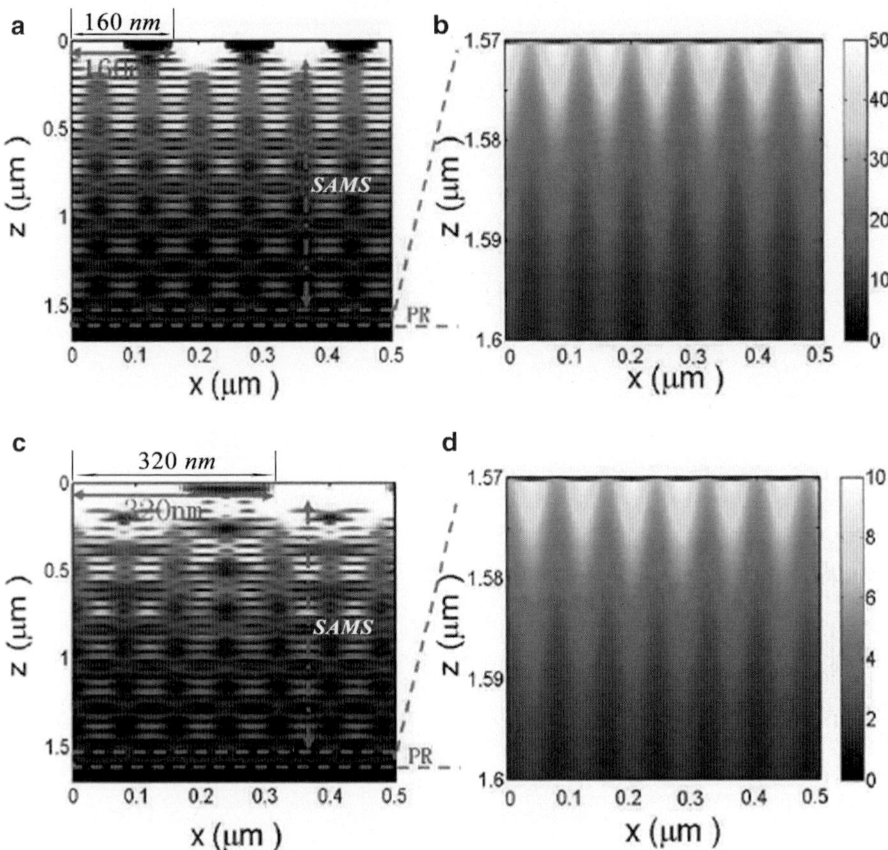

Fig. 6.18 Calculated distributions and cross sections of total electrical field for two configurations: (**a, b**) for the Cr mask with 160 nm periodicity; (**c, d**) for the Cr mask with 320 nm periodicity (Reprinted with permission from Ref. [63], © 2008 Optical Society of America)

same SAMS. From Fig. 6.18a and c, it is clearly shown that the evanescent components scattered from the grating are enhanced and transmitted through the SAMS. Both of the two structures deliver an interference pattern with 80 nm periodicity in the vicinity of the interface between the SAMS and photoresist (Fig. 6.18b and d). If half peak is taken as the feature size, features about 40 nm smaller than 1/11 of exposure wavelength can be achieved.

In 2009, the same group further proposed a dielectric–metal multilayer (DMM) structure to achieve the feature sizes theoretically down to sub-22 nm even to 16.5 nm with diffraction-limited masks at wavelength of 193 nm [64]. As shown in Fig. 6.19, eight pairs of GaN (10 nm)/Al (12 nm) multilayer are designed as a filter allowing only a part of high wave vector k (evanescent waves) to pass through for interference lithography, the mask is 43 nm half-pitch and 40 nm thickness Cr grating, and only the ± 1 order passes through the DMM structure for interference. The simulated total

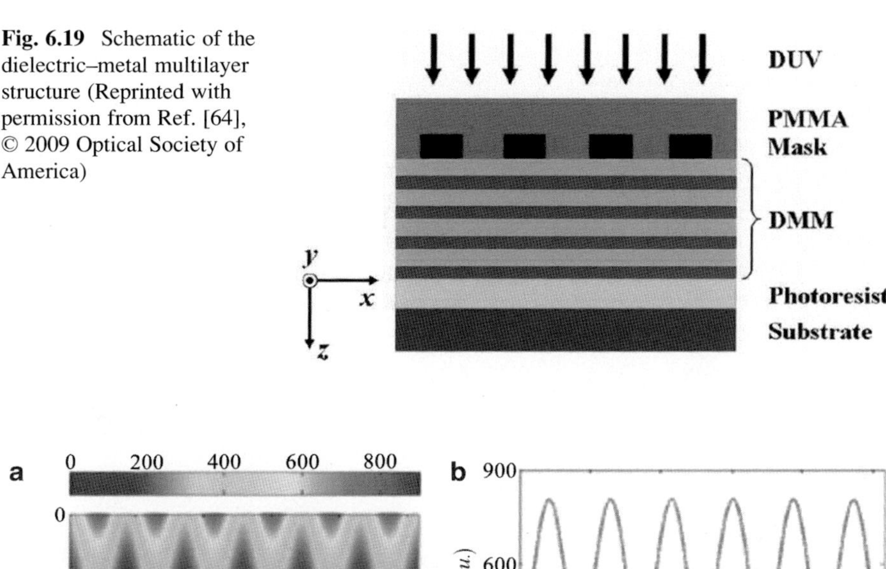

Fig. 6.19 Schematic of the dielectric–metal multilayer structure (Reprinted with permission from Ref. [64], © 2009 Optical Society of America)

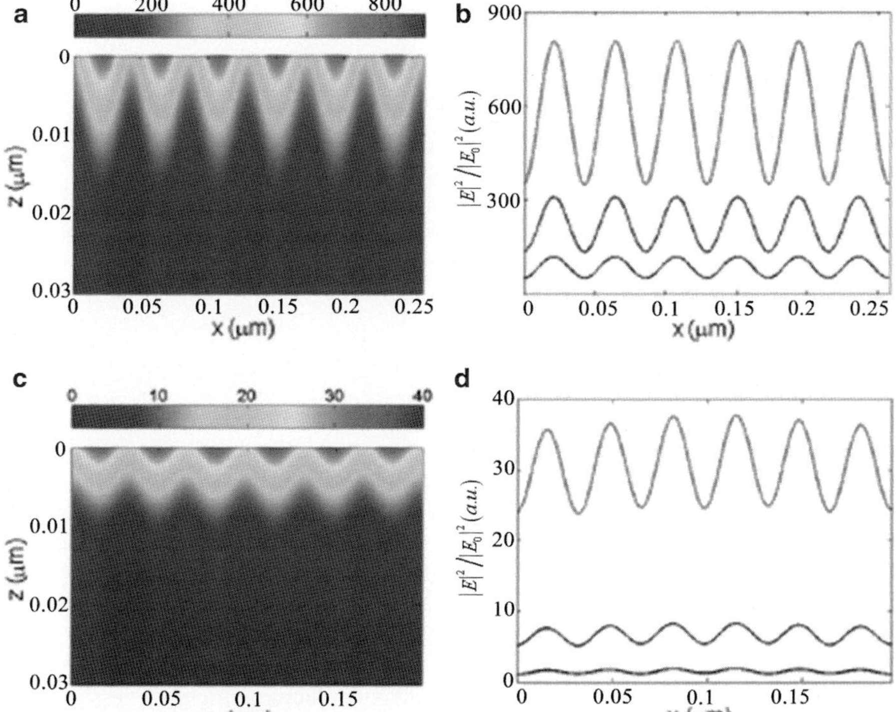

Fig. 6.20 The distribution of E-field intensities: (**a**) and (**b**) for the mask with periods of 86 nm; (**c**) and (**d**) for the mask with periods of 66 nm, respectively (Reprinted with permission from Ref. [64], © 2009 Optical Society of America)

electric field intensity distribution in the photoresist zone is shown in Fig. 6.20a; the feature size of the formed pattern is down to 21.5 nm. By the way, to satisfy the minimum contrast of common negative photoresist, the critical period of the mask can be down to 66 nm for the first order, and the mask is 33 nm half-pitch; thus, as

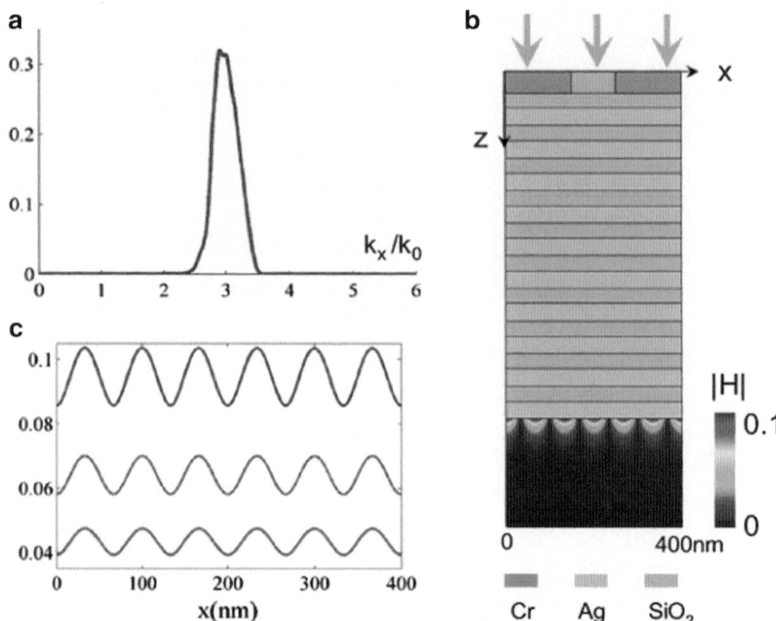

Fig. 6.21 (**a**) The transmission versus the tangential wave vector with TM wave ($\lambda = 405$ nm); (**b**) H-field distribution; (**c**) E-field intensities at 0 nm (*blue*), 10 nm (*green*), and 20 nm (*red*) away from the multilayer-photoresist interface (Reprinted with permission from Ref. [65], ©2008 American Institute of Physics)

shown in Fig. 6.20d, the half-pitch fringes are even down to 16.5 nm. This technique has an exciting potential to print patterns with half-pitch 16 nm and below.

In 2008, Zhang's group proposed a metal–dielectric multilayer structure to generate the subwavelength features projecting from a 1D or 2D diffraction-limited mask [65]. The special material of multilayer can be realized by a metamaterial; thus, the metal–dielectric multilayer has the special properties to split surface plasmon modes due to the interaction of modes on two metal surfaces. Therefore, the designed multilayer can act as an extraordinary spatial filter, i.e., only allows a band of waves with tangential wave vector larger than nk_0 to go through it; n is the refractive index of the medium on the transmission side of the grating. The location and the bandwidth of the passband can be tuned by changing the thicknesses of the metal and dielectric layers.

In this scheme, the width of the spatial frequency passband can be narrow enough to allow only one diffraction order (mth order), and a sub-diffraction-limited pattern with period d/2 m will be formed on the other side of the multilayer as a result of the superposition of the ±mth-order diffraction waves; d is the period of the mask, which is larger than the diffraction limit. Figure 6.21a shows the transmission versus the tangential wave vector for 10 pairs of 40 nm Ag and 35 nm SiO$_2$ multilayer for TM polarization at a wavelength of 405 nm. A 1-D Cr grating mask with 400 nm period is added in front of the multilayer structure and

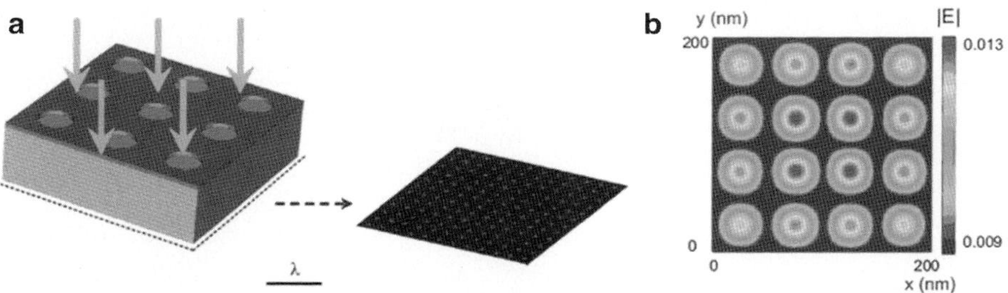

Fig. 6.22 (a) Schematic of the 2D photolithography; (b) E-field distribution at the plane 3 nm after the multilayer with circular polarization incident wave (Reprinted with permission from Ref. [65], ©2008 American Institute of Physics)

illuminated by a light at 405 nm; only the ± 3- order diffraction waves from the mask can go through the multilayers and form a pattern with period six times smaller compared with the mask. The thickness of the Cr mask is 50 nm, and the opening width is 100 nm. This spatial frequency sextupling effect is shown by the simulated H-field distribution in Fig. 6.21b, and the E-field distribution is also shown in Fig. 6.21c at planes 0, 10, and 20 nm.

This method can be easily extended to 2D patterns generation, as shown in Fig. 6.22, the mask is a 2D grating with square lattice on a 50 nm thickness Cr layer, the period is 200 nm, and the diameter of the circular opening is 100 nm. The multilayer is consisted of 12 pairs of 35 nm Ag and 21 nm SiO_2. Only the ± 2-order diffracted waves from the mask that have tangential wave vector $2 \times 405/ 200 \, k_0 \approx 4 \, k_0$ can propagate through the multilayer. As a result, a pattern with period $200/(2 \times 2) = 50$ nm is formed; Fig. 6.22b is the E-field distribution at the plane 3 nm after the multilayer with circular polarization incident wave.

This technique demonstrated a photolithography scheme that can fabricate deep-subwavelength nanometer scale 1D and 2D periodic patterns from diffraction-limited masks, and the period ratio between the mask and the lithographic pattern can be adjusted flexibly by designing appropriate multilayer structure.

6.4.3   Metallic Grating Waveguide Heterostructures

In 2010, Yang et al. proposed a lithography scheme with resonant surface plasmon polaritons for metallic grating waveguide heterostructures (MGWHS) [66]. The schematic of the proposed structure is shown in Fig. 6.23.

A Cr mask, which can be fabricated on a quartz layer by electron beam lithography and a lift-off process, is illuminated from the bottom with p-polarization light of 436 nm. A photoresist is then spun on a metal (silver or Al) layer which clings under the silicon substrates, and the same photoresist is filling in the mask slits. The skin depth of the metal layer in this case is less than 30 nm; thus, the thickness of the metal slab is less than that of the photoresist layer. The thicknesses of the silver layer and

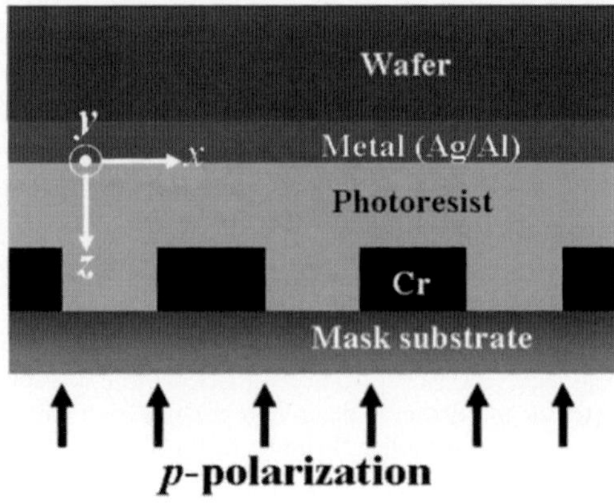

Fig. 6.23 Schematic of the MGWHS for the plasmonic interference lithography (Reprinted with permission from Ref. [66], © 2010 IOP Publishing Ltd)

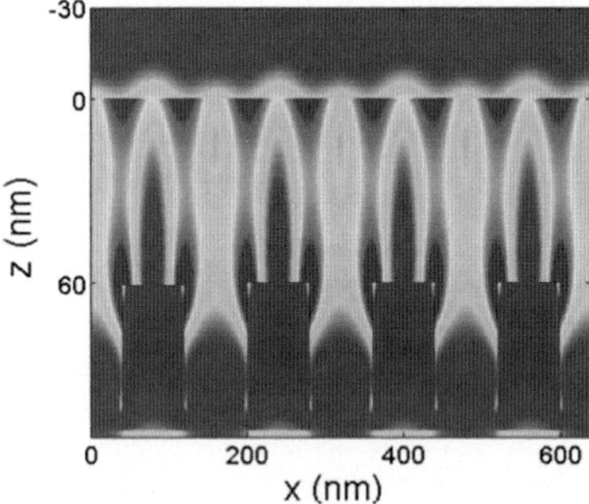

Fig. 6.24 The calculated total E-field distribution with the thicknesses of photoresist and silver are 60 and 30 nm, respectively (Reprinted with permission from Ref. [66], © 2010 IOP Publishing Ltd)

photoresist are chosen as 30 and 60 nm; the silver mask period is 160 nm. Figure 6.24 shows the calculated total E-field distribution, and the relationship between the wave number of the resonant SPPs and the thickness of the photoresist is presented (denoted by +) in Fig. 6.25a. Figure 6.25b shows the calculated total E-field intensity distribution with a phase mask of 136 nm period using the FDTD calculation method. A 34 nm width pattern with high contrast is formed by the interference between the first-order diffracted waves of the mask. The intensity contrast at planes $z = 0$, 10, and 20 nm is 0.489, 0.62, and 0.71, respectively. Similarly, as shown in Fig. 6.25c, a feature size of 25 nm ($\sim\lambda/17$) is formed with a phase mask of 100 nm period. The total E-field intensity distribution of the pattern within a depth of 20 nm along the photoresist/silver interface is also presented (Fig. 6.25c inset). The detailed E-field intensity at planes 0 (red), 10 (blue), and 20 nm (black) below the photoresist/silver interface is shown in Fig. 6.25d. The intensity contrast at planes $z = 0$, 10, and 20 nm is 0.257, 0.574, and 0.358, respectively.

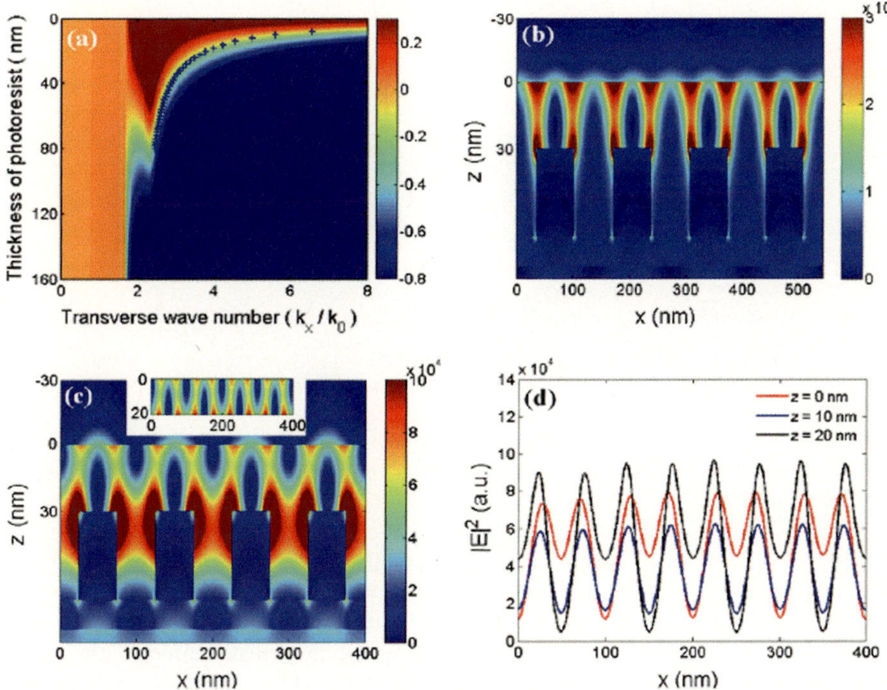

Fig. 6.25 (**a**) The enhancement factor (on a logarithmic scale) versus the thickness of the photoresist (denoted by +), the silver thickness is 30 nm. The calculated total E-field distribution with 30 nm photoresist and the periodic mask of (**b**) 136 nm and (**c**) 100 nm, and the total E-field intensity distribution with 20 nm thickness are also presented (see (**c**) *inset*). (**d**) The electric field of (**c**) at planes 0 (*red*), 10 (*blue*), and 20 nm (*black*) below the photoresist/silver interface (Reprinted with permission from Ref. [66], © 2010 IOP Publishing Ltd)

In 2012, they make a further development to fabricate nanometer scale 1D and 2D periodic patterns with above structure by utilizing the higher diffraction order of the mask and adjusting the incident angle [67]. This approach may provide a cost-effective method for mass production with simple equipment.

6.4.4   SPPs Resonant Cavity Structure

In 2011, Ge et al. reported a tunable and enhanced ultra-deep-subwavelength nanolithography technique using a surface plasmon resonant cavity formed by a metallic grating and a metallic thin-film layer separated by a photoresist layer [68]. The schematic is presented in Fig. 6.26. An SPPs cavity is formed by an upper diffraction-limited metallic grating mask and a lower backing metallic thin film deposited on a fused silica substrate, separated by a photoresist layer or a combination

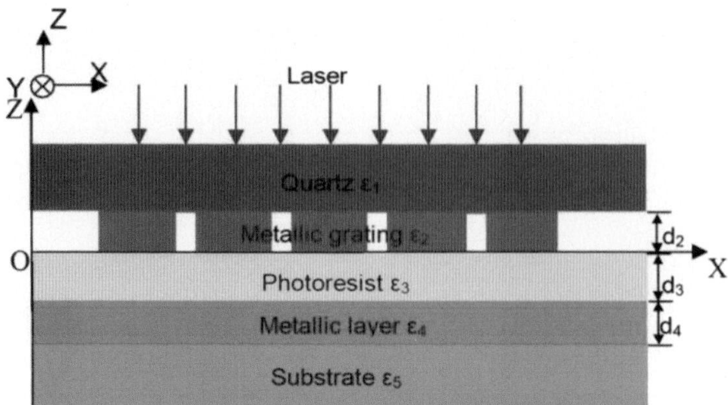

Fig. 6.26 Schematic of a tunable surface plasmon interferometric cavity for ultra-deep-subwavelength photolithography (Reprinted with permission from Ref. [68], © 2011 Optical Society of America)

of photoresist and index-matching fluids. The upper metallic grating and the lower metallic thin film are assumed to be the same silver material. The illumination is incident from the top with a wavelength of 436 nm and p-polarization.

Figure 6.27a, b shows, respectively, the electric field distribution of the cavity structure and a conventional open structure with metallic grating/photoresist/SiO2 substrate (no surface plasmon resonant cavity) as a comparison. In both Fig. 6.27a, b, the period and thickness of the upper Ag grating is assumed to be 600 nm (much larger than the illumination wavelength) and 50 nm, respectively, the slit width of the grating is fixed at 60 nm, and the thickness of photoresist is 50 nm. The Ag thin-film layer deposited on the lower SiO2 substrate in Fig. 6.27a is assumed to be 50 nm. It is seen that the nanopatterns generated in the photoresist layer in Fig. 6.27a, b show different behaviors in terms of resolution and the exposure depth. With the cavity structure, eight pairs of the plasmonic interference patterns are observed within two adjacent open slits of the metallic grating, while only six pairs of patterns are observed in the conventional SPPs open structure under the exact same conditions. It is also seen that the exposure depth in the photoresist layer with the cavity structure is much higher than that of the conventional open structure.

Figure 6.28 shows the electric field distributions of the cavity structure with different photoresist thicknesses. The period and the thickness of the Ag grating are 600 and 50 nm, respectively. The slit width of the grating is fixed at 60 nm, and the thickness of Ag layer on the lower substrate is 50 nm. The thickness of photoresist layer is 50, 30, 20, and 10 nm, respectively, in Fig. 6.28a–d. It is obvious that the number of patterns between the adjacent slits increases with the decrease of thickness of the photoresist layer. The number of pattern pairs increases from 8 in Fig. 6.28a to 10 in Fig. 6.28b, 12 in Fig. 6.28c, and 16 in Fig. 6.28d.

With the cavity technique, the optical resolution can be enhanced about three times that of the conventional open structure, down to about 16.5 nm feature

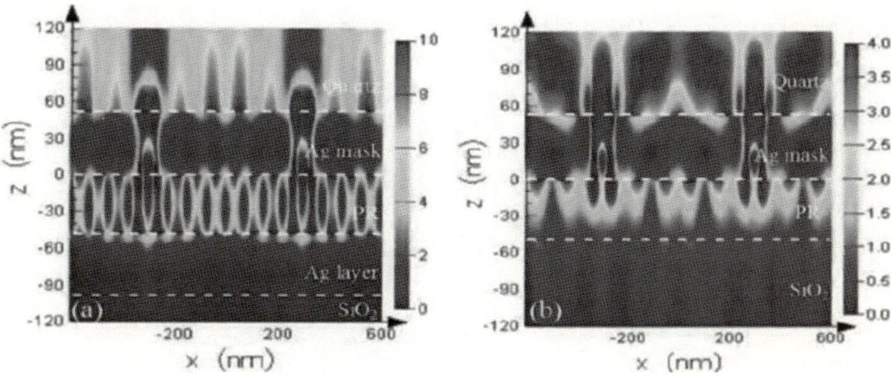

Fig. 6.27 The electric field distributions of (**a**) cavity structure; (**b**) conventional structure (Reprinted with permission from Ref. [68], © 2011 Optical Society of America)

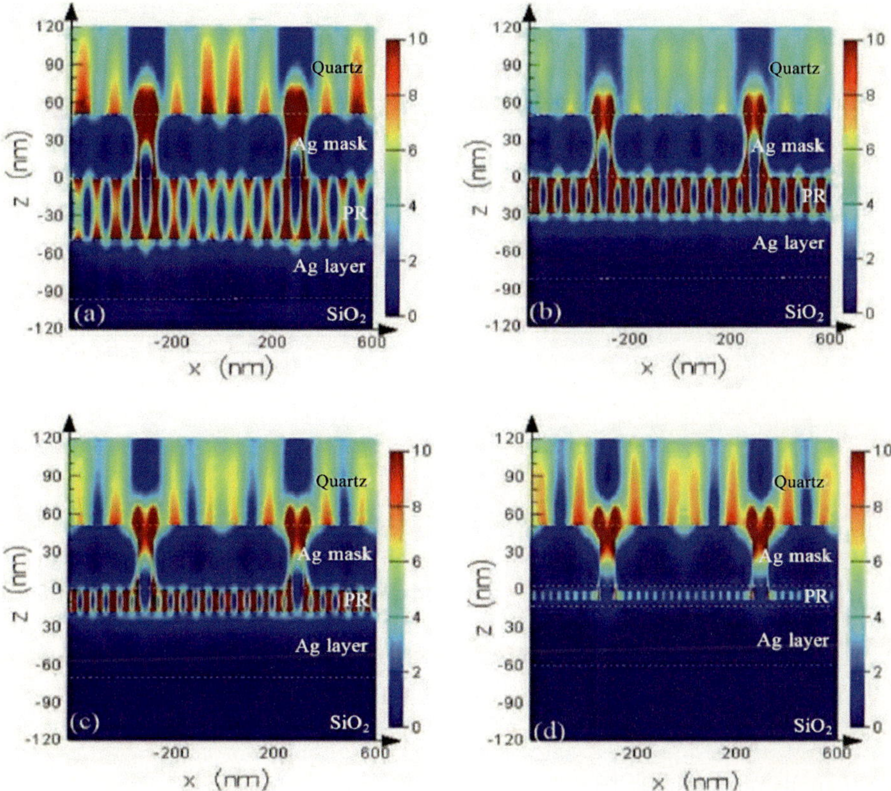

Fig. 6.28 The electric field distributions of the cavity structures with different cavity lengths (i.e., photoresist thicknesses): (**a**) 50 nm, (**b**) 30 nm, (**c**) 20 nm, and (**d**) 10 nm (Reprinted with permission from Ref. [68], © 2011 Optical Society of America)

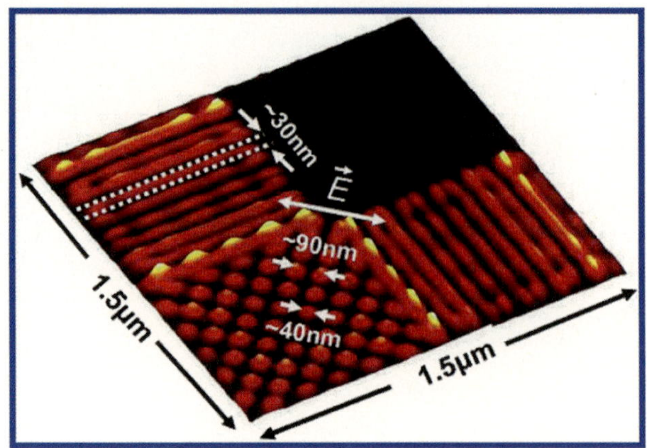

Fig. 6.29 Distribution of the E-field in the photoresist for four perpendicular gratings with 130 nm periodicity (Reprinted with permission from Ref. [69], © 2005 American Chemical Society)

size with a wavelength of 436 nm illumination and a diffraction-limit mask (period = 600 nm). This is comparable to that of using a complicated 30-pair metamaterial structure and 193 nm illumination [64]. With this SPPs cavity technique, the lithography resolution is inherently higher than that of the conventional open SPPs technique. The generated nanopatterns with the cavity technique are of much improved uniformity, contrast, and deep exposure depth compared to the open SPPs technique, which opens a way to generate tunable ultra-deep-subwavelength patterns by using a fixed diffraction-limited mask with capability of large-area, deep exposure depth.

6.4.5   Other Configurations of Grating Coupling

In addition to the aforementioned grating-coupled SPPs lithography techniques, there are some other implementation schemes reported these years.

In 2005, Zhang's group showed that 1D and 2D periodical structures of 40–100 nm features can be patterned using interfering surface plasmons launched by 1D gratings [69]. Multiple 1D gratings are used to convert free-space light into surface plasmon waves, and those waves propagating outside the grating area form an interference pattern when they encounter each other. By using a different number of gratings or surface plasmon waves, various interference patterns such as periodic lines and 2D dot arrays can be obtained. With a 266 nm exposing light polarized along the diagonal direction, four mutually perpendicular 1D Al gratings with 130 nm periodicities yield a square lattice of 2D dot array with a 90 nm periodicity and a 40 nm feature size (Fig. 6.29).

In 2007, Luo's group proposed a localized surface plasmon nanolithography (LSPN) technique to produce patterns with a sub-20 nm line width [70]. The optimized structure shown in Fig. 6.30 is composed of quartz substrate and Al film; the Al film is constructed by grooves (grating) and tapers. The grating can

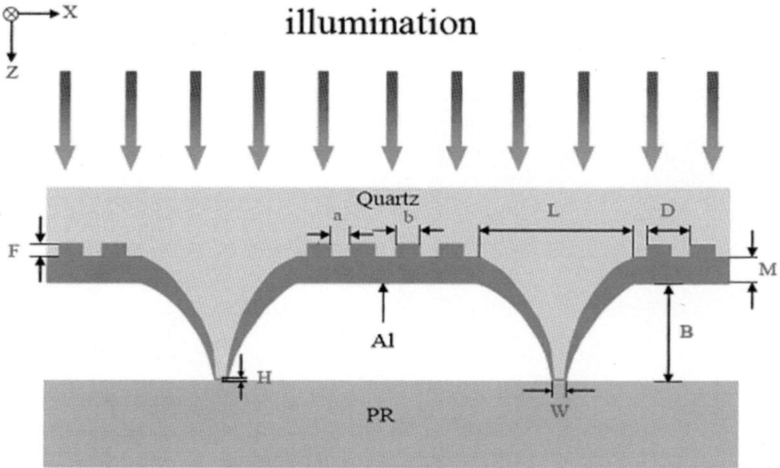

Fig. 6.30 Schematic configuration of the simulated LSPN structure (Reprinted with permission from Ref. [70], © 2007 Optical Society of America)

efficiently excite the SPPs which propagate toward the tip along the tapers surface. Energy concentration is gradually produced in the course of SPPs propagation, and the amplitude of local electric field intensity increases step by step accordingly. At the same time, SPPs are progressively slowed down and stopped at the tip which leads to their accumulation at the tip. Ultimately, a localized electric field with fine distribution and high magnitude forms at the tip. Compared with previous SPP photolithography technique, the LSPN emphasizes its localized characteristic. Figure 6.31 shows a simulated results with the values of $L = 320$ nm, $D = 140$ nm, $W = 15$ nm, $F = 20$ nm, $M = 20$ nm, $B = 100$ nm, and $H = 2$ nm. One spatial electric field peak sharply appears just at the tip position. The electric field intensity profile in the photoresist is shown in Fig. 6.31b. The line width defined as the full width at 0.7 maximum is 19.5 nm.

This method is different from the previous lithographic techniques. The metallic mask is consisted of corrugated metallic structures and tapered structures. The corrugated metallic structures are used to modulate the propagating light and generate SPPs, and the tapered structures act as plasmonic waveguides which cause SPPs propagation and accumulation. The formed patterns depend mainly on the shape of the tip. It is thus believed that it may provide a way for nanolithography of arbitrary patterns with visible or UV light, although a complicated approach has to be solved to design and manufacture the metallic mask.

Murukeshan et al. proposed a metal particle-surface system to pattern periodic nanostructure with excitation of gap modes in 2009 [71]; schematic configuration is shown in Fig. 6.32. As a significant advantage, it can provide a much larger enhanced field to give shorter wavelengths of surface plasmons compared with that achieved with an isolated metal particle or a metal surface configuration alone.

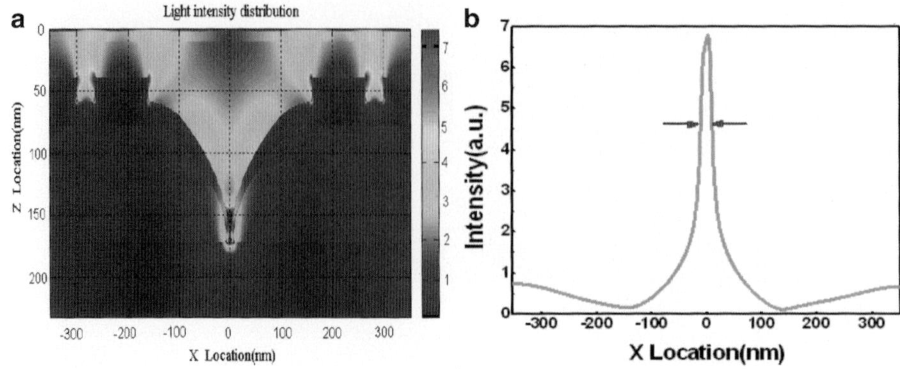

Fig. 6.31 (**a**) Light intensity distribution in the mask and the photoresist; (**b**) E-field intensity profile in the photoresist at the position of 5 nm below the interface of photoresist and tip (Reprinted with permission from Ref. [70], © 2007 Optical Society of America)

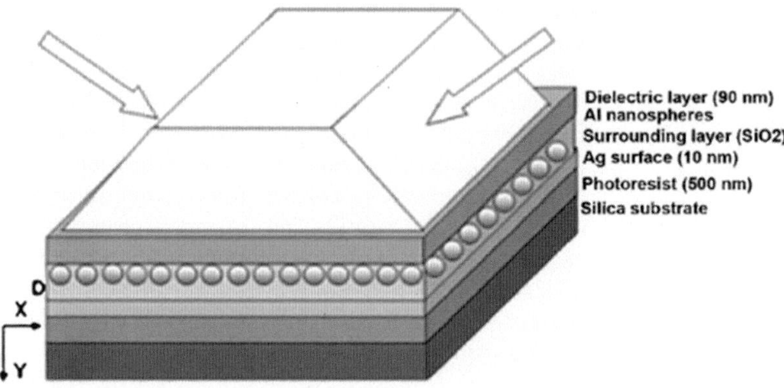

Fig. 6.32 Schematic diagram of the proposed particle-surface system (Reprinted with permission from Ref. [71], © 2009 Optical Society of America)

In 2007, Luo's group proposed a SPPs interference nanolithography based on end-fire coupling method [72]; the physical structure is shown in Fig. 6.33.

The surface-active medium is metal, the middle dielectric layer is set to resist, and the undermost layer is used as the resist substrate whose effects can be neglected when the resist thickness is large enough for SPPs decay length. The p-polarized irradiation light of 365 nm is focused onto one of the end faces of the metal layer but above the metal/resist interface to avoid the local expose of resist. h_0 is the distance between the center of focused light source and the metal/resist interface corresponding to focusing spot offset. Δh is the width of focused light beam. The thicknesses of the metal and resist layers are respectively set to 2 μm and 0.1 μm. Δh is 0.5 μm in view of the diffraction effect of the focused beam and h_0 is half of Δh.

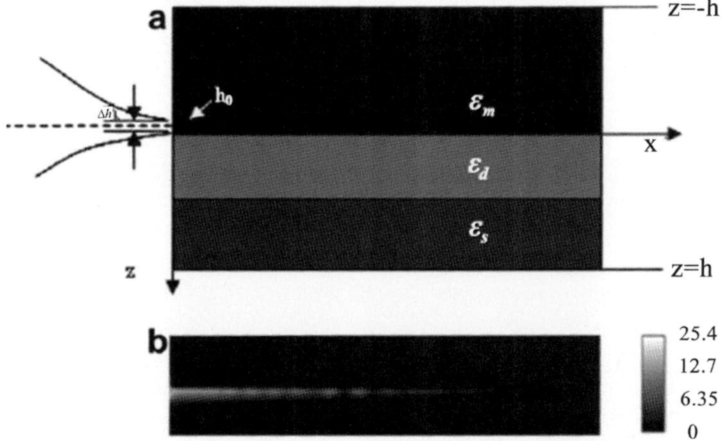

Fig. 6.33 (**a**) Geometry of the end-fire lithography scheme; (**b**) the magnetic field distribution of generated SPPs (Reprinted with permission from Ref. [72], © 2007 Elsevier B.V.)

Fig. 6.34 (**a**) Schematic configuration of two face to face SPPs interference; (**b**) and (**c**) are E_x and E_z components of interference fields, respectively (Reprinted with permission from Ref. [72], © 2007 Elsevier B.V.)

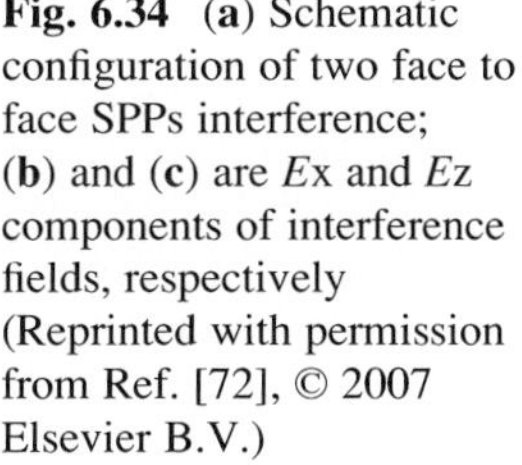

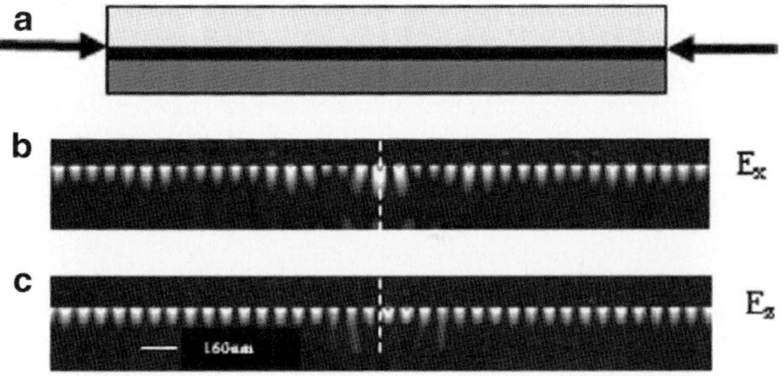

When two mutually coherent plane waves with p-polarizations are incident on each of two end faces of the interface, it will generate two counter-propagating SPPs waves (see Fig. 6.34a). As a result, interference fringes are formed in resist. The electrical field distribution clearly shows that a standing wave or an interference pattern with 80 nm periodicity is formed at the interface shown in Fig. 6.34b, c, which respectively represents E_x and E_z components. If the half-peak width is taken as the feature size, 40 nm features about $\lambda/9$ of the excitation wavelength can be reached.

Furthermore, in 2009, Zhang's group demonstrated similar phenomena, i.e., sharp edge coupling [73]. There are two samples given to demonstrate this method. The first sample is shown in Fig. 6.35b, comprises a 2 μm wide Al strip on quartz

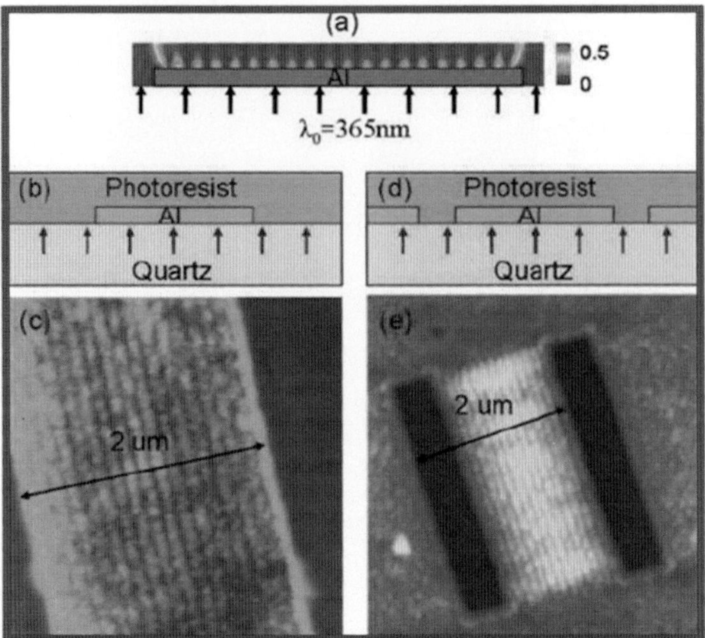

Fig. 6.35 (**a**) Numerical simulation of SPPs interference pattern on top of an Al strip; (**b, d**) schematic experimental configurations of two samples; (**c, e**) AFM image of the exposure pattern using samples (**b**) and (**d**), respectively (Reprinted with permission from Ref. [73], © 2009 American Chemical Society)

substrate. It was fabricated by E-beam lithography followed by a lift-off process. This sample utilizes a sharp Al edge to excite SPPs. The second sample, schematically shown in Fig. 6.35d, was obtained by focused ion beam (FIB) milling (slit openings are about 700 nm wide) in an Al film deposited on the quartz wafer.

The Al thickness is 100 nm in both cases. Subsequently, a 15 nm thick OmniCoat (MicroChem) layer was spun on both samples to increase the adhesion between Al and e photoresist. Finally, a negative near-UV photoresist (SU-8) was spun on the top of the OmniCoat. Both samples were exposed with an exposure dose of 200 mJ/cm^2 by a filtered mercury lamp with a radiation peak at 365 nm.

The exposure results for both samples are presented in panels (c) and (e) of Fig. 6.35, respectively. Clearly, 1D gratings were obtained between the slits/edges in both samples with good contrast and uniformity. The grating periodicities in both cases are about 120 nm, i.e., the SPPs interference pattern line width has reached a 60 nm line width which is equivalent to $\lambda/6$.

It is apparent that this method can be easily extended to 2D geometries; Fig. 6.36a–d shows simulated SPPs interference patterns with four 2D geometries. Obviously, not only periodic but also quasiperiodic and even more complicated 2D

Fig. 6.36 Simulations (**a–d**) and experimental results (**e–h**) of triangle, square, pentagon, and pentagon with rounded corners, respectively. The side lengths are 2, 2, 1.5, and 1.5 μm, respectively, in (**a–d**). The scale bars in all of the AFM images represent 500 nm. The side lengths of the triangle, square, pentagon, and rounded pentagon structures used in the experiment are 2, 2.5, 2, and 1.6 μm, respectively (Reprinted with permission from Ref. [73], © 2009 American Chemical Society)

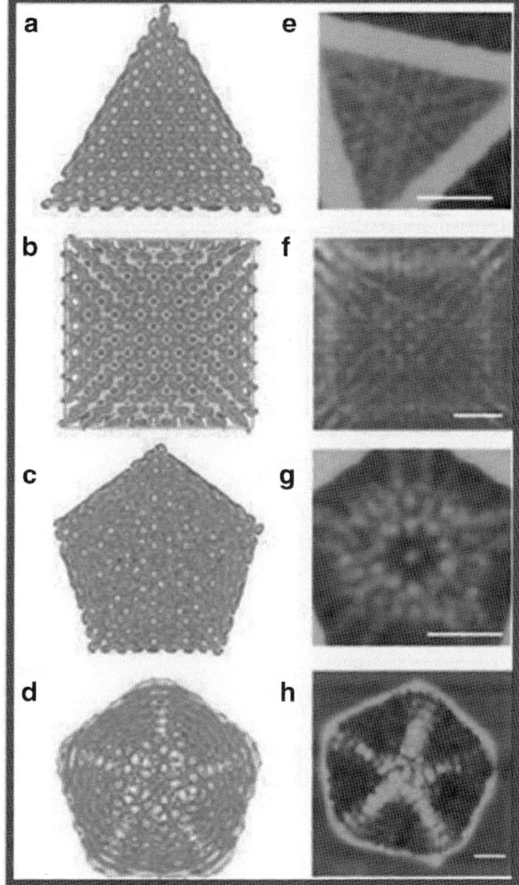

patterns can be realized. The corresponding experimental results are shown in Fig. 6.36e–h with an angular polarized *i*-line beam from a mercury lamp.

It is significant that 2D complex interference patterns can be obtained with different edge/slit geometries by using an edge/slit excitation mechanism. In addition, the patterns can also be manipulated by adjusting incident beam angle, polarization direction, and light frequency; this dynamical manipulation method of surface plasmon patterns will have profound potentials in nanolithography, particle manipulation, and other related fields. Moreover, Ueno et al. reported another innovative lithography system based on two-photon photochemical reactions induced by plasmonic near-field light and the scattering component of the light in a photoresist film[74, 75], and this method is appropriate for fabricating sharp-edged nanodot patterns with nanoscale accuracy.

6.5 Superlens Imaging Nanolithography

6.5.1 Near-Field Superlens Lithography

In 2005, Zhang's group demonstrated sub-diffraction-limited imaging with a 60 nm half-pitch resolution [76]. The setup of their experiment is shown in Fig. 6.37. A set of embedded objects are inscribed onto the 50 nm thick Cr film. Left side is an array of 60 nm wide slots with a 120 nm pitch, and right side is an arbitrary object "NANO". The line width of the "NANO" object is 40 nm, as illustrated in Fig. 6.38a. The Ag film is separated from the objects by a 40 nm thick PMMA layer. A mercury lamp with i-line of 365 nm is employed as the light source. A control experiment was conducted as well for comparison, in which the Ag film is replaced by PMMA. A 120 nm thick negative photoresist (NFR105G) is coated on Ag film to record the near-field image. The substrate is then exposed with flux of 8 mW/cm^2 and an optimal exposure time of 60 s. With this method, the arbitrary pattern of "NANO" and 1D lines with width of 60 nm (about λ/6) was fabricated, as illustrated in Figs. 6.38b and 6.39a respectively. The control experimental results are shown in Figs. 6.38c and 6.39b. To make a comparison, the cross-sectional line width of letter "A" was measured. In the experiment with the superlens, the line width was about 89 nm while the full width at half-maximum line width is 321 $\pm$ 10 nm in the control experiment. It is apparently that the quality of imaging can be greatly improved by using the superlens. The same structure can also realize the imaging of a 50 nm half-pitch object at λ_0/7 resolution [77].

In 2010, Du et al. proposed a photolithographic method to realize a 30 nm resolution by employing a localized surface plasmon (LSP) mask generated by

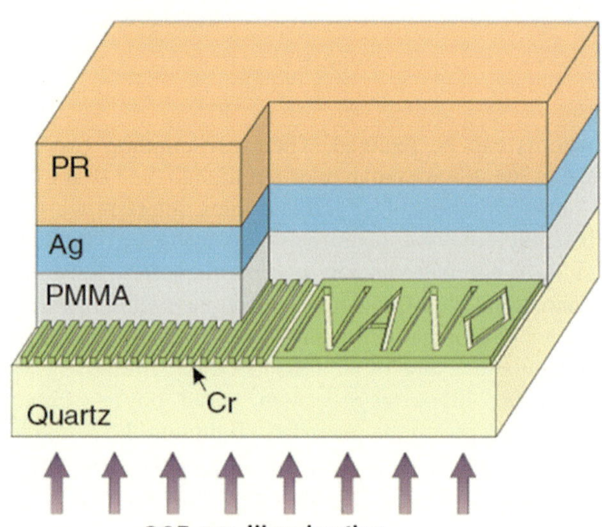

Fig. 6.37 Superlens structure designed by Zhang's group (Reprinted with permission from Ref. [76], © 2005 American Association for the Advancement of Science)

Fig. 6.38 (**a**) Image of "NANO" object; (**b**) AFM image of the exposed pattern with a silver superlens; (**c**) AFM image of the exposed pattern in the control experiment (Reprinted with permission from Ref. [76], © 2005 American Association for the Advancement of Science)

polydimethylsiloxane (PDMS) soft mold and thin Ag film serving as the superlens [78]. The PDMS mask has two unique advantages: high transmittance and intimate contact with the metal film. The thicknesses of Ag and resist (AR-3170) are 18 and 100 nm respectively. The light source is TM-polarized UV light with wavelength of 365 nm. After 15 s exposure and 20 s development, the fine patterns with a width of 30 nm (about 1/12 of the exposing wavelength) and a depth about 20 nm were obtained in resist. The experimental arrangement and the generated patterns are illustrated in Fig. 6.40a, b respectively.

Chaturvedi et al. demonstrated another method using smooth and low loss Ag optical superlens capable of resolving features at 1/12 of the illumination wavelength with high fidelity [79]. The proposed setup is illustrated in Fig. 6.41a; an array of Cr gratings with 40 nm thick and 30 nm half-pitch, which serve as the object, was patterned using a nanoimprint process. Then, a planarization layer with 6 nm thick was deposited on top of the object as a spacer layer to reduce the surface modulation below 1.3 nm. And on top of the spacer layer, a 35 nm thick Cr window layer is patterned to enhance the contrast with dark-field imaging. Subsequently, 1 nm thick Ge and 15 nm thick Ag are evaporated over the window layer followed by coating with a thick layer of photoresist (NOA-73). The Ge layer is introduced to improve the Ag film surface morphology and make the surface roughness less than 0.8 nm. UV light with a wavelength of 380 nm and a power of 80 mW was employed as the light

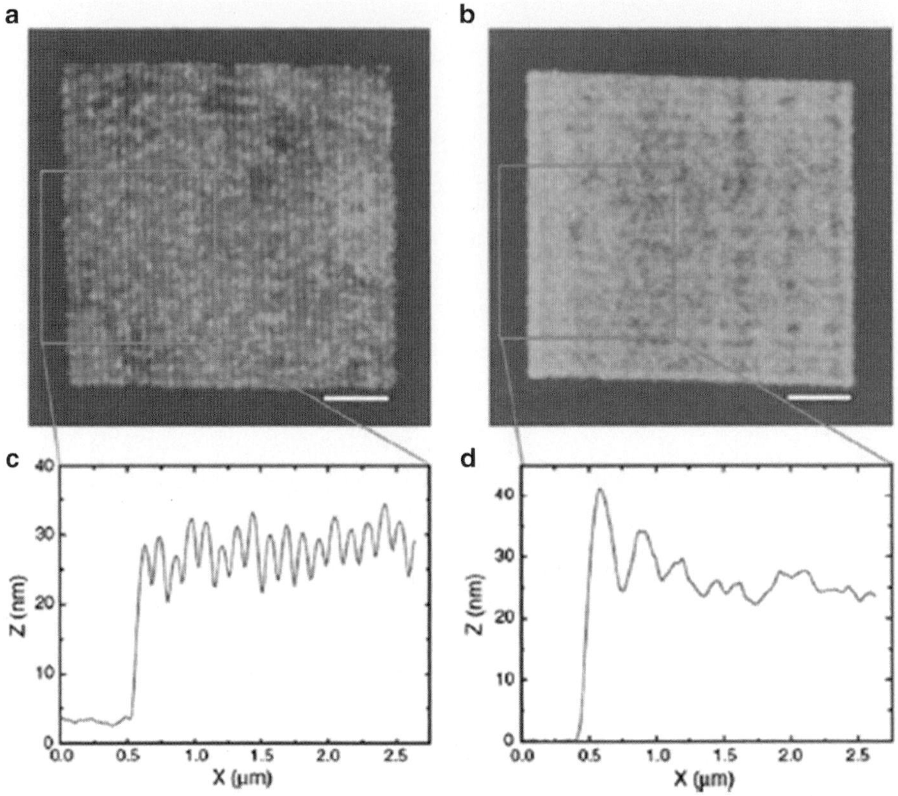

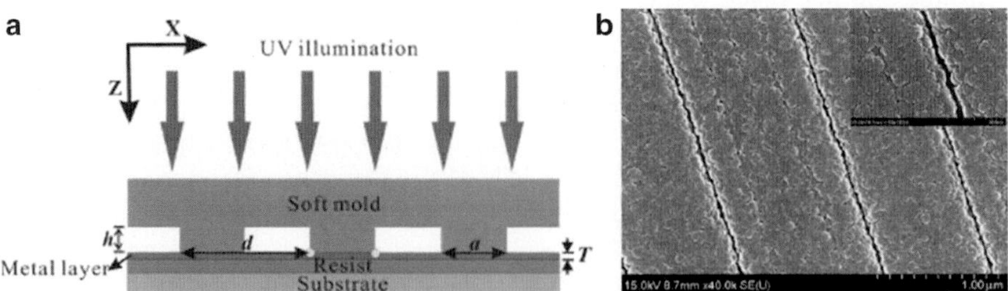

Fig. 6.39 (**a**) AFM image of the exposed patterned with a superlens; (**b**) AFM image of the exposed patterned without a superlens; (**c**) the averaged cross-sectional profile of (**a**); (**d**) The averaged cross-sectional profile of (**b**) (Reprinted with permission from Ref. [76], © 2005 American Association for the Advancement of Science)

Fig. 6.40 (**a**) Experimental arrangement; (**b**) SEM photography of a lithographic result with the PDMS soft mold in 2 μm period and 0.8 μm line width. The *inset* is a magnified image highlighting the achieved minimum feature size of 30 nm (Reprinted with permission from Ref. [78], © 2010 Optical Society of America)

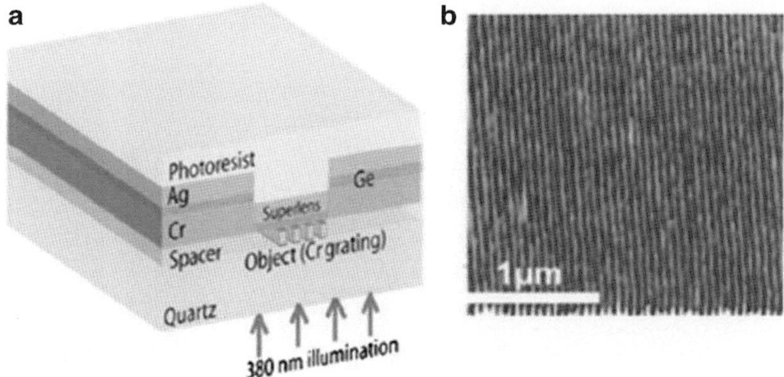

Fig. 6.41 (a) Schematic of a smooth silver superlens designed by Chaturvedi et al. (b) AFM image recorded on the photoresist layer after exposure and development (Reprinted with permission from Ref. [79], ©2010 American Institute of Physics)

source. The experiment result illustrated in Fig. 6.41b showed that 30 nm half-pitch patterns were obtained in photoresist with an exposure time of 120 s.

In 2009, Shi et al. proposed a method to reach the 22 nm lithography node with a 193 nm Al film-based superlens with index-matching layer [80, 81]. In addition, Xu et al. proposed a metal-cladding superlens for projecting deep-subwavelength patterns by adding an additional Ag layer inserted between the photoresist and the substrate [82]. And Blaikie et al. concluded that the choice of light source is crucial for superlens nanolithography experiment [83].

6.5.2 Far-Field Superlens Lithography

The near-field superlens nanolithography is still restricted because the image plane is located in the proximity of the superlens, normally about a few tens of nanometers or even less. Thus, Liu et al. firstly demonstrated a magnifying superlens which can form sub-diffraction-limit image in far field [84–87], as illustrated in Fig. 6.42. The magnifying superlens consists of a curved periodic stack of Ag (35 nm) and Al_2O_3 (35 nm) deposited on a half-cylindrical cavity fabricated on a quartz substrate (Fig. 6.42a); the object inscribed into a 50 nm thick Cr layer was a pair of 35 nm wide lines spaced 150 nm apart (Fig. 6.42b). At last, the magnified image was obtained after the magnifying superlens; once the magnified feature is larger than the diffraction limit, it can then be imaged with a conventional optical microscope. Furthermore, superlens supports the propagation of a very broad spectrum of wave vectors; it can magnify arbitrary objects with sub-diffraction-limited resolution. The recorded image of the letters "ON" shows the fine features of the object (Fig. 6.42d). The experiment has successfully

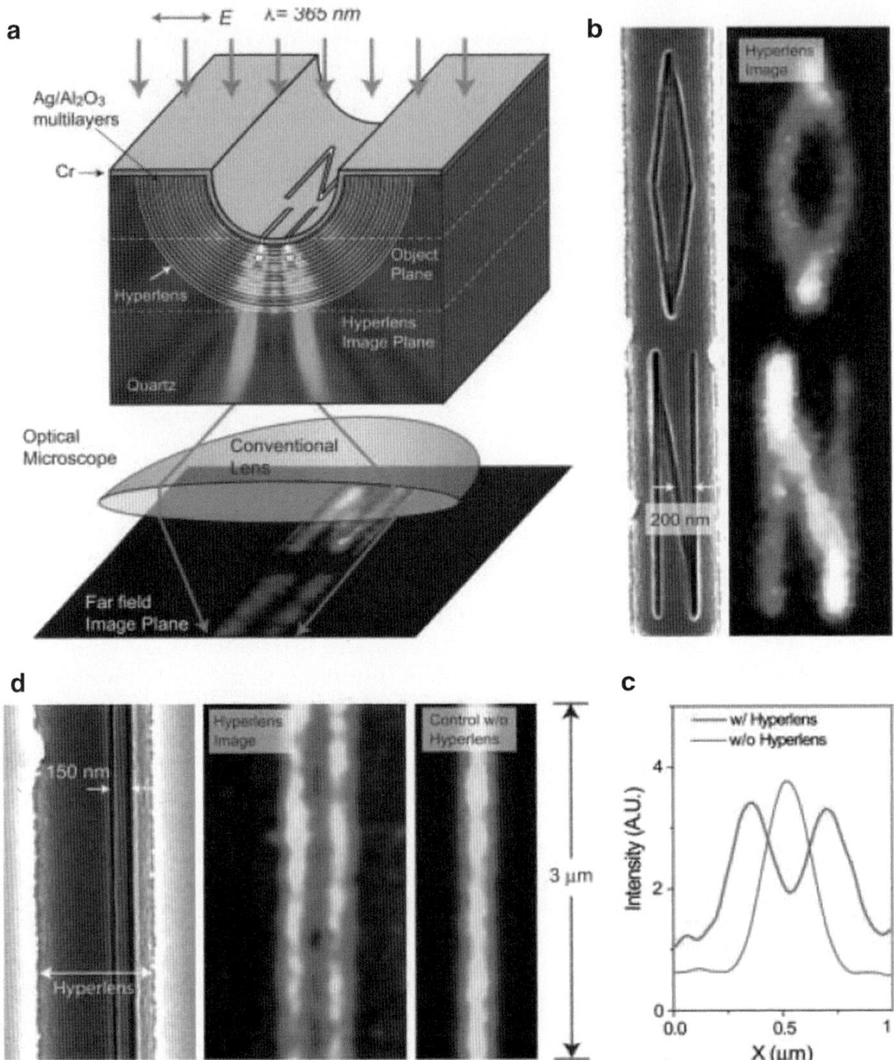

Fig. 6.42 Magnifying optical superlens. (**a**) Schematic of superlens and numerical simulation of imaging; (**b**) superlens imaging of line pair object: From *left* to *right*, SEM image of the line pair object, magnified superlens image with 150 nm spaced line pair, and the image of a control experiment without the superlens. (**c**) The averaged cross section of superlens image of the line pair object (*red*), whereas the image obtained in the control experiment (*green*); (**d**) the object "ON" imaged with sub-diffraction resolution. Line width of the object is about 40 nm. The superlens is made of 16 layers of Ag/Al2O3 (Reprinted with permission from Ref. [84], © 2007 American Association for the Advancement of Science)

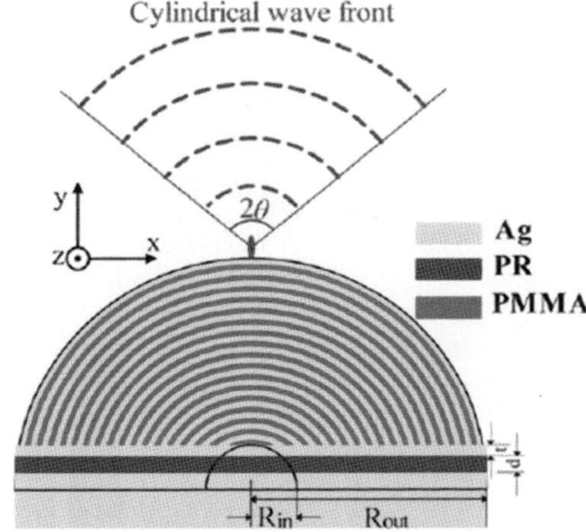

Fig. 6.43 Schematic of the demagnifying superlens (Reprinted with permission from Ref. [88], © 2012 Elsevier B.V.)

demonstrated the imaging ability of the far-field superlens, showing the potential to be applied to nanolithography.

In contrast, Luo's group proposed a demagnifying truncated superlens composed of pairs of metal–dielectric cylindrical multilayers to achieve deep-subwavelength imaging in far field [88]. As illustrated in Fig. 6.43, an optimized superlens with inner radius of $R_{in} = 0.4$ μm and outer radius of $R_{out} = 1$ μm is composed of 30 pairs of 10 nm PMMA/10 nm Ag cylindrical multilayers, which is partially polished at a longitude height of R_{in}. Then a 20 nm thick Ag film was coated on the bottom to avoid the enhancement of local surface plasmon polarizations between the superlens and the 30 nm thick photoresist; in addition, a silver cladding is placed under the photoresist film to improve the lithography quality. The demagnification factor is 2.5, calculated by R_{out}/R_{in}. A 365 nm TM-polarized cylindrical wave with angular aperture 2θ is incident from the far field and converging on the vertex of the superlens. The diffraction-limited spot is focused by far-field converging cylindrical wave.

Figure 6.44a is the simulated result of E-field distribution with oil immersion; Fig. 6.44b is the corresponding intensity along the outer surface of superlens in Fig. 6.44a. The E-field distribution shown in Fig. 6.44b illustrates that a clear focusing spot at a FWHM of 120 nm converges on the vertex of superlens. It is noted that the superlensing effect constrains that the silver and surrounding dielectric have matched permittivity; therefore, a broad band of spectrum components of electric and magnetic fields is continuously coupled into the superlens without distortion.

In contrast to coordinate transformation and conformal transformation theory to flat-to-flat planar superlens design [89–91], this type of truncated superlens can be used for near-field imaging, laser direct writing lithography, optical data storage, and so on.

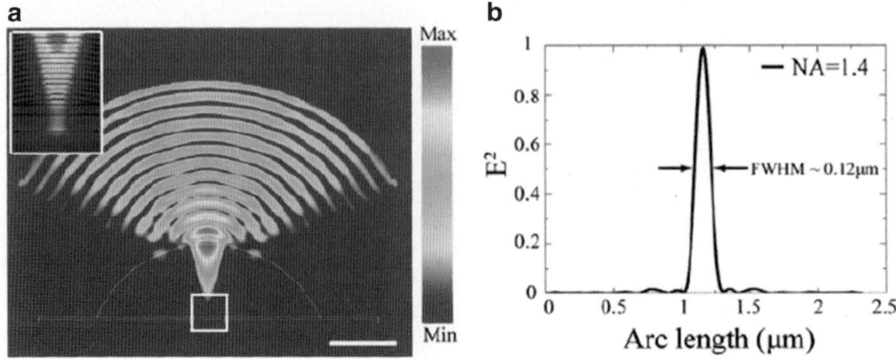

Fig. 6.44 (a) Numerical result of E-field distribution with oil immersion; (b) the corresponding intensity along the outer surface of superlens. The scale bar in (a) is 0.5 µm. The *inset* at the *top left* corner represents the E-field distribution in square zone (Reprinted with permission from Ref. [88], © 2012 Elsevier B.V.)

6.6 SPPs Direct Writing Nanolithography

The fabrication of the mask with nanometer size is complicated, time consuming, and high cost. In addition, mask-based lithography has one obvious disadvantage which is only able to fabricate the specific patterns defined by mask. Fortunately the introduction of direct writing lithography provides a new route, which is more flexible and can be used to fabricate arbitrary patterns.

Zhang's group demonstrated a practical plasmonic scanning optical microscopy system (NSOM) experimentally for near-field lithography for the first time in 2008 [92]. As shown in Fig. 6.45, the conic plasmonic lens consists of a subwavelength aperture at the apex of the cone surrounded by through concentric rings in a thin Al film deposited on a tapered fiber tip. The optimal parameters of the cone angle, aperture diameter, ring periodicity, ring width, and Al layer thickness are 75°, 100, 300, 50, and 80 nm, respectively, with the light source of 365 nm.

During the lithography, the incident light was coupled into the near-field scanning microscope tip to expose the positive photoresist (IX965G) on a Si substrate. With this method, about a 100 nm beam spot was obtained, and the light intensity at the focal point can be 36 times more than the intensity through a single aperture due to constructive interference of SPPs.

Besides, H-shaped, C-shaped, bowtie-shaped apertures, etc., can also be employed as the NSOM probe [93–99], but this type of integration requires a flat surface on top of the probe. As a typical example, bowtie-shaped structure will be introduced below.

Kim et al. reported a plasmonic lithography technique with bowtie-shaped contact probes [98]. As illustrated in Fig. 6.46, the contact optical probe is fabricated as follows: a conical shape solid immersion lens with a flat surface in 30 µm diameter on top was employed. Additionally, FIB (SII SMI3050) milling is

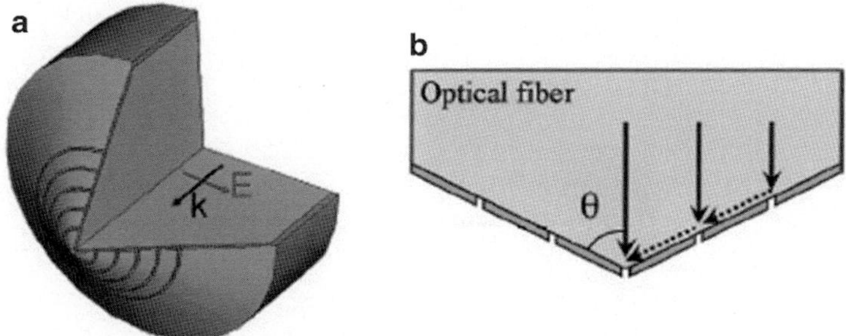

Fig. 6.45 (**a**) Schematic of a conic plasmonic lens; (**b**) the cross section of the conic plasmonic lens (Reprinted with permission from Ref. [92], © 2008 American Chemical Society)

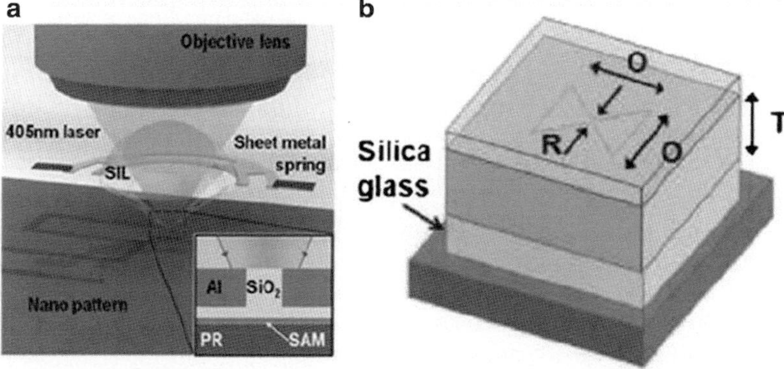

Fig. 6.46 (**a**) The structure of the bowtie-shaped probe. (**b**) Schematic of a bowtie aperture (Reprinted with permission from Ref. [98], © 2009 Optical Society of America)

used for a small contact area with 3 μm diameter top surface. An Al film of 120 nm thickness is coated on the flat surface, and a thin Cr layer of 2 nm thickness is added to ensure the adhesion of the Al film and the glass substrate. In order to obtain the highest transmission of the aperture and the smallest spot size, bowtie apertures with about 140 nm outline dimension were formed on the metal film with focused ion beam (FIB) lithography. Then, a 300 nm thick silica glass film was deposited on the Al film, but with the purpose of controlling the gap between the probe and photoresist and protecting aperture from contamination, the silica glass film thickness was reduced to 10 nm using FIB milling. Finally, a single-atom self-assembled film was coated on the surface of silica film to reduce the friction between the probe and the photoresist during scanning.

In this experiment, the laser source (CrystaLaser, BCL-025-405S) with 405 nm wavelength has a polarization maintaining optical fiber to keep the degree of the polarization at about 0.98. The solid immersion lens is coupled with a high-NA

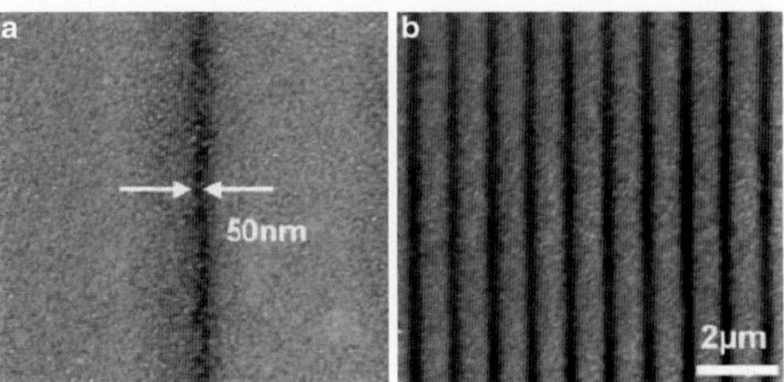

Fig. 6.47 AFM images of the exposed patterns recorded at a speed of 10 mm/s. (**a**) A single line of 50 nm width; (**b**) multiple line pattern of 150 nm width with 1 μm pitch (Reprinted with permission from Ref. [98], © 2009 Optical Society of America)

(0.8) objective lens (Nikon CFI LU Plan Epi ELWD × 100) to focus the laser beam onto the bowtie-shaped aperture, and the focused spot is estimated to be 340 nm. Then the bowtie-shaped aperture excites surface plasmons to expose the photoresist (Shipley S1805), which is 400 nm thick. At last, the patterns can be fabricated in resist by scanning the stage beneath the optical probe. It is noted that, without an external gap control unit, the probe is in intimate contact with resist by applying a certain pressure of $13 \pm 2.5\mu N$ in the scanning range of the resist surface, and the system has a highest throughput of 10 mm/s. The experiment results are shown in Fig. 6.47; the smallest line width of 50 nm illustrated can be realized when the polarized laser light with 405 nm wavelength and a power of 0.5 mW was employed (see Fig. 6.47a). When the laser power is set as 1.25 mW, the lines with a width of 150 nm and a pitch of 1 μm can be fabricated, as shown in Fig. 6.47b.

Srituravanich et al. demonstrated another promising new method with flying plasmonic lens [100]. The schematic of the experiment is illustrated in Fig. 6.48. In this lithography experiment, a UV continuous-wave laser with 365 nm was focused down to a spot of several micrometers onto a plasmonic lens, which further focused the beam to a sub 100 nm spot onto the spinning disk for writing of arbitrary patterns. The laser pulses were controlled by an electrooptic modulator according to the signals from a pattern generator. A "plasmonic flying head" with arrays of plasmonic lenses fabricated on its bottom surface was employed as the direct writing probe. As illustrated in Fig. 6.49, the plasmonic lens consists of a concentric ring grating with a through hole in the center. The plasmonic lens was illustrated in Fig. 6.49. As the subwavelength spots are only formed in the near field, the distance between the plasmonic lens and resist has to be controlled at about 20 nm. To achieve high-speed scanning while maintaining this nanoscale gap, a novel air-bearing slider to fly the plasmonic lens arrays at a height of 20 nm above the substrate at speeds of between 4 and 12 ms^{-1} was adopted. Apart from the rigid distance control, the parallelism of the plasmonic lens and the substrate is also needed to be controlled precisely to ensure

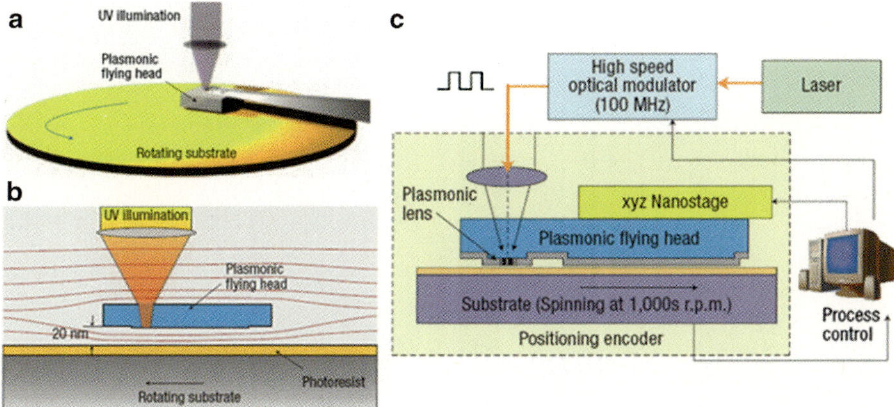

Fig. 6.48 (**a**) The schematic showing the plasmonic lens focusing the light onto the rotating substrate. (**b**) The plasmonic head flying 20 nm above the rotating substrate. (**c**) Schematic of the process control system (Reprinted with permission from Ref. [100], © 2008 Macmillan Publishers Ltd)

Fig. 6.49 SEM image of an array of plasmonic lens (Reprinted with permission from Ref. [100], © 2008 Macmillan Publishers Ltd)

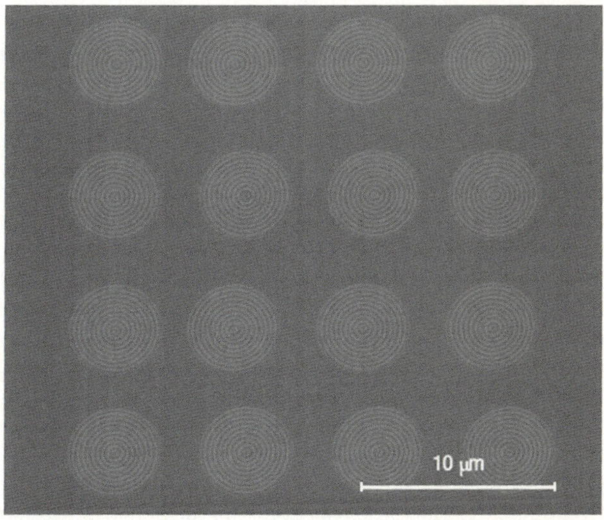

the exposure uniformity. Figure 6.50 shows the experiment results; it is clearly that a 80 nm width line was obtained with the writing speed of 10 m/s (see Fig. 6.50a). Figure 6.50b, c demonstrates the successful patterning of arrays of the acronym "SINAM" with a feature size of 145 nm.

Using large arrays of plasmonic lenses enables parallel writing for high throughput. If a plasmonic flying head carrying 1,000 plasmonic lens could write a 12 in. wafer in just 2 min at a scanning speed of 10 ms^{-1}, that is to say, a high throughput

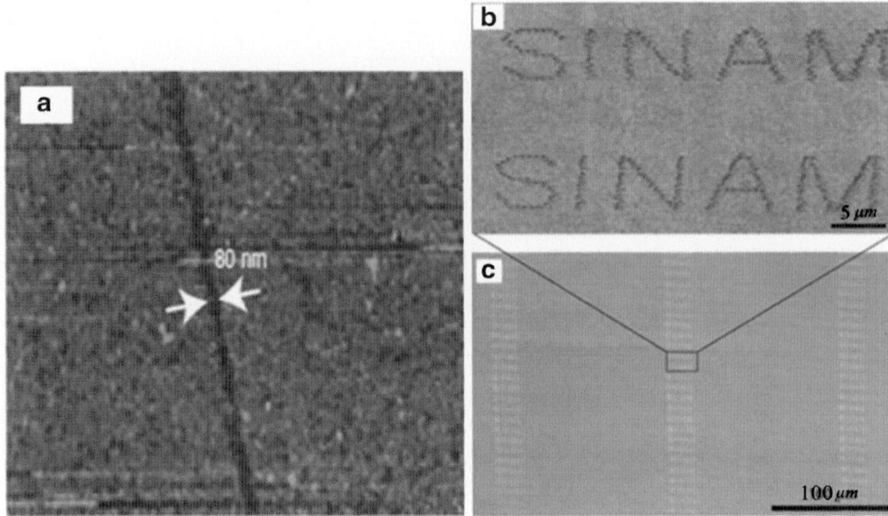

Fig. 6.50 (**a**) AFM image of a pattern with 80 nm linewidth; (**b**) AFM image of arbitrary writing of "SINAM" with 145 nm linewidth; (**c**) optical micrograph of patterning of the large arrays of "SINAM" (Reprinted with permission from Ref. [100], © 2008 Macmillan Publishers Ltd)

of 30 wafer/h can be achieved. Recently, the same group integrated the lithography system with another plasmonic lens [101]. The metallic thin-film structure consisting of two concentric rings with an H-shaped aperture in their center is used as the plasmonic lens; 50 nm lines width can be resolved in the resist with the 355 nm laser beam illumination, at a speed of about 10 m/s.

SPPs direct writing nanolithography is considered as a promising technique in fabricating arbitrary patterns for its flexibility and low cost. In addition, SPPs direct writing nanolithography can be implemented in air, and there is no such rigid environment requirement as any other nanolithography techniques abovementioned.

6.7   Outlooks

Without any doubt, the concept of the SPPs has had a significant impact on science and technology, and it may be the key mechanism to break the optical diffraction limit. The SPPs have many interesting applications, including high-density optical storage, biosensing, photonic crystals, and nanolithography. This chapter reviewed the main aspects of the SPPs-based nanolithography: prism-coupled SPPs nanolithography, grating-coupled SPPs nanolithography, superlens imaging nanolithography, and SPPs direct writing nanolithography. Prism-coupled SPPs nanolithography is expected to provide a convenient route for patterning high-throughput nanometer structures with advantages of maskless and large area. By using different grating masks associated with appropriate structure, grating-coupled SPPs nanolithography can form various interference patterns such as periodic lines and two-dimensional arrays or even more complicated patterns with resolution exceeding the optical diffraction limit. Superlens imaging nanolithography is perfect in "projection lithography" with the extended capability of magnifying or demagnifying arbitrary objects with sub-diffraction-limited resolution. SPPs direct writing nanolithography is a maskless lithography, which has the potential in making arbitrary patterns or masks. Although the SPPs-based nanolithography shows very promising and attractive advantages, much more work needs to be done in terms of the feasibility of the application of the technology into the real nanomanufacturing industry. SPPs-based nanolithography is currently a fast-growing research area in which new ideas and improvements about the technology continue to emerge. It is considered to be the next-generation lithography technology of nano features, large area, fast speed, and cost-effective.

References

1. Rothschild M, Bloomstein TM, Curtin JE et al (1999) 157 nm: deepest deep-ultraviolet yet. J Vac Sci Technol B 17:3262–3266
2. Gwyn CW, Stulen R, Sweeney D et al (1998) Extreme ultraviolet lithography. J Vac Sci Technol B 16:3142–3149
3. Silverman JP (1998) Challenges and progress in x-ray lithography. J Vac Sci Technol B 16:3137–3141
4. Johnson KS, Thywissen JH, Dekker NH et al (1998) Localization of metastable atom beams with optical standing waves: nanolithography at the Heisenberg limit. Science 280:1583–1586
5. Vieu C, Carcenac F, Pepin A et al (2000) Electron beam lithography: resolution limits and applications. Appl Surf Sci 164:111–117
6. Soshnikov IP, Afanas'ev DE, Cirlin GE et al (2011) Fabrication of ordered GaAs nanowhiskers using electron-beam lithography. Semiconductors 45:822–827
7. Luo C, Li Y, Susumu S (2012) Fabrication of high aspect ratio subwavelength gratings based on X-ray lithography and electron beam lithography. Opt Laser Technol 44:1649–1653
8. Manheller M, Trellenkamp S, Waser R et al (2012) Reliable fabrication of 3 nm gaps between nanoelectrodes by electron-beam lithography. Nanotechnology 23:125302–125302
9. Near R, Tabor C, Duan J et al (2012) Pronounced effects of anisotropy on plasmonic properties of nanorings fabricated by electron beam lithography. Nano Lett 12:2158–2164
10. Melngailis J, Mondelli AA, Berry IL et al (1998) A review of ion projection lithography. J Vac Sci Technol B 16:927–957
11. Altun AO, Jeong J-H, Rha J-J et al (2007) Boron nitride stamp for ultra-violet nanoimprinting lithography fabricated by focused ion beam lithography. Nanotechnology 18:465302–465302
12. Wu MC, Aziz A, Witt JDS et al (2008) Structural and functional analysis of nanopillar spin electronic devices fabricated by 3D focused ion beam lithography. Nanotechnology 19:485305
13. Lu K, Zhao J (2010) Focused ion beam lithography and anodization combined nanopore patterning. J Nanosci Nanotechnol 10:6760–6768
14. Oshima A, Okubo S, Oyama TG et al (2012) Nano- and micro-fabrications of polystyrene having atactic and syndiotactic structures using focused ion beams lithography. Radiat Phys Chem 81:584–588
15. Piner RD, Zhu J, Xu F et al (1999) "Dip-pen" nanolithography. Science 283:661–663

16. Mirkin CA (2007) The power of the pen: development of massively parallel dip-pen nanolithography. ACS Nano 1:79–83
17. Basnar B, Willner I (2009) Dip-pen-nanolithographic patterning of metallic, semiconductor, and metal oxide nanostructures on surfaces. Small 5:28–44
18. Nafday OA, Lenhert S (2011) High-throughput optical quality control of lipid multilayers fabricated by dip-pen nanolithography. Nanotechnology 22:225301
19. McAlpine MC, Friedman RS, Lieber DM (2003) Nanoimprint lithography for hybrid plastic electronics. Nano Lett 3:443–445
20. Battaglia C, Escarre J, Soederstroem K et al (2011) Nanoimprint lithography for high-efficiency thin-film silicon solar cells. Nano Lett 11:661–665
21. Higashiki T, Nakasugi T, Yoneda I (2011) Nanoimprint lithography and future patterning for semiconductor devices. J Micro-Nanolith Mems Moems 10:043008
22. Qin F, Meng Z-M, Zhong X-L et al (2012) Fabrication of semiconductor-polymer compound nonlinear photonic crystal slab with highly uniform infiltration based on nanoimprint lithography technique. Opt Express 20:13091–13099
23. Blaikie RJ, McNab SJ (2001) Evanescent interferometric lithography. Appl Optics 40:1692–1698
24. Chua JK, Murukeshan VM, Tan SK et al (2007) Four beams evanescent waves interference lithography for patterning of two dimensional features. Opt Express 15:3437–3451
25. Chua JK, Murukeshan VM (2009) UV laser-assisted multiple evanescent waves lithography for near-field nanopatterning. Micro Nano Lett 4:210–214
26. Sreekanth KV, Chua JK, Murukeshan VM (2010) Interferometric lithography for nanoscale feature patterning: a comparative analysis between laser interference, evanescent wave interference, and surface plasmon interference. Appl Optics 49:6710–6717
27. Ebbesen TW, Lezec HJ, Ghaemi HF et al (1998) Extraordinary optical transmission through subwavelength hole arrays. Nature 391:667–669
28. Weeber J-C, Krenn JR, Dereux A et al (2001) Near-field observation of surface plasmon polariton propagation on thin metal stripes. Phys Rev B 64:045411
29. Shalaev VM, Kawata S (2007) Nanophotonics with surface plasmons. Elsevier, Amsterdam
30. Xie Z, Yu W, Wang T et al (2011) Plasmonic nanolithography: a review. Plasmonics 6:565–580
31. Luo X, Yan L (2012) Surface plasmon polaritons and its applications. IEEE Photonic J 4:590–595
32. Srituravanich W, Fang N, Sun C et al (2004) Plasmonic nanolithography. Nano Lett 4:1085–1088
33. Srituravanich W, Fang N, Durant S et al (2004) Sub-100 nm lithography using ultrashort wavelength of surface plasmons. J Vac Sci Technol B 22:3475–3478
34. Srituravanich W, Durant S, Lee H et al (2005) Deep subwavelength nanolithography using localized surface plasmon modes on planar silver mask. J Vac Sci Technol B 23:2636–2639
35. Barnes WL, Dereux A, Ebbesen TW (2003) Surface plasmon sub-wavelength optics. Nature 424:824–830
36. Ozbay E (2006) Plasmonics: merging photonics and electronics at nanoscale dimensions. Science 311:189–193
37. Grigorenko AN, Geim AK, Gleeson HF et al (2005) Nanofabricated media with negative permeability at visible frequencies. Nature 438:335–338
38. Stefan AM (2007) Plamonics: fundamentals and applications. Springer, Berlin
39. Kretschmann E, Raether H (1968) Radiative decay of non-radiative surface plasmons excited by light. Zeitschrift Fur Naturforschung Part a-Astrophysik Physik Und Physikalische Chemie 23A:2135–2136
40. Otto A (1968) Excitation of nonradiative surface plasma waves in silver by the method of frustrated total reflection. Zeitschrift Fur Physik 216:398–410
41. Ritchie RH, Arakawa ET, Cowan JJ et al (1968) Surface-plasmon resonance effect in grating diffraction. Phys Rev Lett 21:1530–1532

42. Zayats AV, Smolyaninov II (2003) Near-field photonics: surface plasmon polaritons and localized surface plasmons. J Opt A-Pure Appl Opt 5:S16–S50
43. Zayats AV, Smolyaninov II, Maradudin AA (2005) Nano-optics of surface plasmon polaritons. Phys Rep 408:131–314
44. Thio T, Pellerin KM, Linke RA et al (2001) Enhanced light transmission through a single subwavelength aperture. Opt Lett 26:1972–1974
45. Lezec HJ, Degiron A, Devaux E et al (2002) Beaming light from a subwavelength aperture. Science 297:820–822
46. Garcia-Vidal FJ, Martin-Moreno L, Ebbesen TW et al (2010) Light passing through subwavelength apertures. Rev Mod Phys 82:729–787
47. Pendry JB (2000) Negative refraction makes a perfect lens. Phys Rev Lett 85:3966–3969
48. Fang N, Zhang X (2003) Imaging properties of a metamaterial superlens. Appl Phys Lett 82:161–163
49. Brongersma ML, Kik PG (2007) Surface plasmon nanophotonics. Springer, Berlin
50. Guo X, Du J, Guo Y et al (2006) Large-area surface-plasmon polariton interference lithography. Opt Lett 31:2613–2615
51. Xiong W, Du JG, Fang L et al (2008) 193 nm interference nanolithography based on SPP. Microelectron Eng 85:754–757
52. Sreekanth KV, Murukeshan VM (2010) Large-area maskless surface plasmon interference for one- and two-dimensional periodic nanoscale feature patterning. J Opt Soc Am A-Opt Image Sci Vis 27:95–99
53. Sreekanth KV, Murukeshan VM (2011) Multiple beams surface plasmon interference generation: a theoretical analysis. Opt Commun 284:2042–2045
54. Lim Y, Kim S, Kim H et al (2008) Interference of surface plasmon waves and plasmon coupled waveguide modes for the patterning of thin film. Ieee J Quant Electron 44:305–311
55. Guo X, Dong Q (2010) Coupled surface plasmon interference lithography based on a metal-bounded dielectric structure. J Appl Phys 108:113108
56. He M, Zhang Z, Shi S et al (2010) A practical nanofabrication method: surface plasmon polaritons interference lithography based on backside-exposure technique. Opt Express 18:15975–15980
57. Pockrand I (1978) Surface plasma oscillations at silver surfaces with thin transparent and absorbing coatings. Surf Sci 72:577–588
58. Luo XG, Ishihara T (2004) Surface plasmon resonant interference nanolithography technique. Appl Phys Lett 84:4780–4782
59. Luo XG, Ishihara T (2004) Subwavelength photolithography based on surface-plasmon polariton resonance. Opt Express 12:3055–3065
60. Doskolovich LL, Kadomina EA, Kadomin II (2007) Nanoscale photolithography by means of surface plasmon interference. J Opt A-Pure Appl Opt 9:854–857
61. Bezus EA, Bykov DA, Doskolovich LL et al (2008) Diffraction gratings for generating varying-period interference patterns of surface plasmons. J Opt A-Pure Appl Opt 10:095204
62. Bezus EA, Doskolovich LL (2010) Grating-assisted generation of 2D surface plasmon interference patterns for nanoscale photolithography. Opt Commun 283:2020–2025
63. Xu T, Zhao Y, Ma J et al (2008) Sub-diffraction-limited interference photolithography with metamaterials. Opt Express 16:13579–13584
64. Yang X, Zeng B, Wang C et al (2009) Breaking the feature sizes down to sub-22 nm by plasmonic interference lithography using dielectric-metal multilayer. Opt Express 17:21560–21565
65. Xiong Y, Liu Z, Zhang X (2008) Projecting deep-subwavelength patterns from diffraction-limited masks using metal-dielectric multilayers. Appl Phys Lett 93:111116
66. Yang XF, Fang LA, Zeng BB et al (2010) Deep subwavelength photolithography based on surface plasmon polariton resonance with metallic grating waveguide heterostructure. J Opt 12:045001

67. Yang X, Li W, Zhang D (2012) Subwavelength lithography using metallic grating waveguide heterostructure. Appl Phys A-Mater Sci Process 107:123–126
68. Ge W, Wang C, Xue Y et al (2011) Tunable ultra-deep subwavelength photolithography using a surface plasmon resonant cavity. Opt Express 19:6714–6723
69. Liu ZW, Wei QH, Zhang X (2005) Surface plasmon interference nanolithography. Nano Lett 5:957–961
70. Wei XZ, Luo XG, Dong XC et al (2007) Localized surface plasmon nanolithography with ultrahigh resolution. Opt Express 15:14177–14183
71. Murukeshan VM, Sreekanth KV (2009) Excitation of gap modes in a metal particle-surface system for sub-30 nm plasmonic lithography. Opt Lett 34:845–847
72. Guo XW, Du JL, Luo XG et al (2007) Surface-plasmon polariton interference nanolithography based on end-fire coupling. Microelectron Eng 84:1037–1040
73. Liu Z, Wang Y, Yao J et al (2009) Broad band two-dimensional manipulation of surface plasmons. Nano Lett 9:462–466
74. Ueno K, Takabatake S, Onishi K et al (2011) Homogeneous nano-patterning using plasmon-assisted photolithography. Appl Phys Lett 99:011107
75. Ueno K, Takabatake S, Nishijima Y et al (2010) Nanogap-assisted surface plasmon nanolithography. J Phys Chem Lett 1:657–662
76. Fang N, Lee H, Sun C et al (2005) Sub-diffraction-limited optical imaging with a silver superlens. Science 308:534–537
77. Lee H, Xiong Y, Fang N et al (2005) Realization of optical superlens imaging below the diffraction limit. New J Phys 7:1–16
78. Zhang Y, Dong X, Du J et al (2010) Nanolithography method by using localized surface plasmon mask generated with polydimethylsiloxane soft mold on thin metal film. Opt Lett 35:2143–2145
79. Chaturvedi P, Wu W, Logeeswaran VJ et al (2010) A smooth optical superlens. Appl Phys Lett 96:043102
80. Shi Z, Kochergin V, Wang F (2009) Depth-of-focus (DoF) analysis of a 193nm superlens imaging structure. Opt Express 17:20538–20545
81. Shi Z, Kochergin V, Wang F (2009) 193nm superlens imaging structure for 20nm lithography node. Opt Express 17:11309–11314
82. Xu T, Fang L, Ma J et al (2009) Localizing surface plasmons with a metal-cladding superlens for projecting deep-subwavelength patterns. Appl Phys B-Lasers Opt 97:175–179
83. Blaikie RJ, Melville DOS, Alkalsi MM (2006) Super-resolution near-field lithography using planar silver lenses: a review of recent developments. Microelectron Eng 83:723–729
84. Liu Z, Lee H, Xiong Y et al (2007) Far-field optical hyperlens magnifying sub-diffraction-limited objects. Science 315:1686–1686
85. Liu Z, Durant S, Lee H et al (2007) Far-field optical superlens. Nano Lett 7:403–408
86. Liu Z, Durant S, Lee H et al (2007) Experimental studies of far-field superlens for sub-diffractional optical imaging. Opt Express 15:6947–6954
87. Zhang X, Liu Z (2008) Superlenses to overcome the diffraction limit. Nat Mater 7:435–441
88. Yao N, Wang C, Ren G et al (2012) Demagnifying far field focusing spot to deep subwavelength scale by truncated hyperlens for nanolithography. Superlattice Microst 52:63–69
89. Wang W, Xing H, Fang L et al (2008) Far-field imaging device: planar hyperlens with magnification using multi-layer metamaterial. Opt Express 16:21142–21148
90. Han S, Xiong Y, Genov D et al (2008) Ray optics at a deep-subwavelength scale: a transformation optics approach. Nano Lett 8:4243–4247
91. Kildishev AV, Shalaev VM (2008) Engineering space for light via transformation optics. Opt Lett 33:43–45
92. Wang Y, Srituravanich W, Sun C et al (2008) Plasmonic nearfield scanning probe with high transmission. Nano Lett 8:3041–3045

93. Erdmanis M, Viegas D, Hautakorpi M et al (2011) Comprehensive numerical analysis of a surface-plasmon-resonance sensor based on an H-shaped optical fiber. Opt Express 19:13980–13988
94. Sun M, Liu R-J, Li Z-Y et al (2007) Enhanced near-infrared transmission through periodic H-shaped arrays. Phys Lett A 365:510–513
95. Cho E-H, Kang S-M, Leen JB et al (2010) Polymeric light delivery via a C-shaped metallic aperture. J Opt Soc Am B-Opt Phys 27:1309–1316
96. Chen YC, Fang JY, Tien CH (2006) Double-corrugated C-shaped aperture for near-field recording. Jpn J Appl Phys Part 1-Regular Pap Brief Commun Rev Pap 45:1348–1350
97. Zhou L, Gan Q, Bartoli FJ et al (2009) Direct near-field optical imaging of UV bowtie nanoantennas. Opt Express 17:20301–20306
98. Kim Y, Kim S, Jung H et al (2009) Plasmonic nano lithography with a high scan speed contact probe. Opt Express 17:19476–19485
99. Murphy-DuBay N, Wang L, Kinzel EC et al (2008) Nanopatterning using NSOM probes integrated with high transmission nanoscale bowtie aperture. Opt Express 16:2584–2589
100. Srituravanich W, Pan L, Wang Y et al (2008) Flying plasmonic lens in the near field for high-speed nanolithography. Nat Nanotechnol 3:733–737
101. Pan L, Park Y-S, Xiong Y et al (2010) Flying plasmonic lens at near field for high speed nano-lithography. Proc SPIE 7637:763713

Chapter 7
Mass Production of Large-Format Micro-/Nanostructure-Based Optical Devices

7.1 Introduction

It is now more than half a century since the invention of the transistor and the advent of solid-state electronics. The rate at which integrated circuits (ICs) have both improved in performance and declined in cost has been unparalleled, bringing resolutions not only in computing and telecommunications, but actually catalyzing societal change. The ability to fabricate structures from the micro- to the nanoscale with high precision in a wide variety of materials is of crucial importance to the advancement of micro- and nanotechnology and the nanosciences. The days when people thought you could not print features at dimensions smaller than the wavelength of the light being used to do the printing are just a memory. However, nobody can deny that semiconductor lithography continues to become a more and more complex beast with extremely high cost nowadays. As one of the potential successors to 193 nm lithography for 32 nm processing, extreme ultraviolet lithography at 13.5 nm (EUV) is still facing serious technical and economical challenges. The cost of a single EUV exposure tool has been estimated to be above $50 million, and most potential users cannot afford this price [1]. Significant efforts have been made to address the critical issues involved in EUV lithography, such as EUV optics metrology, resist development, and mask defect inspection.

Microoptics and nanophotonics have emerged as exciting new arenas concerned with the interaction of light with micro- and nanostructured materials [2]. By adding a new dimension to microscale science and technology, microoptics and nanophotonics provide challenges for fundamental research and create opportunities for new technologies. Although microoptics and nanophotonics share the technological basis with microelectronics, the structuring techniques may differ considerably. This is caused by the fact that the micro- and nanostructures that are necessary for the functionality of the optical elements can roughly be divided into different classes. The requirements of the profile shape are very different in the different classes, and the specifications and tolerances cannot be generalized for the structures of microoptical elements. If microoptics and nanophotonics are to maximize their impact on the

Q. Liu et al., *Novel Optical Technologies for Nanofabrication*, Nanostructure Science and Technology, DOI 10.1007/978-3-642-40387-3_7,
© Springer-Verlag Berlin Heidelberg 2014

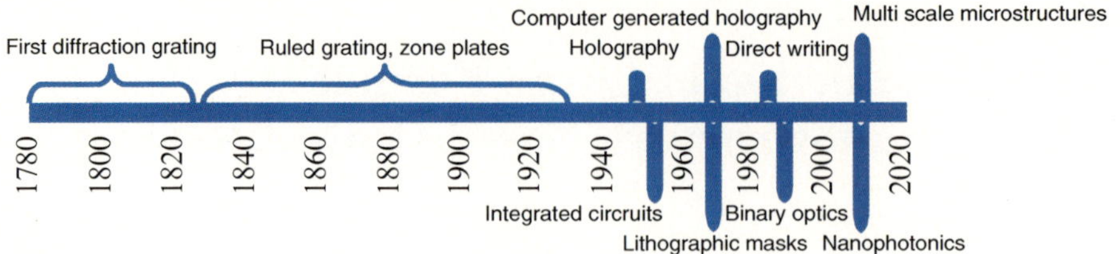

Fig. 7.1 A timeline illustrating key milestones in the history of microoptic fabrication

market and the next generation of technology, those within the fields will need to form a cohesive plan. The timetable for the emergence of new optical platform is dictated by the creation of the necessary infrastructure in circuit design, the adaptation of CMOS processes to optical devices, and the development of new materials.

For the microoptics part, a paper published in 2001 discussed the technical and manufacturing hurdles facing the fabrication of surface relief microoptics in the future [3], shown in Fig. 7.1. It states that *"growing needs and new applications for micro-optical components and systems have driven the need for advances in fabrication and integration technologies for both diffractive and refractive micro-optics. Many capabilities are required, including smaller feature, faster etch rate, precision process control, and many others."* With the advent of the twenty-first century, the design and fabrication of flexible optoelectronic devices, such as functional microlens array film, brightness-enhancement component used in backlight unit, and transparent conductive film, has become hot topics in the relative society. Many novel fabrication approaches, such as roll-to-roll (R2R) process [4], continuous dynamic nanoinscribing [5], and injection mold method [6], have been proposed to satisfy the increasing demand on the requirements of microoptic devices.

For the nanophotonics part, the exploitation of optical phenomena on the nanoscale nanophotonics is opening up a diverse field of study that promises to deliver novel technological solutions. Representatives of several major photonics roadmapping projects met at Photonics West in San Jose, California, on 25 Jan. 2007, to identify emerging common themes and goals for nanophotonics [7]. It has been addressed that the technological needs, low-cost manufacturing platform, integrated packaging solution, and the infrastructure needed are key issues in depicting a nanophotonic roadmap. Although photonic technologies have revolutionized communications and will do the same for the fields of imaging, display, and lighting applications, the market and technology barriers to this health are still significant. In Europe, more than 300 people from industry and academia working together have contributed to the construction of the roadmap that gives insight into the future of materials, equipment, processes, and applications, which has been well known as the "Merging Optics and Nanotechnologies (MONA)" roadmap [8]. This roadmap identifies key equipment and processes for the improvement of the performance of nanophotonics. The processes that will have the highest potential impact on nanophotonics and at the same time have potential for mass production are MOCVD, CNT CVD, colloidal synthesis, nanophosphor fabrication, sol–gel synthesis, OVPD, UV lithography, nanoimprinting, and etching. The types of equipment and processes with the broadest field of applications are MOCVD, MBE, and colloidal chemistry as bottom-up technologies and UV lithography, e-beam lithography, and nanoimprinting lithography (NIL) as top-down technologies. NIL is likely becoming a key mass-production method to prepare surface reliefs at nanoscale cost-effectively.

In this chapter, an efficient method for mass production of large-format micro-/ nanostructure-based optical devices is proposed and discussed. Firstly, interference lithography and spatial light modulator-based lithography fabricating large-format imprinting mold are present. Several types of prototype machine built in-house are introduced. After exposure and development, the micro-/nanostructure on photo-sensitive material is transferred to metal (Ni, Cu, etc.) through a quasi-LIGA process. The obtained metal mold with surface relief structure is wrapped on a roller as master mold for roll-to-roll process. Through this process chain, large-format micro- or nanostructure-based optical devices can be manufactured in an economical way. Several types of flexible roll-to-roll NIL (UV, thermal-embossing, R2R seamless NIL) techniques are discussed in the following sections. We will also investigate several key issues in R2R NIL process, such as imprinting resist, mold–sample separation, fidelity, and resolution. Finally, potential applications of these techniques are briefly reviewed.

7.2   Approaches of Fabricating Large-Format Imprinting Master Mold

Imprinting master mold and its fabrication is one of the key issues of NIL, which is one of the next-generation lithography approaches that have demonstrated great capability for the fabrication of large-scale planar micro- and nanostructures. The mold is analogous to the photomask in the photolithography. Being a direct contact technique, the resolution depends directly on the size and quality of the mold. In principle the master mold in the NIL process can be fabricated by any patterning technologies that are available, including conventional ones like electron beam writing and focused ion beam writing followed by wet or dry etching, as shown in Fig. 7.2. The resolution of the mold, in turn, depends on the mask technology providing the original patterns. The first-generation mold, fabricated by using existing mask technology, is usually called a master stamp.

The era of photolithography started to develop in the 1970s. It was driven mainly by the development of integrated circuits (ICs). The IC industry created a need for high volume, perfect replication of ever smaller patterns on a substrate, at minimal costs. The main method to achieve this is optical lithography, which can be cataloged as mask lithography and maskless lithography. The branch of mask lithography utilizes templates, also known as photomasks, having transparent and opaque areas. Light is shone through the photomask on a substrate which is coated

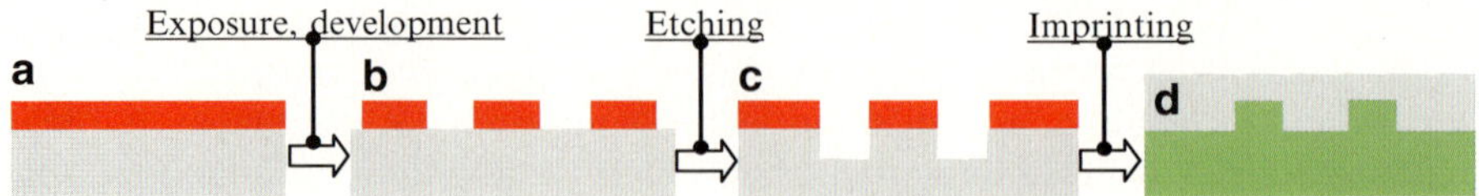

Fig. 7.2 Master mold fabrication by using conventional semiconductor patterning approach. (**a**) Photoresist coating, (**b**) exposure and development, (**c**) dry or wet etching, (**d**) imprinting

with a photosensitive thin film called photoresist. Photoresist areas that are exposed to light will transform to either soluble or insoluble to the photoresist developer. For the maskless lithography case, instead, most commonly, the radiation is focused to a narrow beam. The beam is then used to directly write the image into the photoresist, one or more pixels at a time. An alternative method, developed by Micronic Laser Systems [9] or Heidelberg Instruments [10], is to scan a programmable reflective photomask, which is then imaged onto the photoresist. This has the advantage of higher throughput and flexibility. Both methods are used to define patterns on photomasks [11]. A key advantage of maskless lithography is the ability to change lithography patterns from one run to the next, without incurring the cost of generating a new photomask.

If exposure makes photoresist soluble, it is called positive photoresist and vice versa, and if photoresist becomes insoluble, it is called negative photoresist. Light replicates the patterns from the photomask to the photosensitive film, and further steps are taken to transfer the copied patterns to the substrate. In the early 1970s, the required dimensions for the ICs ranged from 2 to 5 μm. Replication of these patterns was simply achieved by using mercury-arc-based UV light and bringing the photomask and the substrate in close proximity or into contact during the exposure. Systems based on this operation principle are still used today in microfabrication, due to their simplicity, low cost, high throughput, and good process quality. These systems, called UV-contact mask aligners, reach resolutions from a few micrometers to the sub-half-micron level, depending on the exposure wavelength and the contact method. With fully automated systems, the throughput can exceed 100 wafers per hour (wph) and reach an overlay accuracy of 0.25 μm [12] (SUSS MicroTec). An ever increasing demand for smaller and smaller linewidths has required more complex exposure systems. Nowadays, state-of-the-art systems in IC production reach 32 nm linewidths by using deep-UV ArF-light sources at 193 nm wavelength and exposing patterns using immersion scanners, phase shift masks, and double exposure schemes. Exposure is based on an image reduction technique that projects the photomask onto the substrate and simultaneously reduces the size of the patterns many times. This allows the photomasks to be fabricated with looser tolerances than the final pattern. These systems are also very productive and able to pattern more than one hundred and fifty 300 mm wafers per hour and to reach better than 2.5 nm alignment between subsequent patterning steps [13] (ASML).

After exposure and development, the surface relief structure on the photoresist is needed to be transferred to the substrate by wet or dry etching. Wet etching is characterized by where the material is dissolved when immersed in a chemical

Table 7.1 Comparison of wet and dry etching process

	Wet	Dry
Method	Chemical solutions	Ion bombardment or chemical reactive
Environment and equipment	Atmosphere bath	Vacuum chamber
	Low cost, easy to implement	Capable of defining small feature size (<100 nm)
	High etching rate	
	Good selectivity for most materials	
Disadvantage	Inadequate for defining feature size <1 μm	High cost, hard to implement
		Low throughput
	Potential of chemical handling hazards	Poor selectivity
	Wafer contamination issues	Potential radiation damage
Directionality	Isotropic (except for etching crystal-line materials)	Anisotropic

solution, and dry etching is characterized by where the material is sputtered or dissolved using reactive ions or a vapor phase etchant. Wet etch processes can lead to undercutting, resulting in an isotropic etch profile where the vertical etch rate is approximately equal to the horizontal etch rate. As semiconductor manufacturers continue to shrink feature sizes, this undercutting becomes more intolerable, making wet etching a less desirable technique for material removal. Dry etching is one of the most widely used processes in semiconductor manufacturing. Dry etching uses plasma-generated free radicals to remove material only from the area dictated by a photoresist pattern created during the photolithography step. Only dry etching provides the ability to control sidewall anisotropic etching through plasma-generated nonreactive and reactive ions. Usually, the materials used as master mold are quartz, silicon, or silicon nitride. The fabrication can be found everywhere on the Internet or textbooks about semiconductor process. Table 7.1 compares the two types of etching process. Wet etching is cost-effective, but only suitable for fabricating large feature size. Dry etching is capable of defining small feature size with the limits of low throughput and selectivity.

It is expensive and time-consuming and not suitable for preparing large-format master mold by conventional fabrication methods. In order to achieve industrial-level manufacturing using NIL, one of the key issues is to supply sufficient stamps with lowest cost to high-volume manufacturing machines (HVM). It is therefore preferable to produce the maximum amounts of replicated stamps from a master and to ensure that each stamp replica can deliver as many imprints as possible without losing fidelity. In contrast, nickel stamps have high mechanical strength and durability and more importantly can be cost-effectively produced via electroforming process; therefore, they have large potential to replace the fragile stamps (such as Si, quartz) in NIL, especially when used for using industrial high-volume manufacturing machines. The galvano-templates for nickel electroforming are based on resist templates, as shown in Fig. 7.3, photoresist defined the structures

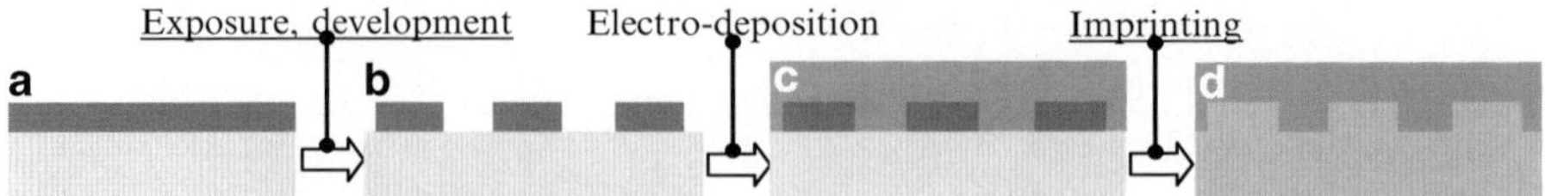

Fig. 7.3 Master mold fabrication by conventional semiconductor patterning approach. (**a**) Photoresist coating, (**b**) exposure and development, (**c**) dry or wet etching, (**d**) imprinting

produced by photolithography. The advantage of this method is the facilitation of easy separation between resist template and electroformed nickel stamp. This method is generic, simple, cost-effective, and promising for mass production of nickel stamp [14].

7.2.1   Interference Lithography

Interference lithography utilizes the interference of two or more coherent laser beams that form an interference pattern on the substrate, which was proposed by Michelson in 1927 [15]. Using the same photoresist material as used in optical mask lithography, the interference pattern can be transferred to the photoresist and subsequently to the other layers on the substrate.

From elementary electromagnetism, one can easily deduce the period p of the interference fringes at a substrate when two plane waves interfere,

$$p = \frac{\lambda}{\sin \theta_1 + \sin \theta_2} \tag{7.1}$$

where λ is the wavelength and θ_1 and θ_2 are the incident angles for the left and the right arms, respectively (Fig. 7.4a). Figure 7.4b illustrates the conventional interference lithography setup. As drawn, the incident angles are assumed equal. A beam splitter based on optical film is used to produce the two interference beams. The splitting beams from a semiconductor ($\lambda = 405$ nm) or argon ion laser ($\lambda = 405$ nm) are spatially filtered before interfering at the substrate. Pinholes here are used as spatial filter to get rid of the undesired spatial frequencies. In practice, however, this setup is sensitive to the turbulence, vibration, and thermal drift limit for the introduction of two separate mirrors. And for fabricating large-format samples, the propagation distance is large. In the MIT setup, this distance is nominally 1 m, which may result in phase nonlinearity [16]. Theoretically, lenses may be used to collimate the beams after the spatial filter and thus eliminate the hyperbolic phase nonlinearity. However, it is questionable whether it is practical to fabricate, align, and keep large optics capable of producing meter-sized gratings with nanometer phase nonlinearity.

In our optical interference head (shown in Fig. 7.4c), it is a grating that splits the incident beam into two beams. Compared to a cube beam splitter, the gating splitter

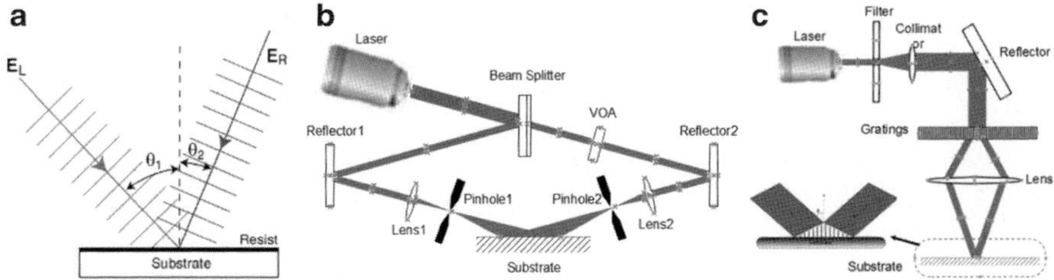

Fig. 7.4 (**a**) Two-beam interference; experimental setup using (**b**) cube beam splitter and (**c**) grating

provides greater tolerance over the laser spatial incoherent length as well as its temporal incoherence. Bi-telecentric lens, which is telecentric in both object and image space, is chosen as interference optical head (as illustrated in Fig. 7.5a). The grating splitter is located in the image plane, and the $\pm$1st-order diffracted beams are recombined on the surface of substrate. The value of numerical aperture (NA) is 0.35 and 0.6 in object space and image space, respectively. The working wavelength is 351 nm and the minimum grating period which can be achieved is 300 nm. If the NA of the imaging lens group is 0.95, the obtained grating period can be further decreased to 195 nm. When the refractive index of fused silica is 1.458, the spot diagram of the proposed bi-telecentric lens is shown in Fig. 7.5b. It can be seen that the root mean square of the radius at normal incidence is 0.99 μm, and those are 0.67 μm at 150 μm and 1.99 μm at 212 μm. The proposed lens system exhibits no parallax error and provides uniform illumination and eliminates vignetting. The optimum lens data and its photo are shown in Fig. 7.5c, d.

Figure 7.6a demonstrates the self-developed digitalized laser direct writing (LDW) system *Holoscan®* based on the two-beam interference. Table 7.2 shows *Holoscan®* specifications. The scanning area is 800 × 600 mm. By using special vector scanning mode, its working speed can reach 600 mm/s. The dimension size of each exposure unit is 300 × 300μm. For the splitting grating can be driven by a motor rotated in its working plane, the orientation of the interference grating on the photoresist is in the range of −90° to + 90°. It makes master mold of large-format diffractive optical variable image possible. Versatile file formats are supported by this system, such as .bmp, .dxf, and self-defined .arr files. Figure 7.6b is the photo of microstructure under microscopy, and Fig. 7.6c, d are the photos of optical variable diffracted image. The visual color is generated by adjusting each interference unit's grating orientation and period. Figure 7.6e shows the interference grating with a period of 400 nm, and Fig. 7.6f is the scanning electronic microscopy (SEM) photo of the nanodots array in which each dot size is 150 nm. Such a nanodot array can be fabricated by twice exposure by rotating the split grating 90°. Compared with the nanoruler developed in MIT, it is for Holoscan to deal with higher efficiency and more supporting data formats because there is no individual mirrors used in the optical path.

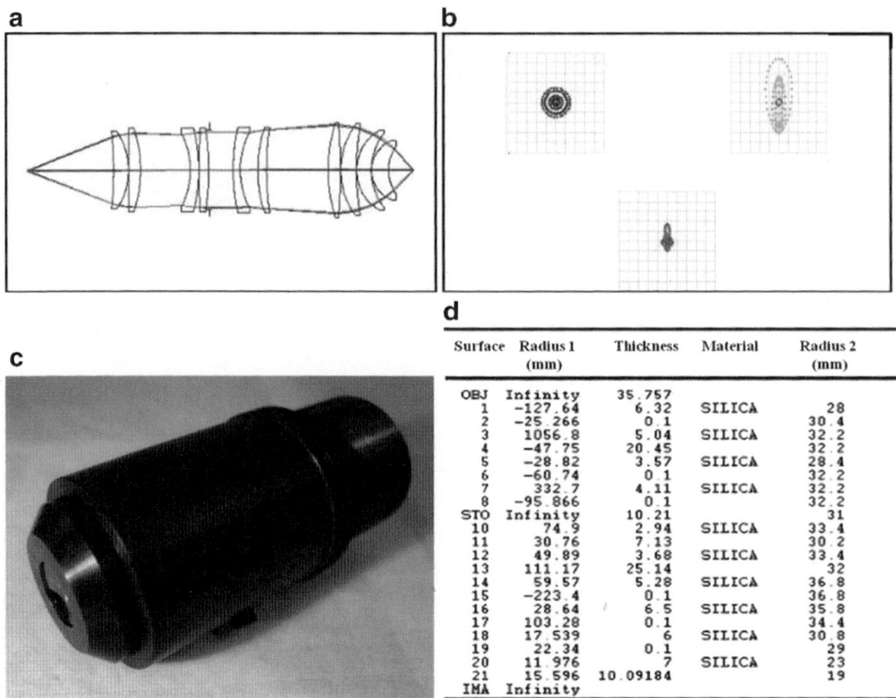

Surface	Radius 1 (mm)	Thickness	Material	Radius 2 (mm)
OBJ	Infinity	35.757		
1	-127.64	6.32	SILICA	28
2	-25.266	0.1		30.4
3	1056.8	5.04	SILICA	32.2
4	-47.75	20.45		32.2
5	-28.82	3.57	SILICA	28.4
6	-60.74	0.1		32.2
7	332.7	4.11	SILICA	32.2
8	-95.866	0.1		32.2
STO	Infinity	10.21		31
10	74.9	2.94	SILICA	33.4
11	30.76	7.13		30.2
12	49.89	3.68	SILICA	33.4
13	111.17	25.14		32
14	59.57	5.28	SILICA	36.8
15	-223.4	0.1		36.8
16	28.64	6.5	SILICA	35.8
17	103.28	0.1		34.4
18	17.539	6	SILICA	30.8
19	22.34	0.1		29
20	11.976	7	SILICA	23
21	15.596	10.09184		19
IMA	Infinity			

Fig. 7.5 (**a**) Bi-telecentric lens system, its spot diagram (**b**) and (**c**) photo, (**d**) design parameters

7.2.2   *Spatial Light Modulator-Based Lithography*

Spatial light modulators (SLMs) are optical devices which can modulate incident light parameters, such as amplitude, phase, polarization, and frequency. Two-dimensional SLMs are commonly used in reality. It can be based on several physical phenomena: linear electrooptic Pöckel effect, quadratic electrooptic Kerr effect, acousto-optic Bragg diffraction, magnetooptical Faraday effect, and different types of electromechanical effects, e.g., micromirrors.

A digital micromirror device, or DMD, is an optical semiconductor invented by Dr. Larry Hornbeck and Dr. William E. "Ed" Nelson of Texas Instruments (TI) in 1987 [17]. It is an important kind of SLM and has widely used in digital light processing (DLP) projection technology. A DMD chip has on its surface several hundred thousand microscopic mirrors arranged in a rectangular array which correspond to the pixels in the image to be displayed. The mirrors can be individually rotated ± 10–$12°$ to an on or off state. In the on state, light from the projector bulb is reflected into the lens, making the pixel appear bright on the screen. In the off state, the light is directed elsewhere (usually onto a heatsink), making the pixel appear dark. To produce grayscales, the mirror is toggled on and off very quickly, and the ratio of on time to off time determines the shade produced (binary pulse width modulation).

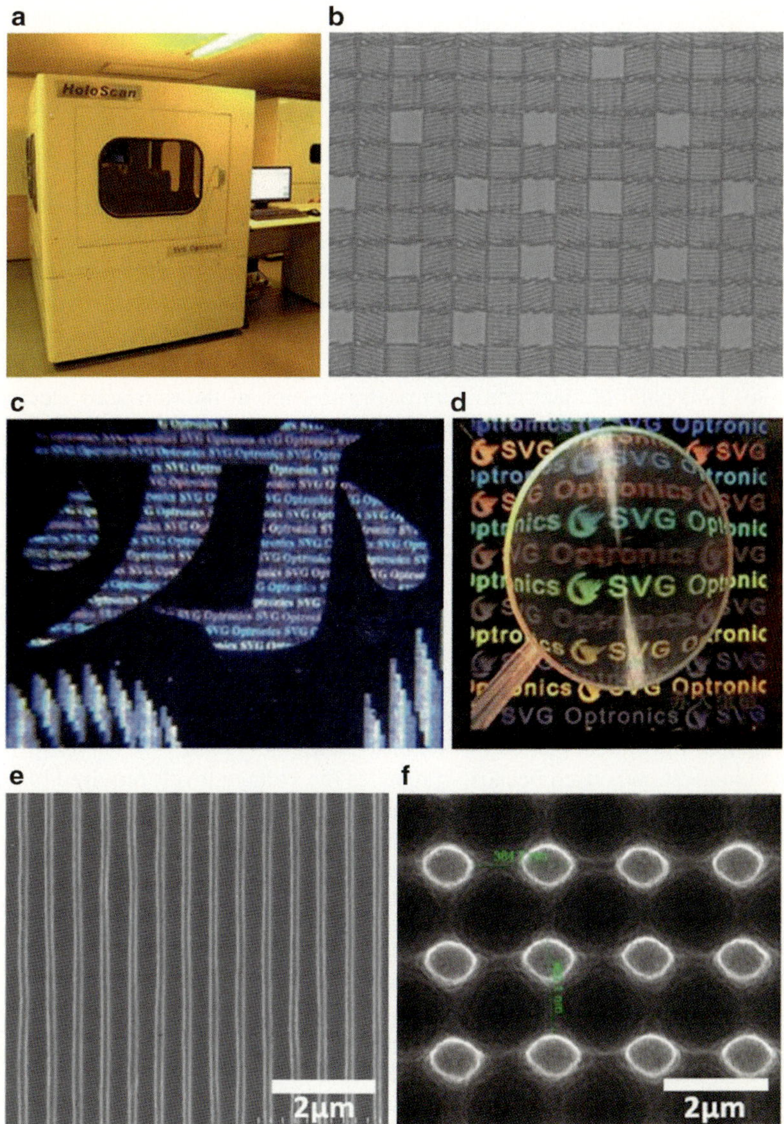

Fig. 7.6 (**a**) Holoscan® LDW system, (**b**) microscopy photo of the fabricated microstructure, (**c–d**) the optical variable diffracted image, (**e**) grating with a period of 400 nm, (**f**) nanodots array fabricated by twice exposure

Contemporary DMD chips can produce up to 1,024 shades of gray (10 bits). The mirrors themselves are made out of aluminum and are around 10–16 μm across. Each one is mounted on a yoke which in turn is connected to two supporting posts by compliant torsion hinges. In this type of hinge, the axle is fixed at both ends and

Table 7.2 Specifications of HoloScan ® LDW system

Format	800 × 600 mm
Efficiency	600 mm/s
Feature size	200 nm (@351 nm)
Repeat accuracy	100 nm
Support data format	Arr, dxf, bmp, etc.
Exposure unit	300 μm^2
Grating orientation	$-90^\circ - + 90^\circ$

literally twists in the middle (shown in Fig. 7.7-a). Because of the small scale, hinge fatigue is not a problem, and tests have shown that even one trillion (10^{12}) operations do not cause noticeable damage. Tests have also shown that the hinges cannot be damaged by normal shock and vibration, since it is absorbed by the DMD super-structure. Two pairs of electrodes control the position of the mirror by electrostatic attraction. Each pair has one electrode on each side of the hinge, with one of the pairs positioned to act on the yoke and the other acting directly on the mirror. The majority of the time, equal bias charges are applied to both sides simultaneously. Instead of flipping to a central position as one might expect, this actually holds the mirror in its current position. This is because attraction force on the side the mirror is already tilted toward is greater, since that side is closer to the electrodes. To move the mirrors, the required state is first loaded into an SRAM cell located beneath each pixel, which is also connected to the electrodes. Once all the SRAM cells have been loaded, the bias voltage is removed, allowing the charges from the SRAM cell to prevail, moving the mirror. When the bias is restored, the mirror is once again held in position, and the next required movement can be loaded into the memory cell.

The bias system is used because it reduces the voltage levels required to address the pixels such that they can be driven directly from the SRAM cell, and also because the bias voltage can be removed at the same time for the whole chip, every mirror moves at the same instant. The advantages of the latter are more accurate timing and a more cinematic moving image. Figure 7.7b shows a type of commercially available DMD-modeled Discovery D4100, and its specifications are listed in Table 7.3, from which we can see the pixel size is 10.8 × 10.8 μm with a very fast refresh frequency 12,000 Hz.

DMD has been used in the pattern generator system [18] for photolithographic mask fabrication or for direct in situ processing on planar substrates. Actually, laser maskless lithography has been developed over more than 20 years. In 1987, the first laser direct writing system CORE-2000 was invented by Etec company. Due to the relative low efficiency compared with optical lithography technique with mask, laser maskless lithography was not put into the industrial fabrication of integrated circuit at that time. However, as the higher resolution are required by the semiconductor society, laser maskless lithography was gradually regarded as an effective method to make masks used in IC exposure. Nowadays, more than 70 % of masks used in IC fabrication are made by laser maskless lithography. Figure 7.8 shows a laser writer system developed by Micronic Mydata from Sweden. The pattern input into the SLM is projected onto the substrate. Thus the optical lens can be seen as a variable mask used in the conventional stepper system [19]

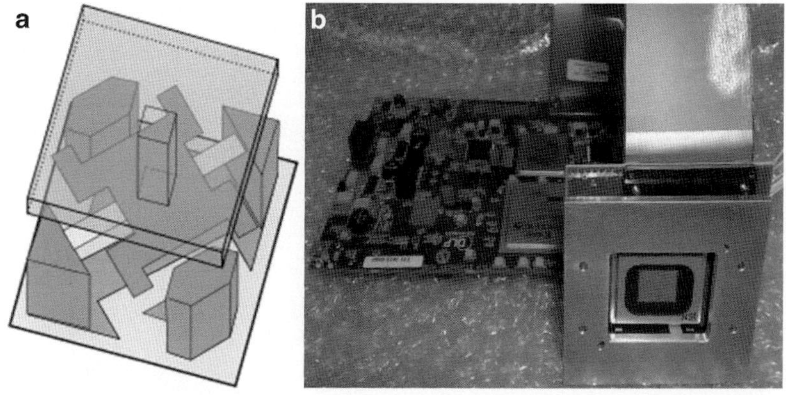

Fig. 7.7 (**a**) Diagram of a digital micromirror showing the mirror mounted on the suspended yoke with the torsion spring running bottom left to top right (*light grey*), with the electrostatic pads of the memory cells below, (**b**) photo of DMD-Discovery D4100

Table 7.3 DMD specifications

Pixel size	10.8 um
Format	0.7/0.55 in.
Matrix	1,024 × 768
Work window	UV/VIS
Refresh frequency	12,000 Hz
Driven software	ALP-4
Switch time	200 ms

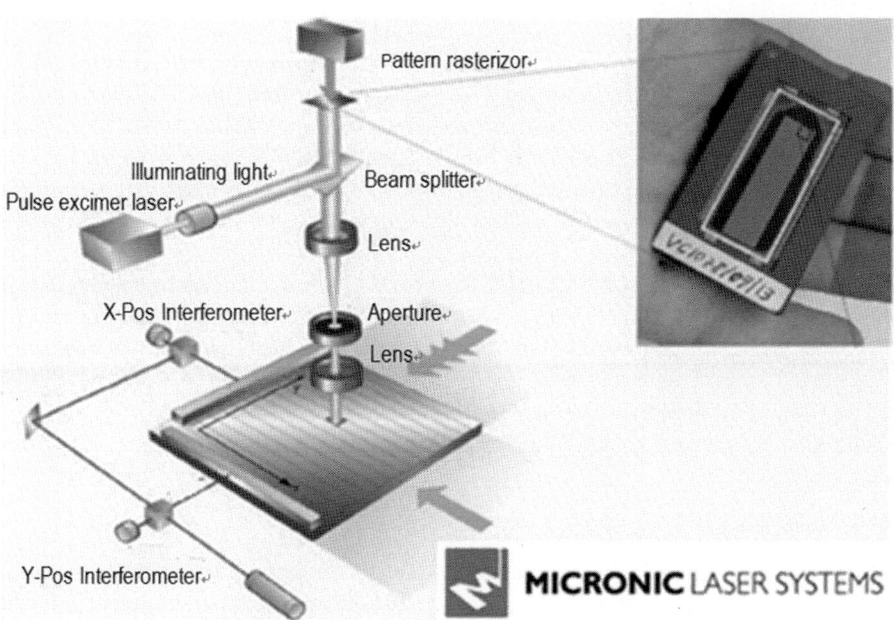

Fig. 7.8 Laser maskless lithography system Sigma developed by Micronic Mydata

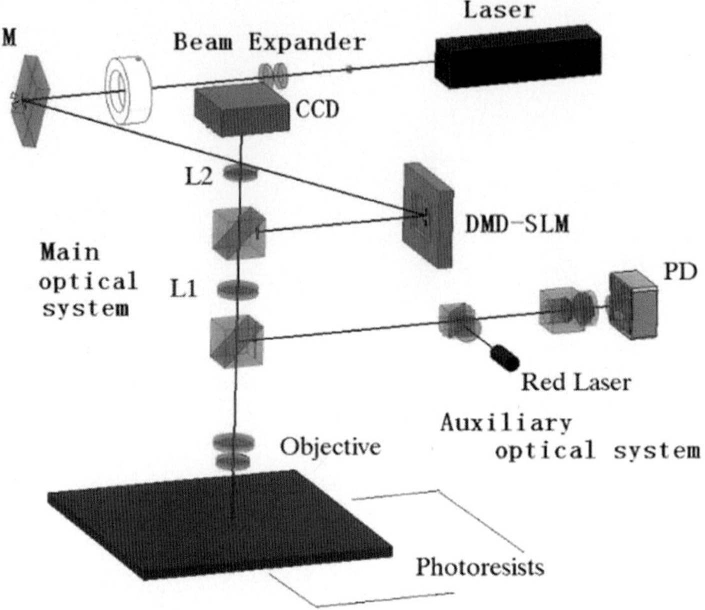

Fig. 7.9 MicrolabA100® LDW system

In order to realize a laser maskless lithography system with high efficiency, high resolution, and low cost, we propose the MicrolabA100® system (as shown in Fig. 7.9). Its specifications are listed in Table 7.4. The source laser beam is conditioned to uniformly illuminate the DMD via an illumination scheme similar to that found in many commercial DLP® projectors. The lens L1 and flat-field objective lens can be used as a bi-telecentric lens system. DMD is on the focal plane of L1 and the surface of photoresist is on the focal plane of objective lens. CCD is aided to observe the projected image. As for the auxiliary optical system, a photodiode detector is used to collect the reflective light signal for tracking the trace of focal spot.

Another LDW system developed (iGrapher®) is characterized by faster writing speed than MicrolabA100® because of the adoption of advanced linear motor and exposure mode, as demonstrated in Fig. 7.10 and Table 7.5. The working rate in this system is 50 mm^2/min with a resolution of 0.5 μm or 300 mm^2/min with a resolution of 1.0 μm, ten times faster than that of MicrolabA100® system. Several key techniques guarantee the iGrapher® LDW system's high performance, which are described as below.

(A) *Data Processing*

Maskless LDW lithography, where image data can be optimally scaled or distorted just prior to writing, addresses the specific needs for registration, as well as for the

Table 7.4 Microlab4A100 specifications

Light source	*Wavelength*	405 nm
	Power	50 mW,
	Mode	TEM00
SLM	DMD 0.55 in.	
Stage	X: 100 mm, Y: 100 mm, θ	
Resolution	0.3 μm(×100)、0.4 μm (×50)	
	1.0 μm(×20)、2 μm(×10)	
Work rate	1.8 mm^2/min @0.4 μm	
	10 mm^2/min @1.0 μm	
Support file	.dxf, .arr, .bmp	

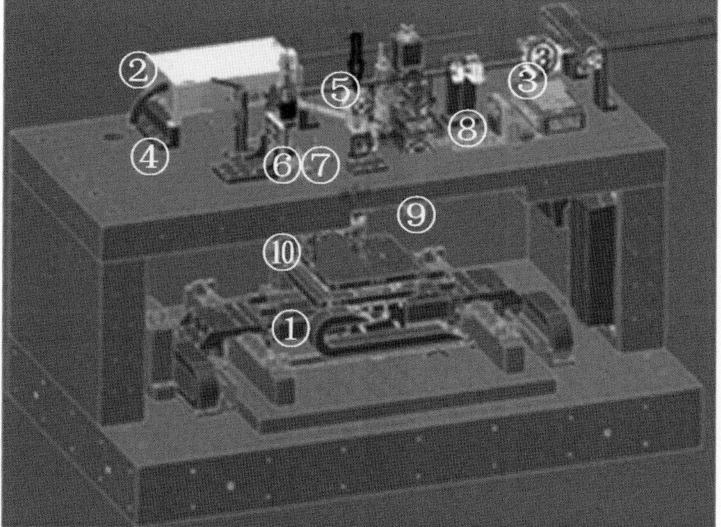

Fig. 7.10 Schematics of iGrapher® LDW system. ① Stage; ② laser; ③ beam shaper;④ beam expander;⑤ collimator; ⑥ aperture;⑦ mirror;⑧ DMD;⑨ projection lens;⑩ detector

Table 7.5 iGrapher specifications

Light source	*Wavelength*	351 nm
	Power	15 ns, 2 mJ/pulse
	Mode	TEM00
SLM	DMD 0.7 in.	
Stage	X: 200 mm, Y: 200 mm, θ	
Resolution	0.5 μm(×50)、1.0 μm (×20)	
Work rate	50 mm^2/min @0.5 μm	
	300 mm^2/min @1.0 μm	
Support file	.dxf, .arr, .bmp, GDSII	

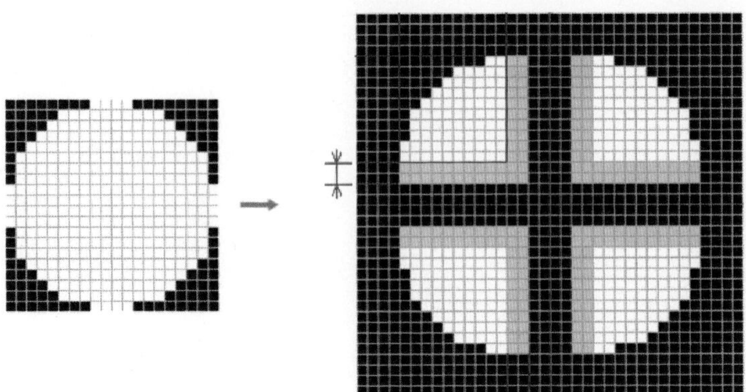

Fig. 7.11 File segmentation and recombination

elimination of process steps and masks that contribute greatly to alignment errors. The first question we met is how to transform the graphic data from computer to a data format recognizable by DMD [20, 21]. Usually, GDSII and DXF are standard file formats for fabricating mask used in IC industry. Both of them are vector image file formats. Vectorial format has the advantages of easy-to-handle, zoom-without-loss. However, for laser LDW tools, in which a pulsed laser beam is raster scanned across the panel substrate to form a pixelated image, we need a raster image. Therefore, in order to achieve high resolution, a big raster image file, maybe several gigabyte, is required to be generated.

Another question must be answered before adopting lithography process is how to fragment file. There are three reasons: (1) Original raster image is too large to be handled by the system. (2) Exposure area of each frame by DMD is limited. For example, for a DMD with a resolution of 1,024 $\times$ 768 and the projection lens with a magnification of $\times 50$, the maximum area in each frame which can be exposed is only about 280×280 μm. (3) Last but not the least, in order to improve the quality of exposure, digital image processing is needed sometimes.

Figure 7.11 shows file fragmentation and recombination. If a raster image is too large to be exposed in one frame, it will be split to several small raster images. In order to get rid of the stitch between each exposure unit, the edge region of each unit is marked by gray level and is to be exposed two times (green area in Fig. 7.11). Figure 7.12 demonstrates how the raster images are encoded and fed into DMD by a self-defined data sequence which refreshes itself with very high frequency.

(B) *Flying Exposure*

For the purpose of increasing the efficiency of LDW lithography, a novel technique called flying exposure is integrated in our system. The main idea of flying exposure is simple. By adopting high-power laser source, the refresh rate of DMD matches that of the input imaging data, then the motion of stage does not need to pause

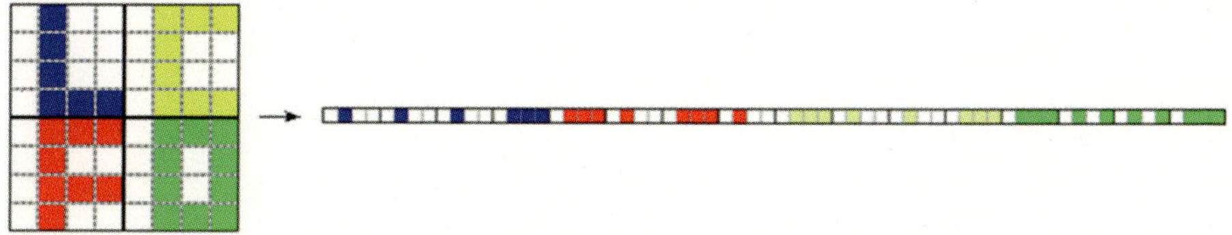

Fig. 7.12   Diagram of data flow fed to DMD

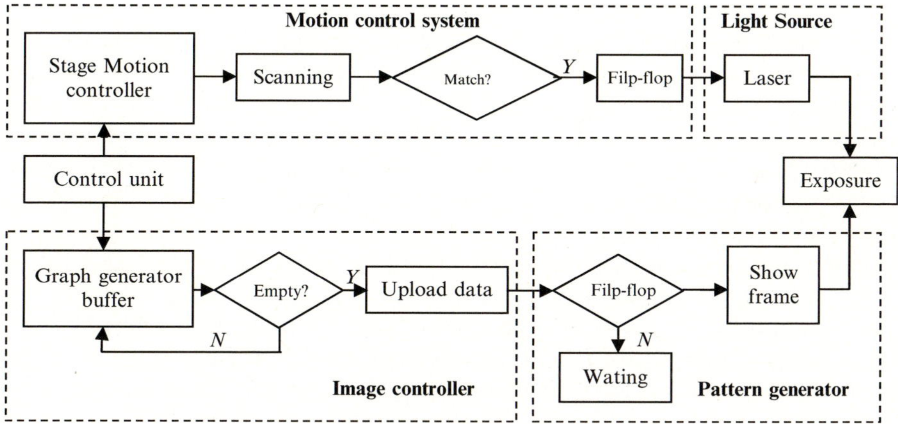

Fig. 7.13   Working principle of flying exposure technique

during the exposure process. It is a kind of position flip-flop mode. The pattern efficiency can be enhanced more than 100 times by using this working mode compared with the traditional one. Figure 7.13 shows the working principle of flying exposure technique adopted in iGrapher® and MicrolabA100® LDW lithography systems. The digital image process controller uploads data to pattern generator and samples the position signal with high frequency (120 MHz) from motion controller as well. Once a phase matching message is received, the pattern generator outputs an image to DMD, then a laser pulse is switched on. Through the optical system, the image now can be projected onto the substrate. The laser pulse width is 10–20 ns and the stage scanning speed is 80–100 mm/s. Therefore, the displacement during the exposure is only about 1–2 nm, which can be neglected in practice.

(C) *Z-Scan Technique*

In order to achieve higher resolution, high numerical aperture (NA) objectives are usually used in LDW system. High NA objective has relatively shorter focal depth. For example, Olympus ×20 objective (NA, 0.46), ×50 objective (NA, 0.8), and

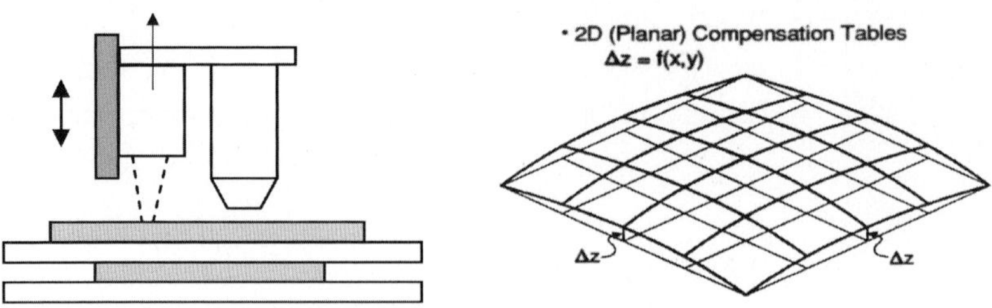

Fig. 7.14 Schematic of z-scan technique

×100 objective (NA, 0.95), their focal depth is about tens of micrometers. The LDW system works at a very fast scanning speed (>100 mm/s), The substrate curvature, un-uniformity of coating photoresist and environmental vibration, etc., can greatly affect the size and shape of the focusing spot, which results in the deterioration of system's resolution and exposure energy. We have developed an active scanning technique (z-scan technique) to overcome the difficulty. The key point here is before processing, surface profile scanning in z-axis is done in advance, and when exposed, the surface profile variation can be compensated by z-axis stage motion controller.

The schematic of z-scanning technique is shown in Fig. 7.14. (1) Laser displacement sensor (SICK OD5-25 T01)is used to record the surface profile by raster scanning, building a relation between Δz and (x, y) coordinate; (2) Filtering redundant fallacious data; (3) Compensate the surface profile data used in exposure process by introduction of least squares method. The range of output analog voltage is between +10 and −10 V, and the z-axis sampling resolution is 0.122 μm. Figure 7.15a indicates the effect of z-scanning technique. The original substrate surface profile, shows large variations, while little variation is observed after z-scan calibration (Fig. 7.15b).

By combination of the abovementioned techniques, the enhanced digital capabilities required for tight registrations and smaller features are provided. The general adoption of digital imaging as a replacement for mask-dependant contact printing will largely depend on the conjunction of several factors, such as acquisition and running costs, print speeds and panel throughput, and registration accuracy. LDW lithography can find wide applications in modern IC, electronics, and optics fields. Figure 7.16a–f illustrates some samples fabricated by our LDW lithography systems iGrapher® and MicrolabA100®. By using high NA (0.95) objective, the minimum linewidth about 0.3 μm on photoresist can be achieved (Fig. 7.16a, by Microlab®). A lithographic mask with dimension size 50 × 50mm can be fabricated by iGrapher® in less than 15 min (Fig. 7.16b, linewidth 2 μm). A photon sieve mask is shown in Fig. 7.16c. Figure 7.16d is a photo of hollow mask with linewidth 100 μm. Figure 7.16e shows the picture of binary optical element, and Fig. 7.16f is its reconstructive image.

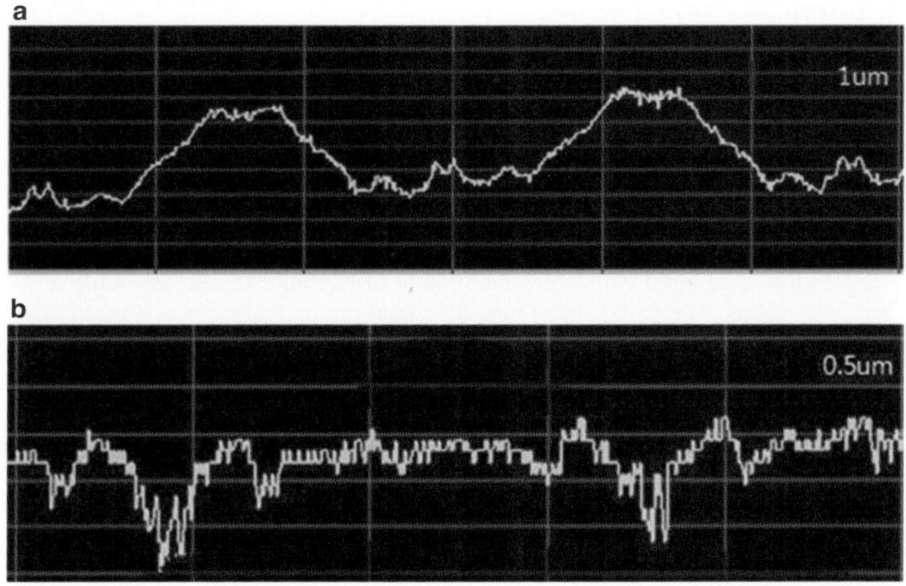

Fig. 7.15 Comparison of the surface profiles without using z-scanning (**a**) and with z-scanning (**b**)

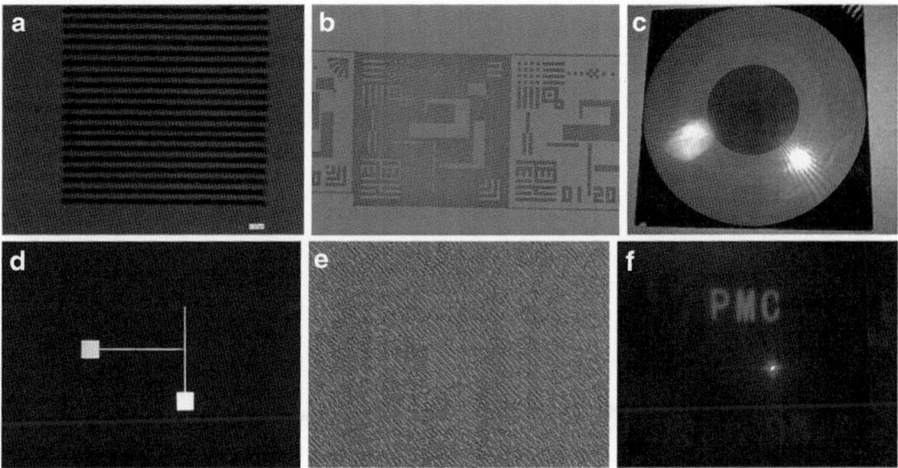

Fig. 7.16 (**a–e**) Samples fabricated by our LDW systems. (**a**) Grating with 300nm linewidth (by MicrolabA100®), (**b**) mask for lithography fabricated by iGrapher®, (**c**) photon sieve mask, (**d**) hollow mask with its linewidth 100 μm, (**e**) microscopy picture of a binary optical element and its reconstructive image (**f**)

7.3 Quasi-LIGA Process

LIGA is the German acronym for X-ray lithography (X-ray lithographie), electro-deposition (galvanofor-mung), and molding (abformtechnik). The process involves a thick layer of X-ray resist (from microns to centimeters) and high-energy X-ray radiation exposure and development to arrive at a three-dimensional resist structure [22]. In quasi-LIGA process, however, not X-ray radiation exposure but conventional UV lithography or LDW lithography is applied. Subsequent electrodeposition fills the resist mold with a metal and, after resist removal, a freestanding metal structure results. The metal shape may be a final product or serve as a mold insert for precision plastic injection molding or as master mold for imprinting, as shown in Fig. 7.3. The quasi-LIGA process is generic, simple, cost-effective, and promising for mass production of micro- and nanostructure-based devices.

After exposure and development, surface relief structure is left on the photoresist layer. Before electrodeposition, it needs to deposit a thin silver seed layer by silver mirror reaction, as shown in Fig. 7.17. The seed layer is always thin enough so it will not affect the micro- or nanostructure.

The silver mirror reaction can be described as:

$$CH_2OH(CHOH)_4-\overset{\overset{\text{O}}{\|}}{C}-H + 2\,Ag(NH_3)_2^{1+} + 3\,OH^{1-} \rightarrow 2\,Ag + CH_2OH(CHOH)_4-\overset{\overset{\text{O}}{\|}}{C}-O + 4\,NH_3 + 2\,H_2O$$

dextrose
(an aldehydic sugar)

- Solutions

  Silver nitrate, $AgNO_3$, 0.10 M (Dissolve 17 g $AgNO_3$ in distilled water to make 1.0 L of solution.)

  Potassium hydroxide, KOH, 0.80 M (Dissolve 22.4 g KOH in distilled water to make 500 mL of solution.)

  Dextrose, $C_6H_{12}O_6$, 0.25 M (Dissolve 11.2 g dextrose in distilled water to make 250 mL of solution.)

  Ammonia, NH_3, concentrated (15 M)

  Nitric acid, HNO_3, concentrated (16 M)

- Procedure (designed for a 125 mL flask)

  1. Place 30 mL of silver nitrate solution in a 150 mL or 250 mL beaker. Add concentrated (15 M) ammonia dropwise, with stirring, until the brown precipitate just dissolves. Add 15 mL of potassium hydroxide solution. If the brown precipitate reforms, add additional ammonia solution, dropwise, until it dissolves.

  2. Pour 3 mL of dextrose solution into the flask. Add the contents of the beaker and stopper the flask. Shake the flask so that the liquid comes in contact with the entire inner surface of the flask. The silver film should begin to form

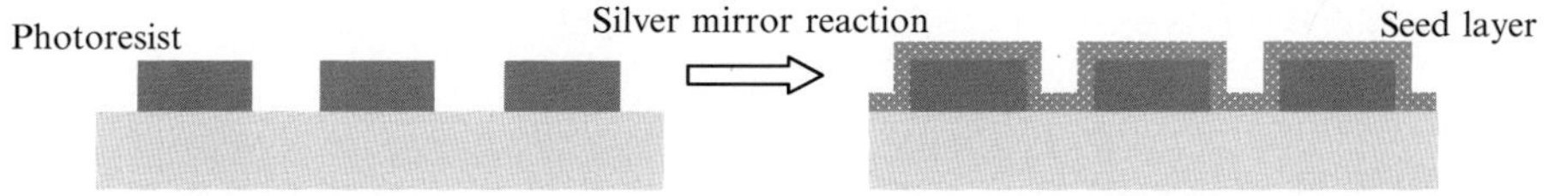

Fig. 7.17 Deposition of a thin silver seed layer on the surface relief by silver mirror reaction

within about 1 min. Continue to shake until the flask has a silver mirror coating (this may take about 5 min).

3. Dispose of the solution and rinse the flask well with water (at least four complete rinsings).

• Notes

1. Any unused reagent should be poured into the appropriate waste container in the laboratory and the container rinsed at least four times with large volumes of water.
2. The silver flask or bottle should be rinsed well with water to prevent formation of explosive silver nitride.
3. Nitric acid is corrosive. Exercise care in handling and work in a hood to minimize fumes. If any gets on the skin, rinse well with water for 10 to 15 min. If reddening or blistering occurs, seek medical assistance. Nitric acid will leave yellow stains on the skin that will persist for several days.
4. Ammonia fumes are irritating and toxic. Work under a fume hood or in a well-ventilated area. If any ammonia solution gets on the skin, rinse well with water.
5. Solutions should be disposed of in the appropriate waste containers in the laboratory.

After the thin silver layer being deposited, a nickel master mold is formed by electroplating the metalized sample. In the electroforming of microdevices with LIGA, the silver-coated substrate, carrying the resist structures, serves as the cathode. The metal layer growing on the substrate fills the gaps in the resist configuration, thus forming a complementary metal structure. The use of a solvent-containing development agent ensures a substrate surface completely free of grease and ready for plating. Most of the plating involves Ni, Cu, Au, etc. The fabrication of metallic relief structures is a well-known art in the electroforming industry. The technology is used, e.g., to make the fabrication tools for records and video discs where structural details in the submicron range are transferred. Electrodeposition is a powerful tool for creating microstructures. Because of the extreme aspect ratio, i.e., several orders of magnitude larger than in the crafting of CDs, electroplating in LIGA poses new challenges.

The nickel electroplating solutions commonly used for electroforming are the Watts and conventional and concentrated nickel sulfamate solutions with and without addition agents. Nickel fluoborate solutions are used, but their popularity appears to be declining. From a commercial perspective, the most important ones

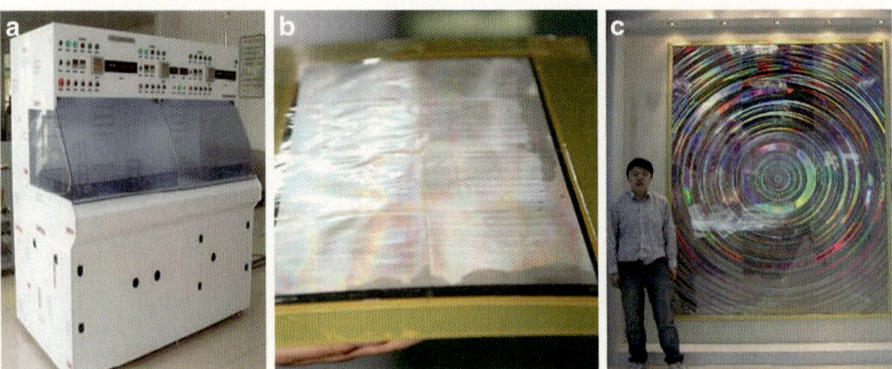

Fig. 7.18 (**a**) The galvanoforming cell with nickel anodes and an auxiliary tank incorporating heating and purification auxiliary equipment, (**b**) shows the nickel plate prepared for imprinting, (**c**) shows the fabricated large-format nickel plate with nanostructure patterning

are those based on nickel sulfamate (Fig. 7.18). The advantages of nickel electroforming from sulfamate solutions are the low internal stress of the deposits and the high rates of deposition that are possible, especially from the concentrated solution (Ni-Speed) [23, 24]. The formulations of these electroforming solutions are given in Table 7.6.

7.4   Nanoimprinting Techniques

7.4.1   Introduction to Nanoimprinting Technique

There exist wonderful demonstrations of nanotechnology-based lasers and other photonic components. However, difficult questions related to the fabrication need to be addressed before such components enter the market. Demonstrations in the scientific literature have relied heavily on the use of direct writing lithographic methods, such as electron beam lithography or focused ion beam lithography. These methods, although excellent for scientific studies, cannot be scaled up to allow cost-effective production of nanophotonics. Lithographic solutions developed for integrated circuits can produce extremely narrow linewidths and deliver high precision but are difficult to transfer to photonics fabrication.

Nanoimprint lithography (NIL) is a nonconventional lithographic technique for high-throughput patterning of polymer nanostructures at great precision and at low costs. Unlike traditional lithographic approaches, which achieve pattern definition through the use of photons or electrons to modify the chemical and physical

Table 7.6 Nickel electroforming solutions and typical properties of the deposits

	Electrolyte composition, g/L		
	Watts nickel	Conventional sulfamate	Concentrated sulfamate
$NiSO_4\ 6H_2O$	225–300		
$Ni(SO_3NH_2)_2\ 4H_2O$		315–450	500–650
$NiCl_2\ 6H_2O$	37–53	0–22	5–15
HBO_3	30–45	30–45	30–45
	Operating conditions		
Temperature (°C)	44–66	32–60	60 or 70
Agitation	Air or mechanical	Air or mechanical	Air or mechanical
Cathode current density A/dm2	3–11	0.5–32	Up to 90
Anodes	Nickel	Nickel	Nickel
PH	3.0–4.2	3.5–4.5	3.5–4.5
	Mechanical properties		
Tensile strength MPa	345–485	415–620	400–600
Elongation %	15–25	10–25	10–25
Vickers hardness 100 g load	130–200	170–230	150–250
Internal stress (MPa)	125–185 (tensile)	0–55 (tensile)	Zero stress can be obtained at various combinations of current density and temperature

properties of the resist, NIL relies on direct mechanical deformation of the resist material and can therefore achieve resolutions beyond the limitations set by light diffraction or beam scattering that are encountered in conventional techniques [25–27]. NIL instead uses a 1:1 template to produce linewidths that have already gone beyond 10 nm mark and foresee virtually no limit in resolution. NIL cannot only create resist patterns, as in lithography, but can also imprint functional device structures in various polymers, which can lead to a wide range of applications in electronics, photonics, data storage, and biotechnology [28–30]. After almost a decade of work and investigation into the technology by several academic and industry groups, NIL finally made itself onto the International Technology Roadmap for Semiconductors and was added to the list of possible candidates for post-optical lithography techniques at the 32 nm node in 2003 [31].

Several approaches toward micro- and nanostructure NIL fabrication have been exploited in the past 17 years, including thermal NIL, ultraviolet NIL (UV NIL), and microcontact printing lithography (μCP) [32]. Whether to choose thermal NIL, UV NIL, or μCP depends on the application. Thermal NIL may be better suited for optical or MEMS applications than semiconductor applications for several reasons. For example, thermal NIL requires that the temperature should be uniform across the substrate surface, which makes it more compatible with full-field templates (and thus less compatible with silicon). Also, because of the high viscosity of the polymer layer in thermal NIL, it requires a fair amount of pressure to get the polymer to flow evenly through the structures, which can cause deformation of the wafer. Because of the required heating and cooling cycles involved in thermal NIL, imprinting an entire 200 or 300 mm wafer is time-consuming. UV NIL, compared with thermal NIL, offers several decisive technical advantages concerning overlay alignment accuracy, simultaneous imprinting of micro- and nanostructures, and tool design due to the absence of high imprint pressures and thermal heating cycles. Its low-viscosity monomer is able to flow through the complex designs and small features of semiconductor chips and requires very low pressure to minimize wafer deformation. Thus, UV NIL offers two approaches for patterning using either rigid stamps (such as quartz) or soft stamps (such as polymer) for structuring of UV-sensitive resists. Promising applications of μCP NIL could be found in the biomedical field. In this case, μCP may make the most sense because the UV light or heat of the other two methods could damage biomaterials. On the other hand, μCP cannot produce the small features that the other methods can because it uses a soft, flexible stamp rather than a hard mold. Besides these three categories of NIL, there are variations in terms of using step-and-repeat tools (much like today's optical steppers) or applying full-wafer techniques.

NIL has been gaining attention and momentum, with more commercial manufacturers and investors joining the game. It is beginning to work its way from academia to commercial industry, although the industrial involvement is still primarily at the R&D level. Some of the biggest players in NIL grew out of university programs – e.g., Molecular Imprints Inc. (MII, Austin, Texas) licensed its technology from work done at the University of Texas, while Nanonex Corp. (Monmouth Junction, N.J.) sprang from research at Princeton University. S. Chou, who founded Nanonex in early 2000 and is still a professor of engineering at Princeton, is considered as a pioneer in NIL. Other major players in the field include EV Group (Schärding, Austria), SUSS MicroTec (Garching, Germany), and Obducat (Malmö, Sweden).

In the abovementioned NIL experiments, an entire flat mold is pressed simultaneously into a polymer cast on a flat substrate, that is to say, plane-to-plane NIL (P2P NIL, shown in Fig. 7.19a). There is an alternative approach to P2P NIL – Fig. 7.19b (plane-to-roll NIL, P2R NIL) and (c) (roller-to-roll NIL, R2R NIL). The roller-based NIL process primarily targets large-area patterning of nanostructures. In the conventional approach, embossing a large area requires a very large force. Huge contact areas between the mold surface and the imprinted nanostructures also produce significant adhesion force, making it difficult or even impossible to separate the mold–sample without damaging the substrates. In thermal NIL, if the mold and substrate are made from materials with different thermal expansion coefficients, such as Si mold and polymer substrate, stress can build up during a thermal cycle of such a magnitude that it can even destroy the Si mold during mold

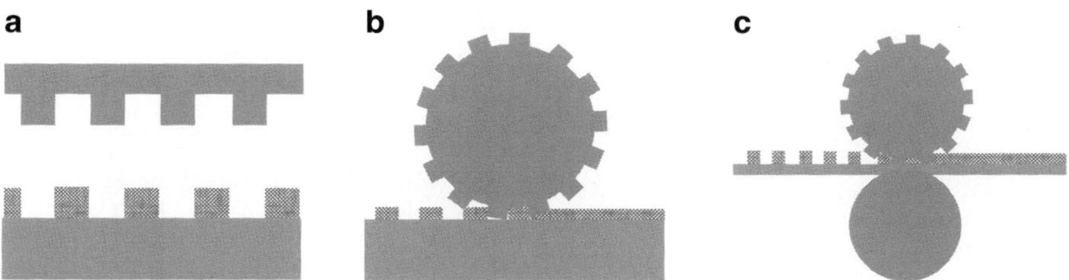

Fig. 7.19 (**a**) P2P NIL, (**b**) P2R NIL, (**c**) R2R NIL

releasing. Roller-based NIL provides a unique solution to these challenges encountered in the conventional wafer-scale NIL process, because imprinting in roller-based NIL proceeds in a narrow region transverse to the web moving direction, thus requiring a much smaller force to replicate the patterns. Also, because the mold used in roller-based NIL is in the form of a roller, the mold–sample separation proceeds in a peeling fashion, which requires much less force and reduces the probability of defect generation [4]. Therefore, compared with P2P NIL, roller-based NIL has the advantage of better uniformity, less force, and the ability to repeat a mask continuously on large-scale substrate.

7.4.2   Flexible Roll-to-Roll Nanoimprinting

R2R NIL can form optical films for flat-panel displays, antireflective coatings for solar cells, and other textured products in mass quantities on large-area pieces and at a high speed. With this method, people can merge nanoimprinting technologies into real-world applications and on an industrial scale. R2R NIL produces nanometer-sized structures of greater complexity using fewer processing steps while minimizing wasted materials. It has evolved from the semiconductor industry's lithography technology to a platform process technology that can be adapted to a wide range of applications.

Previously, roller presses were used in the cement industry to imprint into flexible thin films that can conform to the shape of the roller. However, the pressed feature size in these applications was 1 μm or larger. In this section, the design and construction of roller-based NIL systems, the development of seamless R2R NIL methods, and the demonstration of complex micro- and nanostructure patterns on flexible substrate by R2R NIL are introduced.

Generally, it is difficult to directly make micro- and nanostructure on the surface of roller as imprinting mold. As for our process chain mentioned above, a nickel mold is wrapped on the roller, so inevitably there is a stitching gap. How to realize seamless R2R NIL is still a technical problem in the industry. Figure 7.20a shows the schematic of seamless thermal R2R NIL equipment by using two consecutive complementary rollers and Fig. 7.20b is its photo at working. As illustrated in

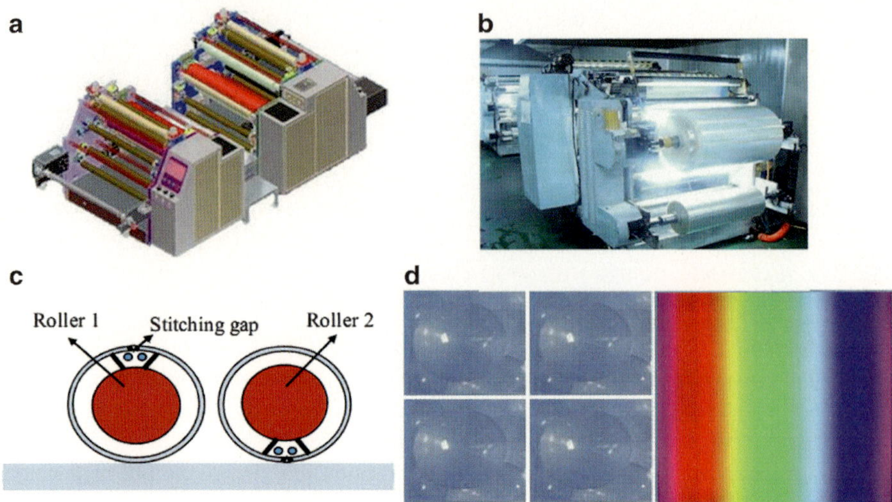

Fig. 7.20 (**a**) Design of seamless thermal R2R NIL with double complementary roller, (**b**) photo of R2R NIL machine, (**c**) schematic of seamless thermal R2R NIL, (**d**) multi-scale micro- (Fresnel lens, *left*) and nanostructure (nanograting, *right*) realized by our thermal R2R NIL with PE pretreatment

Fig. 7.20c, in the proposed seamless thermal R2R NIL process, the stitching gap region on the roller 1 is cooled, and the surface relief structure on the wrapped nickel mold cannot be transferred to the substrate. However, in the corresponding region on the roller 2, it is heated and other parts remain cooled. By carefully choosing the alignment distance, the blank region left by roller 1 can be imprinted by the roller 2 while other regions are unchanged. The main idea here is simple, but it is useful for mass manufacturing of holographic film for security and packaging application. The heating temperature is about 200 °C, greatly above the material's glass transition temperature. The working speed can reach 60 m/min and the length of one foil is 12,000 m. The web width is 800 mm. Another R2R NIL tool in our system is developed by pre-treating substrate with PE polymer, both microstructure and nanostructure can be formed on the resist at one imprinting action. Figure 7.20d is the photo of Fresnel lens (left) and nanograting (right) with feature depth size 10 μm and 200 nm, respectively. The proposed method is efficient in fabricating complex multi-scale micro- and nanostructure on the substrate.

A UV-based R2R NIL was also developed in our institute, as shown in Fig. 7.21. UV technique allows quicker processing because it takes place at room temperature and offers the advantage of fabricating the nanostructures on cross-linkable resins, thus imparting higher thermal and mechanical stability to the imprinted products. The resist film uniformity on the flexible web and the resist thickness are important from both pattern quality and economic point of view. If it is too thin, a film may cause insufficient filling of resist into certain mold regions or cause film fracture during mold releasing. If it is too thick, a resist not only results in accumulation of

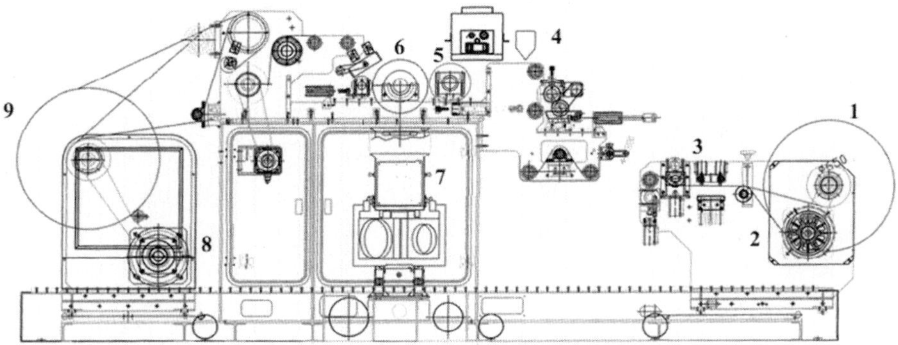

Fig. 7.21 Schematic of UV-R2R NIL system. *1* Unwinder, *2* tension controller, *3* clean roller, *4* slot-die, *5* backup roller, *6* imprinting roller, *7* UV lamp, *8* tension controller, *9* rewinder

resist from successive rolling cycles that hamper further imprinting but also squeezes out excess materials resulting in resist waste. In our UV-R2R NIL system, the coating method by slot-die is used to provide a uniform resin coating. The coating thickness can be controlled by adjusting the speed ratio of the moving web, viscosity, and liquid resist's transfer speed.

7.4.3   Key Issues in Roll-to-Roll NIL

The R2R NIL is a very hot topic in academic and industrial society, including R & D from materials and equipment suppliers to end users. The transition from lab to high-volume manufacturing of micro- and nanostructured films will enable enhanced flexible solar cells, biological applications, plastic electronics, and flat-panel displays. The potential can be seen for photoresists and polymers sustainable manufacturing for micro- and nanolithography. In order to achieve the technical and economic goal as early as possible, there are several critical issues need to be addressed.

1. Reliability

Although NIL can offer extreme high resolution beyond optical diffraction limit, its reliability is still in doubt. The first issue is the acquisition of large-scale mold fabrication. The conventional serial writing by electron or ion beam of patterns with a small feature size and high density is time-consuming, and the writing area is limited to approximately 10^4 μm. All these make large-area (like wafer scale) fabrication almost impossible. The second one is that it is crucial to develop an acceptable standard for evaluating how NIL can satisfy the stringent requirements by the current electronic–photonic integrated solution. Figure 7.22a, c are master molds, and b and d are their replicated patterns with micron feature sizes. The

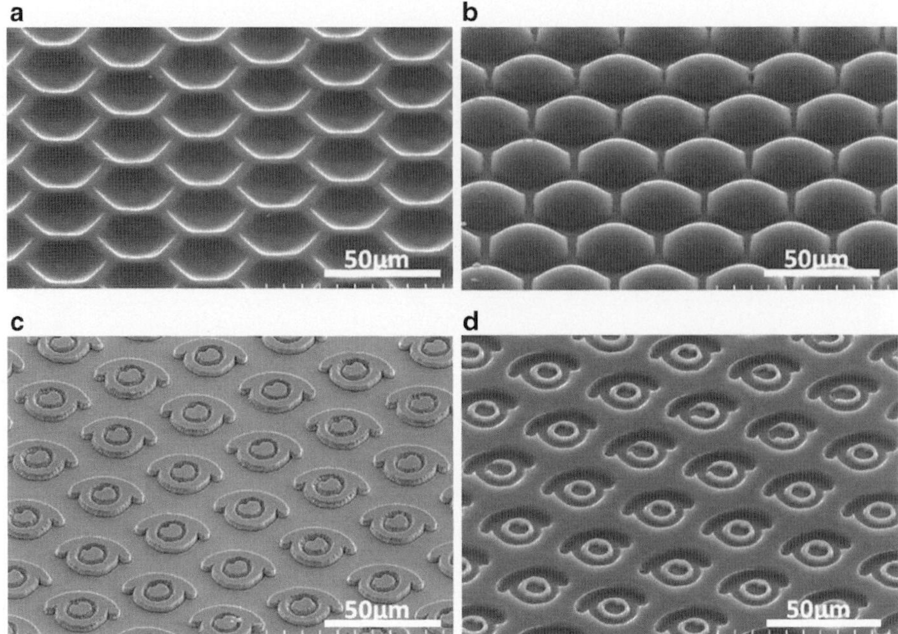

Fig. 7.22   (**a**) Microlens array master mold and its replication on UV resin by R2R NIL (**b**), (**c**) Micro pattern array master mold and its replication on UV resin by R2R NIL (**d**)

fidelity between them needs to be clarified. The NIL infrastructure must be developed as a viable, high-volume, low-cost manufacturing platform during its design, manufacturing, and integration.

2. Materials

In NIL process, the materials of mold and imprinting resist are important for the successful application of NIL. The material for the mold is selected by considering the following properties: hardness, compatibility with conventional microfabrication processes, and the thermal expansion coefficient. A significant difference in thermal expansion between the stamp and the substrate can result in stress or pattern distortion during the cooling period, because thermal imprinting is typically carried out at temperatures of 100–200 °C. Common features of these different molds are high hardness and high mechanical strength. The properties of these hard molds contrast those of elastomeric stamps used in soft lithography and are essential for producing nanoscale features because the protrusion patterns on the mold should not deform, buckle, or collapse during imprinting, even at elevated temperatures. Silicon, dielectric materials (e.g., silicon dioxide or silicon nitride), metals (e.g., nickel), or polymeric materials that have a sufficient Young modulus (e.g., PFPE, PDMS, ETFE) have been used widely for stamp fabrication, because they fulfill the above requirements.

The resist material can be a thermal plastic, thermal setting, or low-viscosity precursor that can be cured either thermally or by UV light. For a fast roll-to-roll process, there are several stringent requirements for the resist materials. First, liquid resists having good coating properties are preferred because they can be continuously and uniformly coated onto plastic substrates and be easily imprinted with low pressure. Second, such liquid resists should have low viscosity before curing for fast imprinting. Third, resists used in R2R NIL need to be quickly cured and with minimum shrinkage after curing. Therefore, conventional resist materials dissolved in solvents and requiring additional baking processes are not suitable for the R2R NIL process. Thermoplastic materials that are typically used in conventional NIL are also not adequate for R2R NIL because they require very high pressures and relatively long processing times to complete the imprinting.

3. Overlay accuracy

The fabrication of micrometer- and nanometer-scale devices in R2R NIL now begins to involve overlay accuracy, such as those in flexible optoelectronic devices and 3D imaging applications. How to guarantee the overlay accuracy in R2R NIL is still a problem to be answered, since conventional alignment methods used in semiconductor are not suitable for flexible substrate. Although efforts have been put to adapt some methods (e.g., moire technique) to NIL, there is a long way to realize the overlay accuracy as its counterpart in IC.

7.5   Application Cases

NIL technology has developed many possible applications in making devices ranging from optics [32–34], patterned media data storage [35], biosensor/ MEMS, microfluidics [36] to molecular electronics [37]. Here we show several successful cases utilizing the technologies (LDW lithography, R2R NIL, etc.) described above.

7.5.1   Optical Security Devices

Nanogratings used as authentic devices are not novel in the optical security systems [38, 39]. Most of these devices are based on guided-mode resonance (GMR) phenomena or surface plasmonics. GMR-based devices are commonly composed of dielectric thin-film layers incorporating zero-order diffractive structures to couple energy diffracted from an incident wave to a leakage mode of the waveguide layer included in the structure. Generally, GMR devices are strongly dependent on wavelengths, incident angles, and polarization states of incident. Surface plasmon-based devices are usually accompanied with thin metal layer which can excite the surface plasmonic wave. With the miniaturization of integrated devices, current

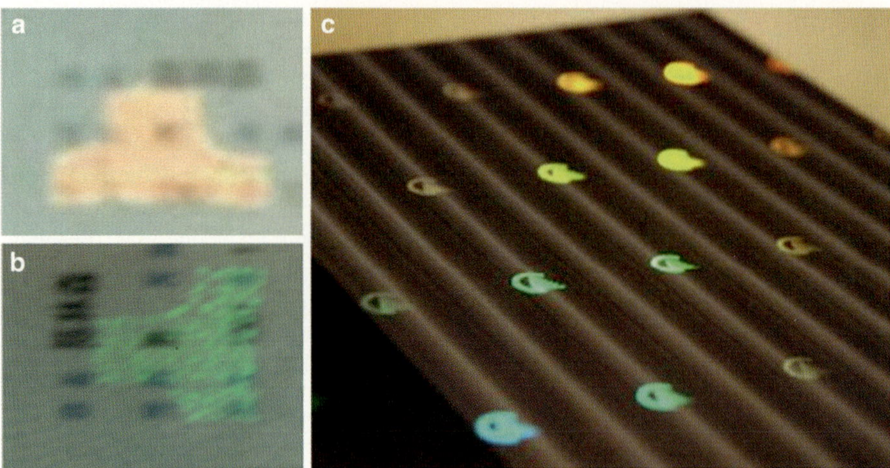

Fig. 7.23 (**a**) and (**b**) are symbols of great wall composed of GMR-based subwavelength grating viewing at different angles, (**c**) micro icon viewing at a large tile viewing angle

research on surface plasmon-based imaging focuses on novel designs aiming at high efficiency, low power consumption, and slim dimension, which poses great challenges to the traditional colorant-based filtering and prism-based spectral splitting techniques. In this context, surface plasmon-based nanostructures are attractive due to their small dimensions and the ability to efficiently manipulate light. Here we show several examples of using nanostructure as security devices. Figure 7.23a, b shows subwavelength grating appearing yellow and green when being viewed from two orthogonal angle. The micro icon in Fig. 7.23c shows color when viewed at a large tilt viewing angle, which is yet invisible at normal incidence. The physical mechanism lies in the fact that the reflective spectrum can be well controlled by using a subwavelength grating buried in a high refractive index layer. The subwavelength grating mold can be fabricated by using interference lithography, and the imprinting and dielectric material deposition can both be involved in R2R process.

7.5.2 Microlens Film Used as Optical Diffuser and Efficiency-Enhanced Component in OLED

An approach for fabrication of microlens arrays (MLA) using spatial light modulator-based lithography method was proposed, and the microlens film was applied to optical diffuser and organic light-emitting device (OLED) for enhancing the extraction efficiency [40–43]. Combined with thermal reflow method, maskless lithography based on a DMD is used to pattern microstructure, and MLA with arbitrary structure and topology can be obtained. The pattern on thick photoresist

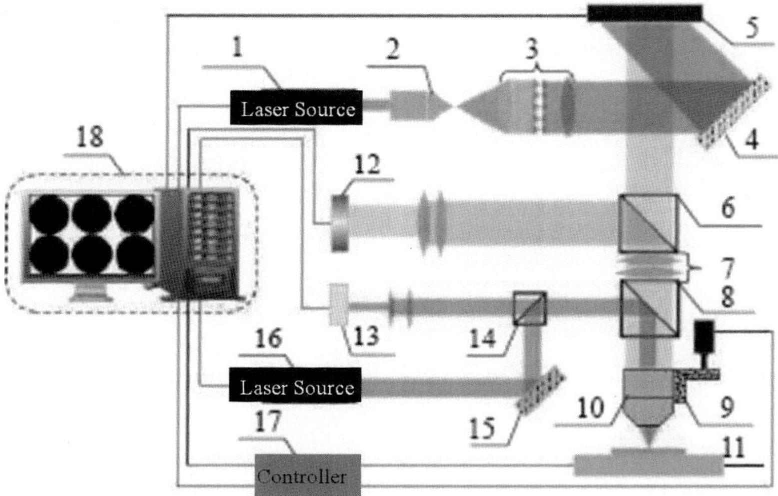

Fig. 7.24 Optical system based on spatial light modulator for fabricating microlens array (Reprinted with permission from Ref. [41]. Copyright 2012, Chinese Journal of Lasers). *1* 405 nm wavelength laser source; *2* beam shaping lens; *3* integrated homogenizing system; *4* mirror; *5* DMD; *6* beam splitter; *7* infinity-corrected optical systems; *8* beam splitter; *9* PZT; *10* objective lens; *11* x-y stage; *12* CCD; *13* four-quadrant detector; *14* beam splitter; *15* mirror; *16* detect laser; *17* controller; *18* computer

layer is projected by a bi-telecentric optical system, as shown in Fig. 7.24. Similar to the optical system depicted in Fig. 7.9, the MLA pattern is input from computer 18 to DMD, and it is projected onto the surface of substrate coated with photoresist. Good surface quality can be realized by thermal reflow method [44]. The process can follow the steps shown in Fig. 7.25. The SEM photos obtained in experiment are presented in Fig. 7.26a–h. The insets are the input patterns in DMD. Arbitrary shape and arrangement can be realized by the digitalized pattern method. For the positive photoresist used in this case, the MLA region is black for blocking light, and the gap between MLA is white for exposure. Compared with classical stereolithography [45] and mask-based exposure lithography, the proposed method has the advantages of low cost and high efficiency. Therefore it is especially suitable for fabricating MLA with feature size ranging from several micrometers to hundreds of micrometers. The obtained microlens array can be transferred to a nickel mold by quasi-LIGA process, which can be used as an imprinting mold. Figure 7.27 is the profile of the imprinted flexible microlens array film. The point line and the real line denote the cross section of measured profile and ideal thermal reflow profile. The experimental error results from the evaporation of the solvent in photoresist during thermal reflow process.

A new type of optical diffuser with cascaded MLAs is proposed and investigated. The diffuser is composed of a layer of MLAs with small aperture and high numerical aperture (NA) on output side (D1) and a layer of MLAs with

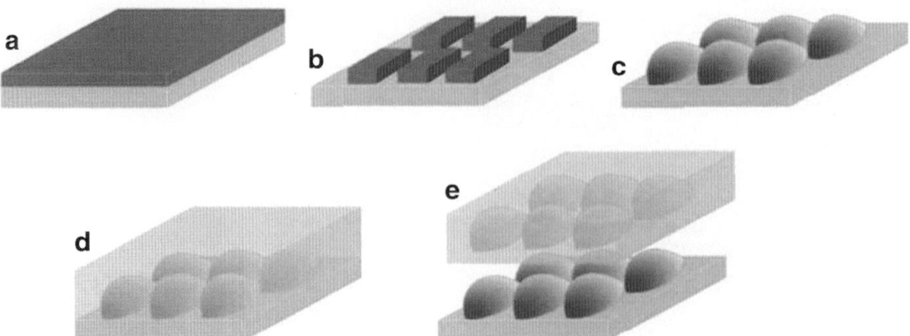

Fig. 7.25 The process chain of the proposed method for fabricating microlens arrays (Reprinted with permission from Ref. [41]. Copyright 2012, Chinese Journal of Lasers)

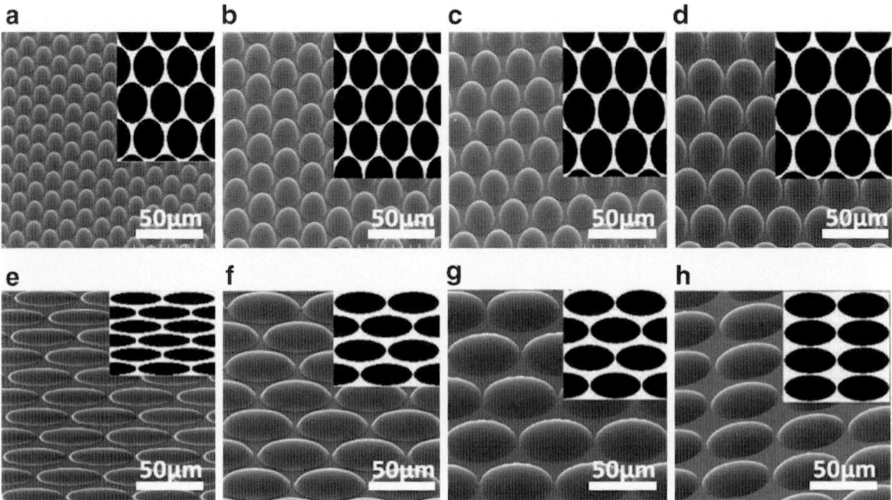

Fig. 7.26 SEM photos of the fabricated microlens array. Aperture: (**a**) 10 μm, circular; (**b**) 30 μm, circular; (**c**) 40 μm, circular; (**d**) 50 μm, circular; (**e**) 50–10 μm, elliptical; (**f**) 50–20 μm, elliptical; (**g**) 50–30 μm, elliptical, hexagonal arrangement; (**h**) 50–30 μm, elliptical, square arrangement (Reprinted with permission from Ref. [41]. Copyright 2012, Chinese Journal of Lasers)

large aperture and low NA on incident side (D2, shown in Fig. 7.28). The SEM photos of microstructure and its surface profile are illustrated in Fig. 7.29a–c. The influence of the parameters (such as sag height and aperture) of cascaded MLAs on the diffusing behavior is investigated by using ray tracing method. The results show that the maximum luminance increases with the ratio of sag height to aperture and an additional 3–5 % optical gain can be accomplished by inserting another layer of MLAs with a height–aperture ratio of 0.01–0.15 on incident side (Fig. 7.30). Experimental results are also demonstrated and the optical gain is measured as

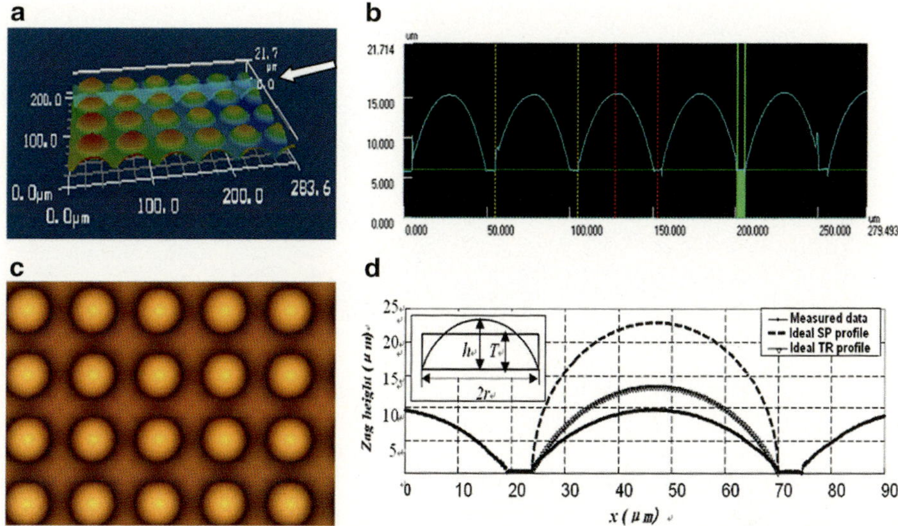

Fig. 7.27 The profile of flexible microlens array film. (**a**) Measured profile; (**b**) cross-section profile; (**c**) the light spots pattern; (**d**) comparison of experimental case and the ideal case (Reprinted with permission from Ref. [41]. Copyright 2012, Chinese Journal of Lasers). Experimental case; ideal case; ideal semi-sphere case. The *inset* is the diagram of thermal reflow)

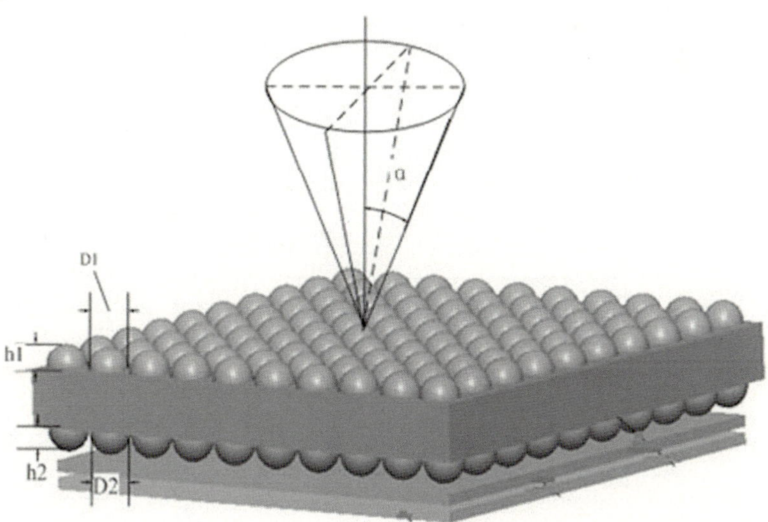

Fig. 7.28 Schematic diagram of the proposed cascaded microlens diffuser (Reprinted with permission from Ref. [42]. Copyright 2010, Acta Optica Sinica)

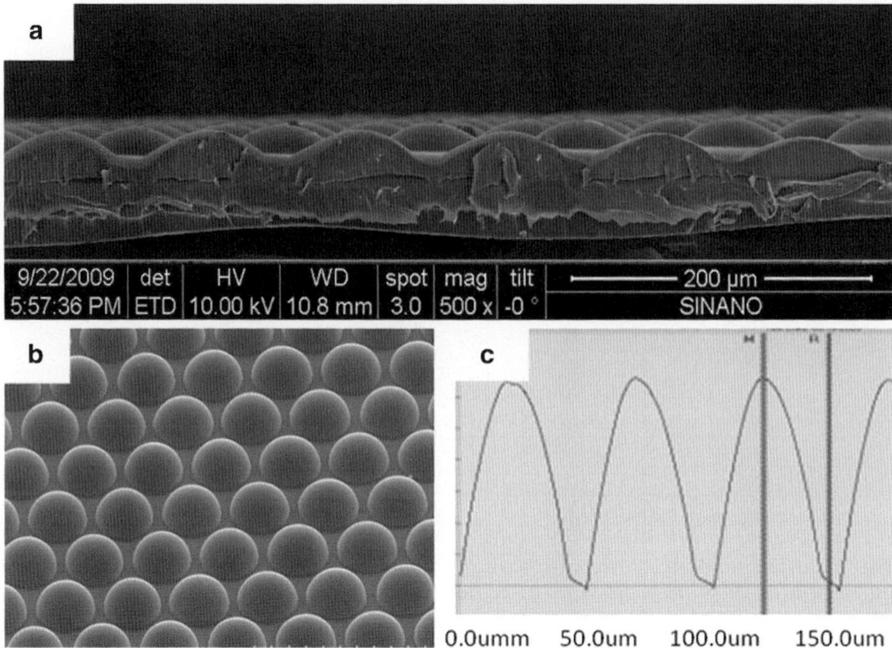

Fig. 7.29 SEM photos of the fabricated cascaded microlens array diffuser. (**a**) Cross section. (**b**) SEM photos of MLA on the output side. (**c**) The surface profile of microlens array on incident side (Reprinted with permission from Ref. [42]. Copyright 2010, Acta Optica Sinica)

40.06 % (Fig. 7.31). Integrating the function of traditional diffuser and brightness-enhancement film, the discussed diffuser is much thinner and more compact.

Organic light-emitting devices (OLEDs) have been attracting considerable interest as a viable alternative to lighting source and display applications due to their high brightness and lower power consumption [46, 47]. Currently, nearly 100 % internal quantum efficiency in OLEDs has been demonstrated by using phosphorescent-emitting materials. However, the external quantum efficiency in a conventional OLED is still limited, since only a small fraction of light generated in the organic layers can be extracted due to total internal reflection (TIR) at the glass substrate/air and indium tin oxide (ITO)/substrate interfaces, induced by refractive index difference. Indeed, the majority of lights generated in the emitting layer are normally trapped, and the light out-coupling efficiency for an OLED with a flat glass substrate is limited to ~20 %.

Many methods have been proposed to enhance light out-coupling in OLEDs, including surface roughness (or random scatters), photonic crystals, microlens arrays, surface plasmons, embedded low-index dielectric grid in organic layers, microcavity structure, etc. [48–53]. However, although a significant enhancement was observed for some reported methods, they have limited applications due to distorted or shifted emission spectrum or complicated fabrication process. On the contrary, microlens array is of great advantage in its simple and reliable processing for large-area and

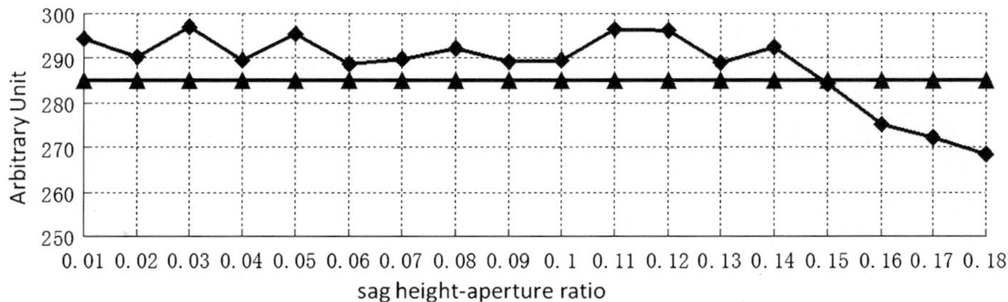

Fig. 7.30 The maximum luminance varies with the ratio of sag height to MLA aperture on the incident side (Reprinted with permission from Ref. [42]. Copyright 2010, Acta Optica Sinica)

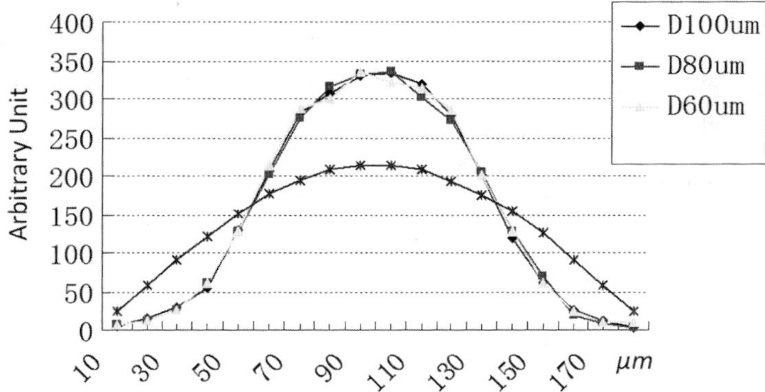

Fig. 7.31 The luminance pattern of the proposed cascaded microlens diffuser (Reprinted with permission from Ref. [42]. Copyright 2010, Acta Optica Sinica)

low-cost fabrication and is especially suitable for reducing the TIR loss at the substrate/air interface with neither spectral variation nor any change on device electrical properties. Previously, microlens arrays with different geometries (e.g., hemispherical, cylindrical, pyramidal, or square) were made of various materials, such as polyethyleneterephthalate (PET), poly-methyl-methacrylate (PMMA), and polydimethylsiloxane (PDMS). However, the performance of microlens arrays with a desired geometrical shape on the light out-coupling enhancement for OLEDs has not yet been studied.

Here, we demonstrate an enhancement of the light out-coupling efficiency by incorporating microlens arrays, which are fabricated by a roll-to-roll UV embossing process on the polycarbonate (PC) film. The influence of microlens shapes on the efficiency of light out-coupling is analyzed experimentally and theoretically. An increase of the light out-coupling by a factor of 1.6 over the unlensed device is realized with an optimized elliptical microlens array.

The structure of an OLED attached with microlens array used in this study is shown schematically in Fig. 7.32a. The organic layers and metal cathode were deposited in high vacuum (base pressure ~5 × 10–6 Torr) by thermal evaporation

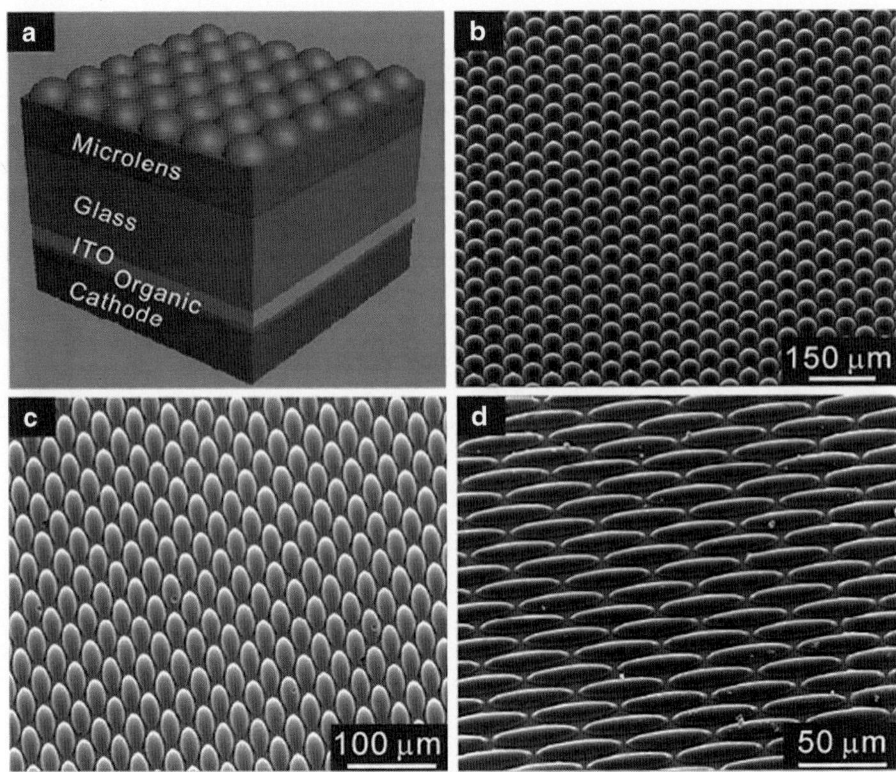

Fig. 7.32 (**a**) Schematic illustration of an OLED attached with microlens array. (**b–d**) A 45° tilted view of SEM images of microlens arrays with (**b**) hemispherical (D50), (**c**) short-elliptical (D50-20), and (**d**) long-elliptical (D50-10) microlenses (Reprinted with permission from Ref. [43]. Copyright 2010, American Institute of Physics)

onto a cleaned ITO glass substrate with a sheet resistance of ~30 Ω/sq. The device structure consists of a 50 nm thick hole transporting layer (HTL) of N, N'-diphenyl-N, N'-bis (1-naphthyl)-(1,1'-biphenyl)-4,4'-diamine (NPB), a 70 nm thick green emitting/electron transporting layer (ETL) of tris (8-quinolinolato) aluminum (Alq3), and a LiF(0.5 nm)/Al (100 nm) bilayer cathode deposited through a shadow mask. Microlens arrays made of the UV hardened epoxy were separately fabricated on the flexible PC film using a roll-to-roll nickel mold transfer and UV curing process. The patterned photoresist (RZJ-390PG) on glass substrate was firstly made by the digital micromirror device (DMD)-based lithography and treated with thermal reflowing process at 120 °C for 5 min to transform them into the desired geometry of microlenses. Then, a nickel mold was formed by electroplating on photoresist surface. Finally, microlens arrays on PC film were made through the roll-to-roll press of the Ni mold on the top of UV hardened epoxy followed by UV curing. The microlens film was then attached to the OLED glass substrate at the air/glass interface with index matching oil. The current density–voltage–luminance

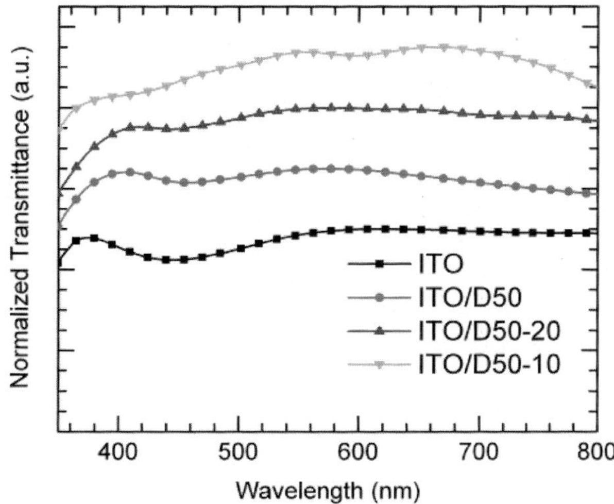

Fig. 7.33 Normalized transmittance spectra of ITO glass substrates attached with various microlens arrays compared to that of the neat ITO substrate (Reprinted with permission from Ref. [43]. Copyright 2010, American Institute of Physics)

(J–V–L) characteristics and electroluminescence spectra of the corresponding devices were measured simultaneously with a programmable Keithley model 2,400 power source and a Photo Research PR 655 spectrometer.

Figure 7.32b–d shows a tilted view of scanning electron microscope (SEM) images of ordered microlens arrays with different microlenses (FEI Quanta 200 FEG). For comparison, three types of microlenses were fabricated, such as hemispherical microlens (50 μm in diameter, referred to as D50) (Fig. 7.32b), elliptical microlens (50 μm in transverse diameter, 20 μm in conjugate diameter, referred to as D50-20) (Fig. 7.32c), and elliptical microlens (50 μm in transverse diameter, 10 μm in conjugate diameter, referred to as D50-10) (Fig. 7.32d). The fill factors, defined as the area ratio of the microlenses to the surface, are estimated to be 0.7, 0.8, and 0.6 for D50, D50-20, and D50-10, respectively.

Optical transmittances of the ITO glass substrate attached with different microlens arrays were measured with a PerkinElmer Lambda 2S UV/visible (VIS) spectrometer and are depicted in Fig. 7.33. It is obvious that microlens arrays exhibit a little difference of transmittance in the visible range of 400–800 nm as compared with that of the ITO glass substrate, except for the D50-10 film. These results indicate that the microlens arrays studied here are wavelength independent and suitable for OLED light extraction.

To verify the enhancement efficiency of the microlens array extracting light into air, OLEDs were fabricated on flat glass substrates. We can plot the luminance efficiency as a function of current density for devices with and without microlens arrays attached to the substrate surface. As evident in Fig. 7.33, the microlenses enhance the light out-coupling. At a current density of 20 mA/cm^2, the light out-coupling enhancement ratios of devices with various microlens arrays relative to unlensed device are 1.5, 1.6, and 1.36 for D50, D50-20, and D50-10 in the normal direction, respectively. This demonstrates that the out-coupling performance of D50-20 is superior to that of D50 and D50-10. It has been reported that when

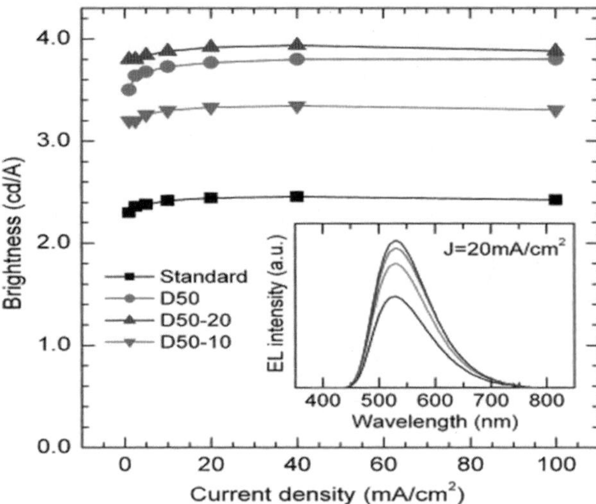

Fig. 7.34 Luminance efficiency of devices with and without microlens arrays versus current density. *Inset*: EL spectra of devices at $J = 20$ mA/cm² (Reprinted with permission from Ref. [43]. Copyright 2010, American Institute of Physics)

integrating the total flux of the light emitted from device, the observed enhancement efficiency is higher than those obtained only from the normal direction, due to more efficient enhancement at larger viewing angles for the devices with microlens array. In addition, as shown in the inset of Fig. 7.34, the electroluminescence (EL) spectra of the corresponding devices at $J = 20$ mA/cm² are independent of the shapes of microlenses, and the microlenses simply redirect the light with negligible color shift.

The optical properties of microlens arrays on the light out-coupling were modeled by Monte Carlo ray tracing calculation with LightTools software, assuming a typical OLED structure as shown in Fig. 7.32a. The microlens layer with refractive index 1.59 is attached to a 1 mm thick glass substrate ($n_{glass} = 1.53$) by using refractive-index-matched oil ($n_{oil} = 1.56$). The glass substrate has a size of 4×4 mm. The OLED device is comprised of a 70 nm thick NPB layer ($n_{NPB} = 1.78$) and a 70 nm thick Alq3 layer ($n_{Alq} = 1.72$), sandwiched between a 100 nm thick ITO anode layer ($n_{ITO} = 1.83$) and a 100 nm thick aluminum cathode. For simplicity, the emitting layer is further simplified as a point lighting source at the interface between NPB and Alq3 at a wavelength of $\lambda = 530$ nm, and the metal cathode is assumed to be an ideal reflector. Since the size of microlenses studied here is much larger than the emission wavelength, refraction rather than diffractive scattering is regarded as the dominant effect. In addition, microcavity effects and waveguide mode at ITO/organic interface are not considered since the microlens array has no significant impact on these processes. Light emitted from the point source is traced through the device considering refraction and reflection at interfaces between layers with different refractive indices. The secondary rays due to total reflection at these interfaces are also taken into account, even though with 5 % energy losses each time (due to the absorption of the glass substrate).

The light out-coupling efficiency of OLEDs can be enhanced by a factor of 1.6 with elliptical microlens arrays attached to the glass substrate. The microlens arrays are made by a roll-to-roll mold transfer process, which is advantageous for the easy

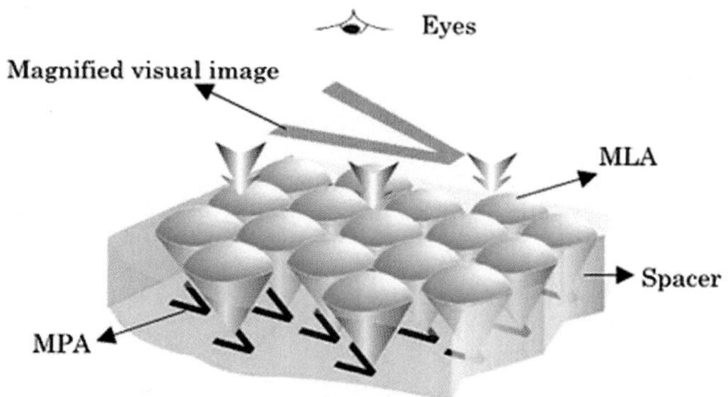

Fig. 7.35 Schematic of the proposed Glass pattern device (Reprinted with permission from Ref. [57]. Copyright 2012, Optical Society of America)

and large-area fabrication through thermal reflow and UV curing process. The performance of microlens arrays with various geometrical shapes for OLEDs has been analyzed experimentally and theoretically, demonstrating that the fill factor of microlenses on the surface is the predominant factor in determining the enhancement of light out-coupling efficiency.

7.5.3   AMOS Devices

The phenomenon of regular moiré magnification occurs whenever an array of lenses is used to view an array of identical objects situated at the focal plane of the lenses. It is not a new phenomenon and has been employed as anti-counterfeiting device used in banknote [54, 55]. However, the random microlens-based moiré phenomenon, called Glass pattern effect [56], has not been investigated. In the Glass pattern case, the magnified moiré pattern is concentrated around a certain point in the superposition, and it gradually fades out and disappears if one goes away from this point. It has been revealed that it is not really required that one has identical dots shapes in both random layers. Generally, the Glass pattern effect is usually realized by making transparent photocopies of a document or of a patterned image. By choosing appropriate dots shapes for the two layers, one can synthesize a Glass pattern of any desired shape and intensity profile.

A superposition of randomly distributed microlens array (MLA) and a corresponding micro pattern array (MPA) consisting of tiny icons are proposed to give rise to particularly integrated dynamic Glass patterns [57]. The schematic view of the discussed device is shown in Fig. 7.35.

A plurality of MLA with arbitrary distribution is disposed on the top surface of the spacer.

The MPA icons are situated at the focal plane of MLA, with the distance between them is exactly the same as the thickness of the spacer. The superposed MLA and MPA are slightly geometrically transformed (scaled, rotated, translated) to each other so as to generate the Glass pattern effect. Similar to the conventional type, a single magnified image is expected to appear in the superposition. The purpose of the introduction of MLA here is to generate the effect of parallax and contributes to sampling the multiple images and to rendering the process more efficient than pinholes. By carefully choosing the geometrical parameters, the synthetically magnified image can show either the parallactic or the orthoparallactic motion effect when viewed from different angles. We name this novel visual motion phenomenon as dynamic Glass pattern.

In order to obtain such dynamic Glass pattern, homogeneously distributed dots patterns are firstly produced by molecular dynamics, combined with a low-discrepancy sequences technique [58, 59]. The low-discrepancy technique used here is Sobol sequences, which utilize a base of two to form successively finer uniform partitions of the unit interval and then render the coordinates in each dimension. It has been reported that the random dots generated by this method are free from visible roughness as well as any moiré patterns when superposed on other regular patterns. The obtained dots position data are input into AutoCAD in order to generate the expected MLA and MPA lithographic mask.

A single synthetically magnified image would appear, centered about a certain superposition point. To build the mathematic model for our Glass pattern device assumes that every microlens is regarded as a pinhole to pick up the intensity of the corresponding focusing point in MPA layer. A piece of dynamic Glass pattern material is in plane view being oscillated or rotated about its vertical axis ($-45°$ to $+45°$). An array of character "V" is used as input micro icon array. If the single synthetically magnified image "V" moves according to parallactic motion, it would appear to be displaced left and right as the material is oscillated around the vertical axis (left bottom inset in Fig. 7.36). However, it is oscillated about the vertical rotational axis, which shows orthoparallactic motion and is counterintuitive to the normally expected parallactic motion (right bottom inset in Fig. 7.36). Random noises appear in the neighboring regions in both cases. This occurs because of the local correlation between the incident light ray and the corresponding intensity in MPA plane. It falls in a non-related manner in the far region.

The magnifying factor, the orientation of the synthetic image, and the dynamic visual movement are determined by the scaling ratio r and the skew angle θ between the MPA and MLA:

$$M(r,\theta) = \frac{r}{(1 + r^2 - 2r\cos\theta)^{\frac{1}{2}}} \tag{7.2}$$

$$\varphi = \arctan\left(\frac{\sin\theta}{r - \cos\theta}\right) \tag{7.3}$$

where $M(r,\theta)$ is the magnifying factor and φ is the angle between the visually

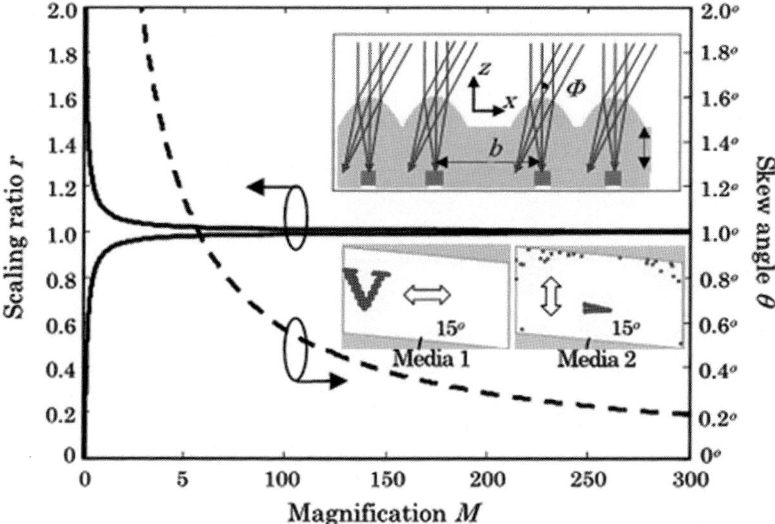

Fig. 7.36 Scaling ratio r (*solid line*) and skew angle θ (*dot line*) versus magnification M. The *top inset* is schematic diagram under tilt viewing angle Φ. The *bottom insets* are single-frame images in at 15° viewing angle, respectively (Reprinted with permission from Ref. [57]. Copyright 2012, Optical Society of America)

synthetic image and the original MPA. As illustrated in Fig. 7.36, $M(r,\theta)$ is very sensitive to the scaling ratio r when MPA and MLA are well aligned ($\theta = 0$, solid line). As $r \rightarrow 1$, $M(r,\theta)$ tends to infinity. When $r = 1$, $M(r,\theta)$ is determined by the skew angle θ (dot line). However, there are two big different visual effects between the two cases. One is that, in the second case ($\theta \neq 0$), the magnified synthetic image is rotated clockwise in MPA plane. The explanation for this effect can be given by Eq. 7.3. It is, nevertheless, possible to confirm that when $r \rightarrow 1$, the angle φ approaches 90°. The other difference is, with the introduction of the skew angle θ, the so-called orthoparallactic motion effect, which can be clearly observed. This novel phenomenon can be explained by the Brouwer fixed point theorem [60], which states given two identical sheets of patterns, if one of the sheets is crumpled and placed on top of the other, then there must be at least one point on the scrupled sheet that is directly above the corresponding point on the bottom sheet. In our case, it means trying to locate the fixed center of the Glass pattern, which can be quantitatively described by the following formulas.

If tilt viewing angle Φ is in x-z plane, the movement distance L of the magnified synthetic image can be calculated by means of integration along the distance L_Φ between two arbitrary icons in the magnified image and the ratio between the focal point shift and the distance b of these two icons in the original pattern:

$$L = \int_0^\Phi L_\Phi dk_\Phi \tag{7.4}$$

where

$$L_\Phi = bM(r,\theta)\sqrt{\sin^2\varphi + \cos^2\varphi \cos^2\Phi} \tag{7.5a}$$

$$k_\Phi = \frac{h\tan\Phi}{b} \tag{7.5b}$$

From Eqs. 7.2, 7.3, 7.4, and 7.5a, b we can obtain the moving distance of visual magnified image from its normal location to that at viewing angle Φ. In the first case ($\theta = 0$), L_Φ equals to $\frac{rh}{2|1-r|}\ln\left(\frac{1+\sin\Phi}{1-\sin\Phi}\right)$, and for $\varphi = 0°$, its direction is along x-axis. In the second case ($\theta \neq 0$), L_Φ $\frac{h\tan\phi}{(1+r^2-2r\cos\theta)^{1/2}}$ is and its azimuthal angle is determined by Eq. 7.3. When $r \rightarrow 1$ and $\theta \rightarrow 0$, φ equals to $(\pi-\theta)/2$ which is almost perpendicular to the x-axis and results in the orthoparallactic motion effect. In both cases, we can see that as the tilt viewing angle increases, the visual magnified image moves away from the concentrated point at a faster speed.

The proposed dynamic Glass pattern device can be fabricated by the steps suggested in Fig. 7.37a–f: formation of MLA photoresist mold using contact lithography and the thermal reflow method, then fabrication of a concave MLA nickel master mold by electroforming and UV nanoimprinting replication from the master to the top surface of 100 µm thickness polyester (PET) film (Toyobo, A4300). The convex MPA nickel plate can be obtained by following similar steps except thermal reflow. The superposition of MPA onto the lower surface of PET can be dealt manually with varying orientations.

During the contact lithography, the thickness of the positive photoresist (RJZ-390, RUIHONG Electronics Chemicals) for MLA is determined by the sag height of microlens, which can be predicted by the volume conservation of photoresist islands in the thermal melting process. The focal length of the microlens equals to the thickness of PET film. The aperture and sag height of each microlens is 50 and 6.0 µm, respectively. To ensure the pigment to be filled into the groove, the microgroove width and depth is chosen as 3.0 µm, respectively, by rule of thumb. The photoresist molds are not suitable for repeated replication because of its fragility and degradation. Instead, rigid nickel master molds are developed for the UV resin replication. A 40 nm thick silver seed layer was coated on the photoresist mold by silver mirror reaction, then the nickel master mold is formed by electroplating the metalized sample (1,500 mA, 2 h), the thickness of which is about100 µm. The photoresist layer is subsequently removed with acetone. Complementary surface microstructure on the photoresist mold can be transferred to the nickel master mold. After the superposition of MPA and MLA, a pigmented coloring material with submicron particles is filled into the MPA voids by a gravure-like doctor blading process.

Figure 7.38a shows the microstructure image of the fabricated MLA and MPA measured by a 3D laser scanning microscope (Keyence, VK 9700). The aperture of randomly distributed MLA is 50 µm and its sag height is 6 µm. The micro icon

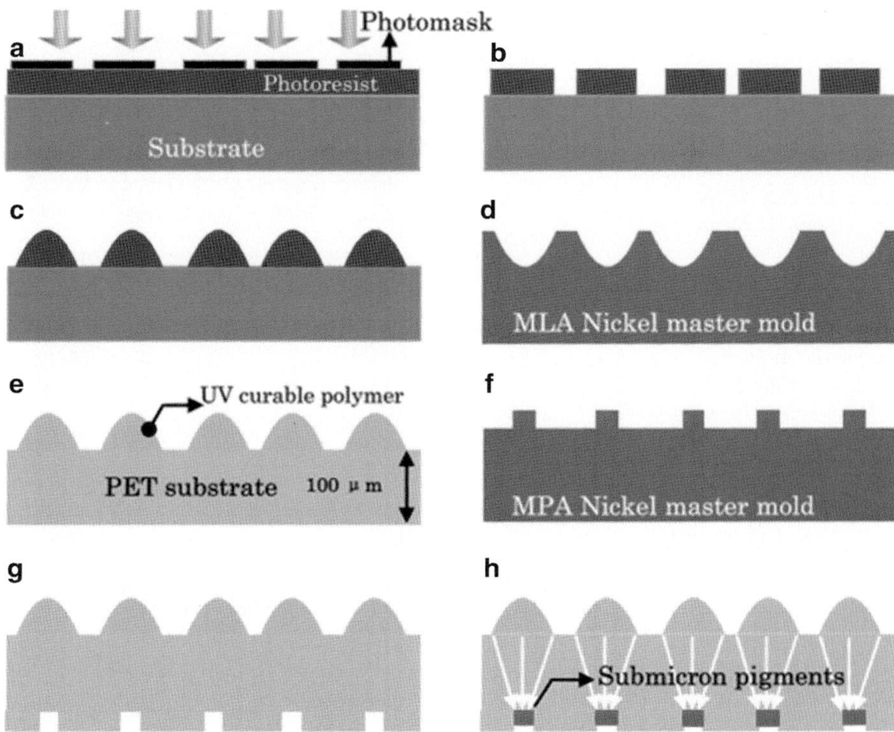

Fig. 7.37 Fabrication process of dynamic Glass pattern device: (**a**) contact exposure on the photoresist through a photomask, (**b**) development, (**c**) thermal reflow to form MLA, (**d**) MLA nickel master mold by electroplating, (**e**) MLA replication by UV nanoimprinting, (**f**) MPA nickel master mold with similar process, (**g**) MPA replication on the other side of PET substrate with alignment, (**h**) submicron particle pigment filled in the grooves of MPA (Reprinted with permission from Ref. [57]. Copyright 2012, Optical Society of America)

character "v," the linewidth and dimension size of which is 3 μm and 40×40 μm, respectively, is on the back side of PET. The scaling ratio of MPA to MLA is 1.0 in this case. The macro visual image is shown in Fig. 7.38b. The dimension of the magnified character "v" is 5×5 mm, which means a magnification factor of 125 is achieved for this sample ($\theta = -0.45°$). As predicted, the fabricated sample shows orthoparallactic motion effect and the macroimage rotates clockwise. The microlenses sample the MPA and fill the aperture of the lenses with information relating to the icon at the focal points. There are randomly distributed noises in the neighbor region, but thanks to the filled pigment which enhanced the image contrast, we can still clearly observe the Glass pattern realized by microlens array.

Fig. 7.38 (a) Surface profile of randomized MLA and MPA based on a piece of PET substrate. The aperture of the microlens is 50 µm and its sag height is 6 µm. The dimension size of the micro icon "v" is 40×40 µm, (b) picture of fabricated sample. The dimension size of the magnified character "v" is 5×5 mm (Reprinted with permission from Ref. [57]. Copyright 2012, Optical Society of America)

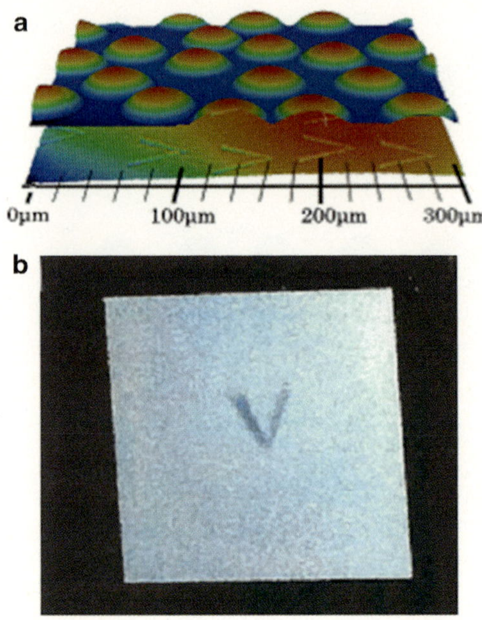

7.6   Outlook

Microoptics and nanophotonics are really wide research fields that cover many interesting applications, branching from cutting-edge sciences, including plasmonics, metamaterials, and cavity quantum electrodynamics in high-Q cavities, all the way to applied sciences, such as silicon nanophotonics for on-chip optical interconnections, single-frequency semiconductor light sources, optical films used in flat-panel display, and organic optoelectronic devices for lighting and solar energy. Many practical devices have utilized micro- or nanostructure patterned surfaces which consist of microlens array, nanogratings, photonic crystals, waveguides and metal structures, etc. Without doubt, the intellectual and material resources for achieving this will require coordination along the entire technology supply chain. Whereas in most industries coordination occurs through market mechanisms, the microelectronics industry provides a prime example of how coordination can be improved intentionally through an industry technology roadmap. For over 10 years, researchers and engineers in our group have been focusing on the application of micro- or nanostructure to promote the performance of optoelectronic devices. Vast experience in laser direct writing and R2R process laid the solid foundation for our application-oriented research. We believe that a bright future is promised for the realization of mass production of large-format micro-/nanostructure-based optical device. Perhaps most importantly, these technologies can extend human's capabilities to help people live a better life.

References

1. Wikipedia: the Free Encyclopedia (2013) http://en.wikipedia.org/wiki/Photolithography. Accessed 22 Jan 2013
2. Jahns J, Brenner KH (2004) Microoptics: from technology to applications. Springer, Dordrecht
3. Suleski TJ, Kolte RDT (2001) A roadmap for micro-optics fabrication. Proc SPIE 4440:1–15
4. Ahn SH, Guo LJ (2008) High speed roll-to-roll nanoimprint lithography on flexible plastic substrates. Adv Mater 20:2044–2049
5. Ok JG, Park HJ, Kwak MK, Hernandez CAP, Ahn SH, Jay Guo L (2011) Patterning of nanogratings by nanochannel-guided lithography on liquid resists. Adv Mater 23 (38):4444–4448
6. Lee BK, Kim DS, Kwon TH (2004) Replication of microlens arrays by injection molding. Microsyst Technol 10:531–535
7. Kirchain R, Kimerling L (2007) A roadmap for nanophotonics. Nat Photonics 1:303–305
8. MONA: Merging Optics & Nanotechnology (2005) http://www.ist-mona.org/home.asp. Accessed 22 Jan 2013
9. Micronic mydata (2012) http://www.micronic.se/. Accessed 22 Jan 2013
10. Heidelberg Instrument (2012) http://www.himt.de/en/home/. Accessed 22 Jan 2013
11. Wikipedia: the Free Encyclopedia (2013) http://en.wikipedia.org/wiki/Maskless_lithography. Accessed 22 Jan 2013
12. SUSS MicroTec (2012) http://www.suss.com. Accessed 22 Jan 2013
13. ASML (2013) http://www.asml.com/asml/show.do?lang=EN&ctx=277&rid=543. Accessed 22 Jan 2013
14. Zhou Y, Luo G, Asbahi M, Eriksson T, Keil M, Ring J, Carlberg P, Jiawook R, Heidari B (2011) A method for metallic stamp replication using nanoimprinting and electroforming techniques. Microelectron Eng 91:112–120
15. Michelson AA (1927) Studies in optics. University of Chicago, Chicago. Reprinted by Dover, Dover (1995)
16. Chen CG (2003) Beam alignment and image metrology for scanning beam interference lithography fabricating gratings with nanometer phase accuracy. Massachusetts Institute of Technology, Cambridge
17. Wikipedia: the free encyclopedia (2013) http://en.wikipedia.org/wiki/Digital_micromirror_device. Accessed 22 Jan 2013
18. Wilsher T (2007) The mask handbook. Routledge, New York
19. Ljungblad U, Askebjer P, Karlin T (2005) A high-end mask writer using a spatial light modulator. Proc SPIE 5721:43–52
20. Voss SH, Talmi M, Saniter J, Reisig A, Heinitz J, Haugeneder E (2006) High-speed data storage and processing for projection mask-less lithography systems. Microelectron Eng 83:976–979
21. Liu H, Richards B, Zakhor A, Nikolic B (2010) Hardware implementation of Block GC3 lossless compression algorithm for direct-write lithography systems. Proc SPIE 7637:763716
22. Hak MG (2001) MEMS handbook. CRC Press, Boca Raton
23. Wearmouth WR, Belt KC (1979) Electroforming with heat-resistant sulfur-hardened nickel. Plat Surf Finish 66(10):53–57
24. Hammond RAF (1970) Nickel plating from sulphamate solutions. Part 3 – structure and properties of deposits from conventional solutions. Metal Finish J 16(188):234–243
25. Guo LJ (2007) Nanoimprint lithography: methods and material requirements. Adv Mater 19:495–513
26. Chou SY, Krauss PR, Renstrom PJ (1996) Imprint lithography with 25-nanometer resolution. Science 272:85–87
27. Merino S, Retolaza A, Juarros J, Schift H (2008) The influence of stamp deformation on residual layer homogeneity in thermal nanoimprint lithography. Microelectron Eng 85:1892–1896

28. Wu W, Hu M, Ou FS, Li Z, Williams RS (2010) Cones fabricated by 3D nanoimprint lithography for highly sensitive surface enhanced Raman spectroscopy. Nanotechnology 21:255502
29. Kim SH, Le KD, Kim JY, Kwon MK (2007) Fabrication of photonic crystal structures on light emitting diodes by nanoimprint lithography. Nanotechnology 18:055306
30. Gao H, Liu Z, Zhang J, Zhang GM, Xie GY (2007) Precise replication of antireflective nanostructures from biotemplates. Appl Phys Lett 90:123115
31. ITRS: International Technology Roadmap for Semiconductors (2009) www.itrs.net. Accessed 22 Jan 2013
32. Homburg O, Hauschild D, Lissotschenko V (2008) Manufacturing and application of micro-optics: an enabling technology for the 21st century. Optik and Photonik 4(48):52
33. Lee KD, Ahn SW, Kim SH, Lee SH, Park JD, Yoon PW, Kim DH, Lee SS (2006) Nanoimprint technology for nano-structured optical devices. Curr Appl Phys 6(S1):149–153
34. Chaix N, Landis S, Gourgon C, Merino S, Lambertini VG, Durand G, Perret C (2007) Nanoimprinting lithography on 200 mm wafers for optical applications. Microelectron Eng 84:880–884
35. Ouchi T, Arikawa Y, Homma T (2008) Fabrication of CoPt magnetic nanodot arrays by electrodeposition process. J Magn Magn Mater 320(22):3104–3107
36. Mills CA, Martinez E, Bessueille F, Villanueva G, Bausells J, Samitier J, Errachid A (2005) Production of structures for microfluidics using polymer imprint techniques. Microelectron Eng 78:695–700
37. Jung GY, Ganapathiappan S, Li X, Ohlberg DAA, Olynick DL, Chen Y, Tong WM, Williams RS (2004) Fabrication of molecular electronic circuits by nanoimprint lithography at low temperatures and pressures. Appl Phys A Mater Sci Process 78(8):1169–1173
38. Wu ML, Hsu CL, Lan HC, Huang HI, Liu YC, Tu ZR, Lee CC, Lin JS, Su CC, Chang JY (2007) Authentication labels based on guided-mode resonant filters. Opt Lett 32 (12):1614–1616
39. Kolle M, Cunba PMS, Scherer MR, Huang F, Vukusic P, Mahajan S, Baumberg JJ, Steiner U (2010) Mimicking the colourful wing scale structure of the Papilio blumei butterfly. Nat Nanotechnol 101:2010
40. O'Neill FT, Sheridan JT (2002) Photoresist reflow method of microlens production part II: analytic models. Optik 113(9):405–419
41. Shen S, Pu DL, Hu J, Chen LS (2012) Fabrication of microlens arrays using spatial light modulator based lithography method. Chn J Lasers 39(3):0316003
42. Zhuang XL, Zhou F, Shen S, Chen LS (2010) Characteristics of diffusers with cascaded microlens arrays. Acta Opt Sin 30(11):3306–3310
43. Yang JP, Bao QY, Xu ZQ, Li YQ, Tang JX, Shen S (2010) Light out-coupling enhancement of organic light-emitting devices with microlens array. Appl Phys Lett 97:223303
44. Lu Y, Chen S (2008) Direct write of microlens array using digital projection photopolymerization. Appl Phys Lett 93:041109
45. Tang CW, VanSlyke SA (1987) Organic electroluminescent diodes. Appl Phys Lett 51:913
46. Reineke S, Lindner F, Schwartz G, Seidler N, Walzer K, Lüssem B, Leo K (2009) White organic light-emitting diodes with fluorescent tube efficiency. Nature 459:234–239
47. Adachi C, Baldo MA, Thompson ME, Forrest SR (2001) Nearly 100% internal phosphorescence efficiency in an organic light-emitting device. J Appl Phys 90:5048–5051
48. Schnitzer I, Yablonovitch E, Caneau C, Gmitter TJ, Scherer A (1993) 30% external quantum efficiency from surface textured, thinfilm light emitting diodes. Appl Phys Lett 63:2174
49. Do YR, Kim YC, Song YW, Lee YH (2004) Enhanced light extraction efficiency from organic light emitting diodes by insertion of a two-dimensional photonic crystal structure. J Appl Phys 96:7629–7633
50. Möller S, Forrest SR (2002) Improved light out-coupling in organic light emitting diodes employing ordered microlens arrays. J Appl Phys 91:3324–3328

51. Feng J, Okamoto T (2005) Enhancement of electroluminescence through a two-dimensional corrugated metal film by grating-induced surface-plasmon cross coupling. Opt Lett 30:2302–2304
52. Sun Y, Forrest SR (2008) Enhanced light out-coupling of organic light-emitting devices using embedded low-index grids. Nat Photonic 2:483–487
53. Nakayama T, Itoh Y, Kakuta A (1993) Organic photo- and electroluminescent devices with double mirrors. Appl Phys Lett 63:594–596
54. Hutley MC, Hunt R, Stevens RF, Savander P (1994) The moiré magnifier. Pure Appl Opt 3:133–142
55. Stenblik RA, Hurt MJ, Jordan GR (2008) Micro-optic security and image presentation system, US PatentNo. 0037131
56. Glass L (1969) Moiré effect from random dots. Nature 223:578–580
57. Shen S, Lou YM, Hu J, Zhou Y, Chen LS (2012) Realization of glass patterns by a microlens array. Opt Lett 37(20):4248–4250
58. Chang JG, Lee CT (2007) Random-dot pattern design of a light guide in an edge-lit backlight: integration of optical design and dot generation scheme by the molecular-dynamics method. J Opt Soc Am A 24(3):839–849
59. Kang MW, Guo KX, Liu ZL, Zhang ZH, Chen B, Wang RZ (2010) Dot pattern designing on light guide plate of backlight module by the method of molecular potential method. J Disp Tech 6(5):166–169
60. Amidror I (2007) The theory of the moiré phenomenon: aperiodic layers. Springer, Dordrecht

Printed by Printforce, the Netherlands